Molecular and Cell Biology
of Sexually Transmitted Diseases

Molecular and Cell Biology of Human Diseases Series

Series Editors

D. J. M. WRIGHT MD FRCPath
Reader in Medical Microbiology,
Charing Cross and Westminster Medical School, London, UK

L. C. ARCHARD PHD
Senior Lecturer in Biochemistry,
Charing Cross and Westminster Medical School, London, UK

The continuing developments in molecular biology have made possible a new approach to a whole range of different diseases. The books in this series each concentrate on a disease or group of diseases where real progress is being made in understanding the pathogenesis, diagnosis and management. Experts on aspects of each area provide a text accessible to scientists and clinicians in a form which records, advances and points to the potential application of these advances in a clinical setting.

Other titles in the series

Molecular and Cell Biology of Opportunistic Infections in AIDS
Edited by Steven Myint and Alan Cann

Molecular and Cell Biology of Muscular Dystrophy
Edited by Terry Partridge

Molecular and Cell Biology of Neuropsychiatric Disease
Edited by Frank Owen and Ruth Itzhaki

Molecular and Cell Biology
of Sexually Transmitted Diseases

EDITED BY

David Wright and Leonard Archard

Charing Cross and Westminster Medical School, London

CHAPMAN & HALL

London · Glasgow · New York · Tokyo · Melbourne · Madras

Published by Chapman & Hall, 2–6 Boundary Row, London SE1 8HN

Chapman & Hall, 2–6 Boundary Row, London SE1 8HN, UK

Blackie Academic & Professional, Wester Cleddens Road,
Bishopbriggs, Glasgow G64 2NZ, UK

Chapman & Hall, 29 West 35th Street, New York NY10001, USA

Chapman & Hall Japan, Thomson Publishing Japan, Hirakawacho Nemoto
Building, 6F, 1–7–11 Hirakawa-cho, Chiyoda-ku, Tokyo 102, Japan

Chapman & Hall Australia, Thomas Nelson Australia, 102 Dodds Street, South
Melbourne, Victoria 3205, Australia

Chapman & Hall India, R. Seshadri, 32 Second Main Road, CIT East, Madras
600 035, India

First edition 1992

© 1992 Chapman & Hall

Typeset in 10/12 Sabon by Falcon Typographic Art Ltd, Fife, Scotland
Printed in Great Britain at the University Press, Cambridge

ISBN 0 412 36510 3

A catalogue record for this book is available from the British Library

Library of Congress Cataloging-in-Publication data available

∞ Printed on permanent acid-free text paper, manufactured in accordance
with the proposed ANSI/NISO Z 39.48–199X and ANSI Z 39.48–1984

'The pleasure is momentary, the position
ridiculous and the expense damnable'
Fourth Earl of Chesterfield
(1694–1773)

And the cost of promiscuity is the risk of infection.

It is to those with such infections
that we dedicate this book.

Contents

Contributors

John F. Alderete
The University of Texas
Health Science Center at San
Antonio
7703 Floyd Curl Drive
San Antonio
Texas 78284–7758
USA

Leonard C. Archard
Department of Biochemistry
Charing Cross and Westminster
Medical School
Fulham Palace Road
London W6 8RF
UK

Rossanna Arroyo
The University of Texas
Health Center at San Antonio
7703 Floyd Curl Drive
San Antonio
Texas 78284–7758
USA

Michael F. Barile
Laboratory of Mycoplasma
Center for Biologics Evaluation
and Research
FDA
Bethesda
Maryland 20892
USA

David E. Barker
Department of Medicine
University of Chicago
910 East 58th Street
Chicago
Illinois 60637
USA

Neil W. Blake
Department of Biochemistry
Charing Cross and Westminster
Medical School
Fulham Palace Road
London W6 8RF
UK

Alain Blanchard
Laboratory of Mycoplasma
Institut Pasteur
Paris 75724
France

Jeffrey J. Cream
Department of Dermatology
Charing Cross Hospital
Fulham Palace Road
London W6 8RF
UK

Don C. Dailey
The University of Texas
Health Science Center at San
Antonio
7703 Floyd Curl Drive
San Antonio
Texas 78284–7758
USA

Angus Dalgleish
St Georges Hospital
Blackshaw Road
London SW17
UK

Geoffrey M. Dusheiko
University Department of
Medicine
Royal Free Hospital School of
Medicine (London)
Rowland Hill Street
London NW3 2PF
UK

Jean Engbring
The University of Texas
Health Science Center at San
Antonio
7703 Floyd Curl Drive
San Antonio
Texas 78284–7758
USA

Tim J. Harrison
University Department of
Medicine
Royal Free Hospital School of
Medicine (London)
Rowland Hill Street
London NW3 2PF
UK

Marjorie B. Kovler
Department of Medicine
University of Chicago
910 East 58th Street
Chicago
Illinois 20892
USA

Mohammed A. Koshnan
The University of Texas
Health Science Center at San
Antonio
7703 Floyd Curl Drive
San Antonio
Texas 78284–7758
USA

Michael W. Lehker
The University of Texas
Health Science Center at San
Antonio
7703 Floyd Curl Drive
San Antonio
Texas 78284–7758
USA

A. D. B. Malcolm
Department of Biochemistry
Charing Cross and Westminster
Medical School
Fulham Palace Road
London W6 8RF
UK

Jim M. McKay
The University of Texas
Health Science Center at San
Antonio
7703 Floyd Curl Drive
San Antonio
Texas 78284-7758
USA

Marjorie A. Monnickendam
Formerly Institute of
Ophthalmology
Judd Street
London WC14 9QS
UK

Lyn D. Olsen
Laboratory of Mycoplasma
Center for Biologics Evaluation
and Research
FDA
Bethesda
Maryland 20892
USA

Colin D. Porter
Division of Cell and Molecular
Biology
Institute of Child Health
University of London
30 Guildford Street
London WC1N 1EH
UK

Bernard Roizman
Kovler Viral Oncology
Laboratories
The University of Chicago
910 East 58th Street
Chicago
Illinois 60637
USA

Leo M. Schouls
Unit Molecular Microbiology
National Institute of Public
Health and Environmental
Protection
PO Box 1
3720 BA Bilthoven
The Netherlands

H. Steven Seifert
Department of Microbiology –
Immunology
NorthWestern University Medical
School
303 East Chicago Avenue
Chicago
Illinois 60611
USA

David R. Soll
Department of Biology
University of Iowa
Iowa City
Iowa 52242
USA

David Wilks
Department of Medicine
Addenbrooke's Hospital
Cambridge
UK

Preface

The appearance of acquired immune deficiency syndrome in the early 1980s rescued immunology as a science in search of a disease. In the same way we anticipate that molecular biology will secure for the HIV–infected patient, as for those with other sexually transmitted diseases, the means of rapid diagnosis and possibly the basis of a vaccine. The combined immuno-molecular approach is reflected in topics chosen for this book.

The current basic methods are reviewed in scientific but not resource terms. The United Kingdom is perhaps the only country to provide a comprehensive free service for the treatment of sexually transmitted disease and these techniques, applied to diagnosis of venereal disease, could act as a catalyst in transferring molecular biology to general routine microbiology. The modern enlightened infectious disease consultant will grasp these methods and incorporate them into his or her training, so as to enter the next century familiar with current usage.

The concern with HIV and associated opportunistic infections, has led to overspecialization: no other microbe since the days of leprosy and tuberculosis has had academic departments, units or even hospitals devoted to it. However, the socio-medical motivation for these developments should not obscure the study of other venereal diseases, many of which occur concurrently with sexually transmitted retroviral infections. It is to revitalize interest in those other sexually transmitted diseases, now relatively neglected, that we have directed the major portion of the book, particularly the application of cell and molecular biology to the understanding of their pathogenesis, diagnosis and immunological treatment.

Pride of place must go to the bacterial diseases, syphilis, gonorrhoea and chlamydial infections but mycoplasma has now come of age. These are becoming increasingly important, especially as a possible co-factor in inducing AIDS. *Haemophilus (in sensu strictu Pasteurella) ducreyi* and *donovanosis* have not been discussed since knowledge of their molecular biology is still, at present, too rudimentary to merit more than a paragraph or two.

The selection of viral diseases has been made on the basis of lack of recent definitive reviews. We have therefore included sections on herpes, molluscum contagiosum and hepatitis B infection. Other primary liver viruses have been omitted as their sexual transmission as a major cause of spread, remains speculative. Discussion of HIV infection is excluded since specialist works are now appearing on the subject. However, there is a place for research which proposes a novel pathogenesis of AIDS, illustrating immunology at a molecular level. Similarly there is no chapter on the molecular biology of papilloma virus infection and cervical carcinoma. This topic has become a molecular subspeciality and merits a distinct treatise and not just a short review, which is all that could be allotted if the balance of the diverse subjects in this book was not to be disturbed.

A chapter on *Candida* is included with concentration on modern typing, a prerequisite for the molecular analysis which could lead to better understanding of the role of specific *Candida* species in human disease. The section on trichomonads presents variation as part of a general biological phenomenon.

David Wright and Leonard Archard
Charing Cross Westminster Medical School
1991

Introduction: Detection and diagnosis of sexually transmitted diseases by molecular biological techniques •

ALAN D.B. MALCOLM

Although venereal diseases such as syphilis and gonorrhoea have diminished significantly in developing countries, they have been replaced by others such as HIV and genital warts infection, chlamydia, etc. (Näher *et al.*, 1989, Havlichek *et al.*, 1990). Of even greater concern is that strains of gonorrhoea with increased resistance to antibiotics such as penicillin, are now emerging (Easmon, 1990; Jacoby and Archer, 1991). Although barrier methods of contraception offer significant protection against the spread of such diseases, it is clear that early diagnosis and mass screening for many of these agents may offer the best hope of controlling their spread. It is therefore crucial to consider the possible assays by which such widespread analyses could be established.

Historically, there have been three widely used techniques for the identification of strains of micro-organisms. Various agents secrete different enzymes into their growth medium and an enzyme profile can often be used to provide information about the nature of the infection (Howard *et al.*, 1986). Tests based on the antigenicity of surface markers of micro-organisms, employing radioimmunoassay or ELISA (Williams *et al.*, 1985), have been prominent recently because of the simplicity of the test and the development of automation. The classical method of identifying an organism's characteristics by growth on selective media is slow and waning in popularity, although it is potentially capable of detecting a single organism.

This introduction highlights how nucleic acid-based methods may be applied to the investigation of sexually transmitted diseases. There are several fundamental reasons why the use of nucleic acids is likely to dominate laboratory tests over the next few years (Desselberger, 1990; Mayrand *et al.*, 1990; Bains, 1991). Although it may seem trivial that we know the general structure of DNA and that such knowledge is limited to a very small number of enzymes and other proteins, this knowledge has allowed us to define clearly the simple base pairing rules of recognition by which one strand of a nucleic acid is able to recognize another. The result is that knowing the sequence and structure of one strand of a nucleic acid molecule allows us to predict the sequence of the complementary strand which will recognize and bind specifically to the original molecule. We are much less familiar with the rules governing interactions between enzyme and substrate or between antibody and antigen and are not yet in a position to make any such generalized predictions about these molecules.

Another useful aspect of working with nucleic acids is that they have similar structures and therefore follow a stereotypic behaviour. Nucleic acid molecules are long, thin, negatively-charged cylinders whose behaviour varies only according to the length (molecular weight) and nucleotide of the individual molecule sequence. A consequence is that the range of laboratory techniques which the practitioner needs to master is very straightforward: having learnt how to handle one nucleic acid molecule, the scientist is able to handle them all. Nucleic acids can be fractionated readily, either by electrophoresis through an appropriate medium (usually polyacrylamide or agarose), by ion exchange chromatographic methods or, more recently, by fast protein liquid chromatography or high pressure liquid chromatography. Painfully this is not the case with enzymes and proteins, all of which have individual pH optima, temperature stabilities, etc., rendering such skills close to an art form.

DNA is physically a robust molecule and can stand temperatures up to about 80°C without denaturation. Even more importantly, denaturation, when it occurs, is reversible on lowering the temperature and this is the basis for hybridization technology. Proteins and enzymes frequently lose their biological activity at temperatures significantly below 80°C, and very rarely regain it on lowering the temperature again. DNA is also chemically robust being fairly resistant to alkali and depurinating only slowly in dilute acid. Proteins are much less tolerant of such changes in their chemical environment.

The stability of a double helix of nucleic acids is influenced by temperature (high temperature breaking the hydrogen bonds between the strands) and by salt concentration. This arises because the two negative strands of nucleic acids repel each other electrostatically. Inside a cell, the organism overcomes this repulsion by coating the nucleic acid with positively-charged proteins or polyamines: in the laboratory the biochemist is able to manipulate this repulsion by changing the salt concentration as high salt reduces the electrostatic repulsion between the two DNA strands. This combination of temperature and salt concentration in a hybridization experiment is referred to as stringency and manipulation of this allows us to vary the stability of the interaction between reagent and analyte virtually at will. In contrast, it is not possible to vary the binding constant of substrate to enzyme or the affinity constant between antigen and antibody in a predictable manner.

The organisms requiring investigation in an STD clinic have DNA or RNA as their genetic material, and therefore can all be investigated by the same technique. Variation in stringency allows us to detect minute differences between different strains of an organism. As little as a single base pair change can be detected in what may be many millions of base pairs in their genomes. Variations in the micro-organism can be characterized by using allele-specific oligonucleotide hybridization in which a single base change can be detected by the change in the stringency under which a complementary oligonucleotide will hybridize to the particular region of the genome: this has been used to identify the origin of AZT-resistant mutations in HIV.

Finally, DNA replicates itself and repairs itself *in vivo* using the enzyme DNA polymerase. This has been available in purified form for many years, enables nucleic acid molecules to be sequenced readily (Sanger *et al.*, 1977; Sanger and Coulson, 1978) and recently has allowed the development of the polymerase chain reaction (PCR) (Saiki *et al.*, 1986; Saiki *et al.*, 1988). It is this latter technique which virtually guarantees that nucleic acid-based technology will take over from conventional methods in the next few years.

The polymerase chain reaction, examples of which will be seen in several of the following chapters, allows us to amplify specifically a defined nucleic acid sequence from a complex mixture. This is possible even where the molecule under investigation may be present in low concentration. It relies on two synthetic oligonucleotides (between 15 and 30 bases long is usual) which are complementary to opposite strands on the genome of interest and separated by between 100 and 1000 base pairs: such oligonucleotides can now be prepared on automated equipment. After denaturation of the DNA under investigation, these oligonucleotides hybridize to the two strands and act as primers for DNA polymerase to synthesize the sequence complementary to the template strand. Because the DNA polymerase is capable of travelling in only one direction along the DNA molecule, the only sequence to be replicated on every single cycle is the sequence between the two oligonucleotides. Each cycle doubles the number of copies of the sequence and the heating and cooling cycle takes approximately ten minutes. Thus, 20 cycles can be achieved in about three hours, leading to a predicted million-fold amplification. The DNA polymerase is able to introduce into the copied DNA either ^{32}P-labelled nucleotides or biotin-labelled nucleotides (Syvänen *et al.*, 1988). In the first case, the product can be identified by electrophoresis and autoradiography: in the second, the biotin-labelled product can be separated on streptavidin columns or beads and identified either by radioactive techniques or, increasingly, non-radioactive techniques involving colour reactions or luminescence. The polymerase chain reaction is automated and can be performed easily on 100 samples at a time. The separation of the product can be performed on a semi-automatic basis using conventional equipment. Subsequent detection depends on the method of labelling, but again technology exists to handle up to 100 samples at a time. PCR has already been used to detect papillomaviruses in cervical mucus

(Tidy *et al.*, 1989; Ward *et al.*, 1990) and HIV in blood samples (Laure *et al.*, 1988; Loche and Mach, 1988).

The polymerase chain reaction is so sensitive that even when it gives accurate results, the level of 'infection' detected may be too low to be of clinical significance. It seems likely that many body orifices (nasal passages, ears, mouth, throat and the ano-genital region) are a more or less permanent reservoir of sub-clinical infections by viruses and micro-organisms. Clearly it will take time to identify appropriate cut-off levels above which clinical intervention is required.

As the sensitivity for the polymerase chain reaction is so great, questions have arisen about the possibility of contamination giving rise to false positive results. Possible contaminants include plasmid vectors and previous PCR products: the sources of these artefacts include pipettes, fingers, reagents, other samples, centrifuges and other equipment, together with the possibility of samples becoming contaminated in the original consulting room. Scrupulous laboratory technique including physical separation of samples, the use of disposable pipettes and pipette tips, aliquoting of reagents before the start of a series of assays and a good choice of both positive and negative controls can help to allay some of these fears. Additionally, techniques have been developed using nested PCR and anti-contamination primers (Finkler *et al.*, 1990; Porter-Jordan *et al.*, 1990; Williamson and Rybicki, 1991) which control many of these potential artefacts.

The barrier to the introduction of these powerful tests is the lack of trained personnel, required both to perform the tests and to interpret the results in a manner which will benefit the patient. It is essential that doctors and pathology laboratory staff become familiar with these techniques as well as the theoretical aspects of the subject. It is hoped that this volume goes some way to fulfil this aim.

References

Bains, W. (1991) Simplified format for DNA probe-based tests. *Clin. Chem.*, 37, 248–53.

Desselberger, U. (1990) Molecular techniques in the diagnosis of human infectious diseases. *Genitourin. Med.*, 66, 313–23.

Easmon, C.S.F. (1990) The changing pattern of antibiotic resistance of *Neisseria gonorrhoea. Genitourin. Med.*, 66, 55–6.

Finkler, E.-B., Pfister, H. and Girardi, F. (1990) Anti-contamination primers to improve specificity of polymerase chain reaction in human papillomavirus screening. *Lancet*, 335, 1289–90.

Havlichek, D.H., Mauck, C., Mummaw, N.L., Moorer, G., Rajan, S.J. and Mushahwar, I.K. (1990) Comparison of chlamydial culture with chlamydiazyme assay during erythromycin treatment of chlamydia genital infections. *Sexually Transmitted Diseases*, 17, 48–50.

Howard, L.V. Coleman, P.F., England, B.J. and Herrmann, J.E. (1986) Evaluation of Chlamydiazyme for the detection of genital infections caused by *Chlamydia trachomatis. J. Clin. Microbiol.*, 23, 329.

Jacoby, G.A. and Archer, G.L. (1991) New mechanisms of bacterial resistance to antimicrobial agents. *New Engl. J. Med.*, 324, 601–12.

Laure, F., Rouziox, C., Veber, F., Jacomet, C., Courgnaud, V. *et al.* (1988) Detection of HIV1 DNA in infants and children by means of the polymerase chain reaction. *Lancet*, ii, 538–40.

Loche, M. and Mach, B. (1988) Identification of HIV-infected seronegative individuals by a direct diagnostic test based on hybridisation to amplified viral DNA. *Lancet*, ii, 418–21.

Mayrand, P.E., Hoff, L.B., McBride, L.J., Bridgham, J.A., Cathcart, R. *et al.* (1990) Automation of specific human gene detection. *Clin. Chem.*, 36, 2063–71.

Näher, H., Niebauer, B., Hartmann, M., Söltz-Szöts, J. and Petzoldt, D. (1989) Evaluation of a radioactive rRNA:cDNA-hybridisation assay for the direct detection of *Chlamydia trachomatis* in urogenital specimens. *Genitourin. Med.*, 65, 319–22.

Porter-Jordan, K., Rosenberg, E.I., Keiser, J.F., Gross, J.D, Ross, A.M. *et al.* (1990) Nested polymerase chain reaction assay for the detection of cytomegalovirus overcomes false positives caused by contamination with fragmented DNA. *J. Med. Virol.*, 30, 85–91.

Saiki, R.K. Bugawan, T.L., Horn, G.T., Mullis, K.B. and Erlich, H.A. (1986) Analysis of enzymatically amplified β-globin and HLA-DQ DNA with allele-specific oligonucleotide probes. *Nature*, 324, 163–6.

Saiki, R.K., Gelfand, D.H., Stoffel, S., Scharf, S.J., Higuchi, R. *et al.* (1988) Primer-directed enzymatic amplification of DNA with a thermostable DNA polymerase. *Science*, **239**, 487–91.

Sanger, F., Niklen S. and Coulson A.R. (1977) DNA sequencing with chain terminating inhibitors. *Proc. Natl. Acad. Sci. USA*, **74**, 5463–7.

Sanger, F. and Coulson, A.R. (1978) The use of acrylamide gels for DNA sequencing *FEBS Lett.*, **87**, 107–10.

Syvänen, A.-C., Bengtström, M., Tenhunen, J. and Söderlund, H. (1988) Quantification of polymerase chain reaction products by affinity-based hybrid collection. *Nucleic Acid Res.*, **23**, 11327–38.

Tidy, J.A., Parry, G.C.N., Ward, P., Coleman, D.V., Peto, J. *et al.* (1989) High rate of human papillomavirus type 16 infection in cytologically normal cervices. *Lancet, ii*, 434.

Ward, P., Parry, G.C.N., Yule, R., Coleman, D.V. and Malcolm, A.D.B. (1990) Comparison between the polymerase chain reaction and slot blot hybridization for the detection of HPV sequences in cervical scrapes. *Cytopathology*, **1**, 19–23.

Williams, T., Maniar, A.C., Brunham, R.C. and Hammond, G.W. (1985) Identification of *Chlamydia trachomatis* by direct immunofluorescence applied in specimens originating in remote areas. *J. Clin. Microbiol.*, **22**, 1053.

Williamson, A.-L. and Rybicki, E.P. (1991) Detection of genital human papillomaviruses by polymerase chain reaction amplification with degenerate nested primers. *J. Med. Virol.*, **33**, 165–71.

Molecular mechanisms of antigenic variation in *Neisseria gonorrhoeae*

H. STEVEN SEIFERT

1.1 INTRODUCTION

Neisseria gonorrhoeae is the causative agent of the sexually transmitted disease gonorrhoea. Gonorrhoea has been recorded in man throughout the medieval and modern eras and possibly since biblical times (Cheng, 1988). This long association between *N. gonorrhoeae* and humans can be attributed to: (1) an efficient mode of transmission; (2) a high percentage of asymptomatic carriers; (3) the rare occurrence of lethality, and (4) the ability of the bacterium to avoid an immune response. This Gram-negative diplococcus (the gonococcus) does not colonize any other ecological niche besides the human body. A consequence of this limited host range is a lack of a valid animal model. Therefore, a majority of facts concerning gonococcal pathogenesis result from *in vitro* studies, supported by information gained by occasional experiments in human volunteers. The focus of this chapter is to examine mechanisms used by the bacterium to evade the immune response. Selected topics on other aspects of gonococcal pathogenesis will be presented but for wider analysis the reader is referred to review articles about gonorrhoea by Cannon and Sparling (1984) or Brooks and Donegan (1985) and specifically about antigenic variation by Seifert and So (1988), Swanson and Koomey (1989), or Meyer *et. al.* (1990).

The most closely related bacterium to *N. gonorrhoeae* is the Gram-negative diplococcus, *N. meningitides* (the meningococcus). The only notable microbiological differences between these two organisms are the presence of a polysaccharide capsule on the meningococcus and its utilization of sugars other than glucose. Despite the similarities in gross structure and surface proteins, these organisms cause very different diseases. The meningococcus

Molecular and Cell Biology of Sexually Transmitted Diseases
Edited by D. Wright and L. Archard
Published in 1992 by Chapman and Hall, London ISBN 0 412 36510 3

normally colonizes the upper respiratory tract. The antiphagocytic properties of the meningococcal capsule allows this micro-organism to survive in the blood stream with resulting bacteraemia. This persistence within the blood stream in turn can provide a greater chance of access to the central nervous system, with the possibility of meningitis. Therefore, the presence of the capsule on the meningococcus potentiates its ability to infect sites distinct from the area of initial colonization. As a sexually transmitted disease agent, the gonococcus normally colonizes the genital tract. Although the gonococcus can infect the nasal pharynx, prolonged colonization by the gonococcus and continued transmission to this site is not usually found. Other virulence factors specific to each organism need to be discovered before the separate virulence properties of each pathogen can be explained. This chapter will concentrate on *N. gonorrhoeae* pathogenesis but all of the molecules discussed, are expressed on both organisms and each are regulated in a similar way.

1.2 GONOCOCCAL PATHOGENESIS

The human immune system is a formidable predator of micro-organisms and the gonococcus has evolved a number of mechanisms to circumvent it. When considering the degree of virulence of an infectious agent, growth within the host is of major importance. The availability of host iron in a form metabolizable to the bacteria is potentially growth limiting. Most iron in mammals is complexed with transferrin or lactoferrin and unavailable for use by micro-organisms. Many bacterial pathogens produce low molecular weight organic compounds, called siderophores, that compete efficiently with transferrin and lactoferrin for iron and deliver it to the bacterial cell. Although the gonococcus does not produce soluble siderophores, it expresses high affinity receptors for transferrin and lactoferrin on its cell surface (Norrod and Williams, 1978; West and Sparling, 1985). Iron needed for growth is released from bound transferrin or lactoferrin at the cell surface. That pathogenic Neisseriae can bind human transferrin and lactoferrin, but not homologous proteins isolated from other mammalian species, may account for part of their host range limitation (Lee and Schryvers, 1988).

Gonorrhoea is presented differently in males and females. Infection of the male urethra often results in painful urination and a purulent discharge from the penis but physical damage to the genital epithelium rarely results from infection. A low level of asymptomatic carriers (1–7%) assures treatment for a majority of infected males (Donegan, 1985). In contrast, females are often asymptomatic (40–60%), and can develop pelvic inflammatory disease (PID) following tissue damage to the uterus or fallopian tube epithelium. Tissue damage in PID may result from cell invasion by the gonococcus or by the action of toxic substances released by the bacterium (Melly *et al.*, 1981). The high numbers of asymptomatic carriers and the risk of PID, make gonorrhoea a potentially more serious disease in females than males.

In a small subset of the infected population, the bacterium leaves the genital tract and becomes systemic. Disseminated gonococcal infection (DGI) relates to a specific set of phenotypic and genotypic alterations in the bacterium. These include resistance to the bactericidal effects of human serum (Schoolnik *et al.*, 1976) and expression of specific auxotrophic mutations (Knapp and Holmes, 1975). A causal relationship between mutations in amino acid metabolism and development of DGI, would seem to be tenuous. Instead, the auxotrophic mutations are probably carried by a subtype of gonococcal strains that have a greater capacity to become systemic. The most serious complication of DGI is bacterial arthritis following colonization of synovial joints by the bacterium.

1.3 DIAGNOSIS AND TREATMENT OF GONORRHOEA

Positive diagnosis of *N. gonorrhoeae* infection relies on the cultivation of viable organisms from infected sites. Patients who show symptoms of gonorrhoea or have infected sexual partners will often be tested for gonorrhoea. A swab of the urethra (and the cervix in females) is taken for Gram-staining to detect gram-negative diplococci associated with polymorphonuclear leukocytes. The absence of diplococci by gram-staining is not sufficient to rule out infection. Therefore, the genital swab material is cultured routinely on gonococcal growth media that selects against other normal genital flora. The most common media (Thayer-Martin media) contains vancomycin, colistin and nystatin to select against other Gram-negative bacteria, Gram-positive bacteria and yeasts, respectively. Confirmation of gonococcal infection is the growth of Gram-negative diplococci that are oxidase positive and can only utilize glucose as a carbon source.

Once a positive diagnosis of gonorrhoea is confirmed, treatment regimens are fairly straightforward. Many clinicians still use high dosages of benzyl penicillin to treat gonorrhoea but the increasing prevalence of penicillinase producing *N. gonorrhoeae* (PPNG) makes this a less useful treatment. The most common alternate treatment regimens are tetracycline, spectinomycin or amoxicillin. The recent acquisition of tetracycline resistance on a conjugal plasmid by the Neisseriae suggests that tetracycline may soon be of diminished usefulness (Morse *et al.*, 1986). The continued selection of antibiotic-resistant strains supports the need to develop a vaccine to combat this disease.

1.4 THE SURFACE OF *N. GONORRHOEAE*

Most of the research into gonococcal pathogenesis has focused on molecules and structures expressed on the cell surface, since these molecules mediate many interactions with the host. The functions encoded by these molecules include: adhesion; adaptation to changing environments and avoidance of the immune system. The underlying concept of this chapter is that the gonococcus has evolved complex systems for changing its surface that can affect all these functions. We will examine how different forms of the gonococcal pilus and protein II (PII) are expressed and how this influences pathogenicity. Before mechanisms of antigenic variation can be examined, an overview of the gonococcal cell surface is required.

There are three major outer membrane proteins that are known to contribute to gonococcal pathogenesis; protein I (PI), protein II (PII) and protein III (PIII). The PI proteins are also defined as porin proteins (POR), while the PII proteins are also designated as opacity proteins (OPa). The properties of the PII protein will be discussed in section 1.8 PI functions as a porin allowing small molecules passage through the outer membrane, between the exterior of the cell and the periplasm. Two major forms of this protein exist in nature, PIA and PIB. The presence of each PI type and the antigenic subtype of the expressed protein constitute the basis of serotyping based on panels of monoclonal antibodies that define individual serovars of each protein (Bygdeman *et al.*, 1983; Knapp *et al.*, 1984). Although antigenic heterogeneity exists within the natural population, enough similarity exists between different PI types to suggest that a vaccine could be developed to elicit bactericidal antibodies to PI. However, the PIII protein is closely associated with PI in the outer membrane and co-purifies with PI (McDade and Johnston, 1980; Lytton and Blake, 1986). PIII antibodies formed against the contaminating protein block the bactericidal activity of antibodies against PI (Joiner *et al.*, 1985). Recently, a PIII⁻ mutant has been developed that may induce sufficient PI antibodies to prevent disease without generating blocking antibodies (Wetzler *et al.*, 1989). There are several other outer membrane-associated proteins that may also contribute to pathogenicity but no role in virulence has been established for these proteins.

The outer leaflet of the outer membrane is composed of lipooligosaccharide (LOS). This structure differs from the lipopolysaccharide (LPS) of enteric bacteria by the lack of a somatic O-antigen. LOS is antigenically variable between strains and may contribute to pathogenesis. The molecular basis of LOS variation has not been determined. LOS has an important role in pathogenesis besides its structural role in the outer membrane. Like other lipid A containing molecules, gonococcal LOS may act as an endotoxin, stimulating production of cytokines in mammalian cells (Gregg *et al.*, 1981). This endotoxin activity of LOS contributes to the development of gonococcal arthritis during a disseminated infection (DGI) and may act to induce epithelial cell damage that results in PID. Several studies have demonstrated that the effect of LOS on eukaryotic cells is enhanced when expression of cell surface adhesins (pili or PII) concentrates the bacterium near epithelia (McGee *et al.*, 1981; Virji and Everson, 1981).

An interesting facet of gonococcal outer membrane structure is the formation of outer membrane blebs (Dorward and Garon, 1989). During growth of the gonococcus, small outer membrane vesicles, carrying the normal complement of outer membrane proteins, bleb out of the cell surface. In theory it is possible that blebs act as antigen decoys absorbing antibodies directed against outer membrane constituents and preventing reactivity to the mother cell. In addition, a role for blebs in transmitting the effects of LOS to eukaryotic cells cannot be discounted, since the effect of LOS occurs even when bacteria are not attached to the damaged cells. Finally, blebs mediate plasmid transfer between gonococcal cells (Dorward *et al.*, 1989). This finding suggests that blebs contribute to the spread of antibiotic resistance, and may also mediate other genetic exchanges between cells.

1.5 IMMUNE RESPONSES IN GONORRHOEA

Patients with uncomplicated gonococcal infection display a wide range of antibody reactivities against the bacterium. Serum antibodies against the invading organism are often produced, but their efficacy in combating disease is doubtful considering the mucosal location of the bacterium. However, the presence of serum antibodies may limit disseminated infections by stimulating complement fixation. Strains that cause DGI demonstrate high levels of resistance to the bactericidal effects of serum. At the surface of the genital mucosa, secretory IgA would be expected to have the greatest effect in preventing colonization, and IgA reactive to most outer membrane antigens can be detected in vaginal secretions (Tramont, 1977). Gonococci express an extracellular protease that cleaves IgA1 antibody and renders it inactive (Plaut *et al.*, 1975). Why the bacterium produces a protease that only recognizes IgA1 is a puzzle since equivalent or larger amounts of IgA2, against which the protease is inactive, is produced at the mucosa. One possibility is that different amounts of IgA1 and IgA2 are produced against individual gonococcal antigens and the protease selectively inactivates antibody directed against these targets. Only if most neutralizing or bactericidal antibodies were of the IgA1 subclass would the protease be efficient in enhancing disease.

The importance of a humoral immune response in combating gonococcal infections is well established but the role of specific cellular immune responses is less certain. Macrophages ingest and effectively kill gonococci, but the interactions of gonococci with neutrophils have been a source of considerable contention. Reports differ as to how well gonococci survive phagocytosis by neutrophils (Thomas *et al.*, 1973; Densen and Mandell, 1978; Casey *et al.*, 1979; Rest *et al.*, 1982; Britigan and Cohen, 1985; Casey *et al.*, 1986). Piliated gonococci may resist phagocytosis but the ability to survive phagocytosis by neutrophils, relates more closely to PII expression (reviewed in Rest and Shafer, 1989). The mechanisms for this differential survival are unknown but the recognition of specific receptors on the surface of neutrophils by some PII proteins may contribute to the difference. Cells expressing this one subset of PII proteins may remain bound to the surface of the polymorphs or, alternatively, may be directed to a different cellular compartment than those cells not expressing PIIs or expressing other PII variants.

The human immune system is efficient at recognizing gonococcal antigens and has several strategies to kill the organism or inhibit infection. The molecular mechanisms used by *N. gonorrhoeae* to change its surface and avoid attack by the immune system will now be described.

1.6 MECHANISMS OF ANTIGENIC VARIATION

Antigenic variation can be defined as a change in antigenicity in a molecule expressed on the surface of an organism that occurs at a rate much higher than the normal mutation rate (Seifert and So, 1988). The term has been applied to a wide variety of processes including antigenic variation, phenotypic variation and antigenic drift. A broad definition of antigenic variation encompasses any mutational event that alters the antigenic property of a molecule. Assuming that mutation and natural selection are required processes for all biological systems, this definition includes all evolution when natural selection is provided by an immune system. This definition is too broad to be useful. In this work antigenic variation describes systems that contain the following properties. (1) Antigenic changes that occur at a frequency considerably higher than the normal mutation rate. (2) Changes within a domain of a surface-exposed molecule that is recognized by an immune system. (3) Antigenic changes that can occur multiple times to provide a panel of antigenic types. Well defined systems that fall within this definition are neisserial PII and pili, borrelial VMP variation (Barbour, 1990), and antigenic variation of trypanosome variable surface glycoproteins (Van der Ploeg, 1987).

Before examining gonococcal antigenic variation, consideration of the relevance of phase variation is appropriate. Phase variation is a reversible change of phenotype that varies between two states or phases. The phases usually result by changing the expression of a single protein. Expression can switch between on or off phases, or between two alternate phases. Both gonococcal pili and PII undergo reversible on/off phase variation. In both cases, phase variation results from mechanisms similar or identical to those controlling antigenic variation. Other well understood systems of phase variation include *E. coli* type I pili (on/off), *Salmonella* flagellin switch (H1/H2) and *Moraxella bovis* pili variation (α/β). Each of these systems relies on an invertible DNA segment to vary the expression of the different phases, a mechanism which is quite different from the mechanisms used to

6

produce gonococcal PII and pilus phase variation (reviewed by Glasgow *et al.*, 1989).

The gonococcal pilus and PII phase variation produces four colony morphotypes. T1 and T2 colonies usually express pili on their surface (Swanson *et al.*, 1971) and T3 and T4 are usually nonpiliated. The T1/T2 colonies are smaller than T3/T4 colonies, have more distinct edges and are more raised or domed than the T3/T4 colonies (Kellogg *et al.*, 1963). The T1 and T3 phenotypes represent PII+ clones while the T2 and T4 are often PII⁻ '(James and Swanson, 1978). The T1/T3 (PII+) colonies are opaque demonstrating a refractile phenotype when transmitting light, while the T2/T4 (PII⁻) colonies are translucent. Each of these four colonial types can produce the other morphotypes by pilus and/or PII phase variation.

1.7 PILI

Pili are long proteinacious structures (also called fimbriae) found on most Gram-negative bacteria that mediate bacterial adherence. The primary role of pili in attachment is to overcome the electrostatic repulsion between cell surfaces by placing specific adhesins away from the bacterial cell surface. These adhesins can be accessory proteins contained within the pilus fibre, or may be the major subunit, pilin. The only known component of the gonococcal pilus is pilin, although other proteins have been suggested to be copolymerized within the pilus fibre (Muir *et al.*, 1988; Johnson *et al.*, 1991).

The gonococcal pilin protein exhibits variations in molecular weight from 16 to 24 Kd (Lambden, 1982). Amino acid sequencing of variant gonococcal pilins has shown that the N-terminal portion of the protein is constant among strains, while the C-terminal portion of the molecule varies. There is also N-terminal sequence homology between pilins from *N. gonorrhoeae*; *N. meningitidis*; *Moraxella bovis*; *Bacteroides nodosus*; and *Pseudomonas aeruginosa* (Hermodson *et al.*, 1978; McKern *et al.*, 1983; Sastry *et al.*, 1983; Marrs *et al.*, 1986). These pilins form a family that share an unusual 6 or 7 amino acid leader sequence and a methyl-phenylalanine as the first residue in the mature protein, called N-met-phe or Type IV pili. All of these pilin proteins are produced as a preprotein that carries a six or seven amino acid leader peptide that is cleaved during pilus assembly.

1.7.1 Pilus functions

The gonococcal pilus has been the focus of many studies concerning gonococcal pathogenesis. The lack of a valid animal model for the disease has made it necessary to address most questions concerning gonococcal pathogenicity *in vitro*. However, the pioneering studies of Kellogg and coworkers (1963, 1968) linking changes in colonial morphology (caused by changes in piliation) with virulence of the gonococcus in laboratory volunteers, were instrumental in demonstrating the essential role of the pilus in virulence. More recent human volunteer studies conducted by the US Army and collaborators supported the Kellogg findings linking piliation and virulence (Swanson *et al.*, 1987). While many other virulence factors are believed to contribute to pathogenicity, the pilus is the only determinant that has been proven to be essential for virulence.

The pilus mediates adherence of the bacterium to a variety of eukaryotic cells *in vitro* (reviewed in Stephans, 1989). These target cells include: nonciliated cells of the columnar epithelium in the fallopian tube organ culture model, a variety of cell types in tissue culture (including: HEC-I-B cells, HEp2 cells, HeLa cells, cervical cells, buccal epithelial cells, and Chang conjunctival cells) and human erythrocytes. Adherence mediated by the pilus appears to be the first step in cell and tissue invasion. Pilus mediated adhesion may also localize the bacterial cells near the epithelium where diffusible toxic substances (such as LOS) can promote inflammation and disrupt the mucosa. Pilus antibody blocks the adherence of gonococci in all of these systems, if the antiserum is raised to the homologous pilus type (Virji *et al.*, 1982; Tramont, 1977, Ashton *et al.*, 1977). Pilin specific monoclonal antibodies can also block adherence supporting a direct role for pilin in adherence (Virji and Heckels, 1984; Nicolson *et al.*, 1987). In addition, a single gonococcal strain expressing different pilin proteins binds to eukaryotic cells with different affinities (Lambden *et al.*, 1980). Although the relative contribution of minor constituents of the pilus to adhesion remains to be determined, the effect of pilin antigenic variation on binding supports a direct role for pilin.

1.7.2 Genetic organization of pilin loci

The pilin protein is produced from a locus called *pilE* or the expression locus (Figure 1.1). The strain that yielded the first pilin clones (MS11) carries two *pilE* loci; *pilE1* and *pilE2* (Meyer *et al.*, 1982; Meyer *et al.*, 1984). It is now generally accepted that the existence of two expression loci in MS11 was inadvertently selected by isolating colonies with stable piliation phenotypes during passage. Unless a clinical isolate is found that contains two expression loci prior to *in vitro* passage, the existence of two expression loci in MS11 will only be historically significant. Comparison of variant *pilE* gene sequences provides an overall picture of the pilin protein in which the first third carries constant amino acid sequences, the middle third contains semivariable amino acid sequences and the last third encodes hypervariable sequences (Figure 1(a): Hagblom *et al.*, 1985; Segal *et al.*, 1986). This is a generalized picture of the pilin protein since, conserved as well as variable sequences are contained in the last two thirds of the protein (Figure 1(a)).

8

Other chromosomal pilin loci, designated *pilS*, are silent and encode partial gene segments carrying the two thirds of the coding region, representing the carboxy terminal portion of the protein (Figures 1.1 and 1.2). These loci have from one to six copies of variant pilin genes. Silent gene copies start at either pilin amino acid 29, 31 or 44. None of the silent copies encode the amino terminal amino acids, nor do they encode any transcriptional or translation start signals. However, each silent copy carries carboxy-terminal codons corresponding to the last or next to last amino acid residue of pilin. Both the silent partial gene segments and the expression locus contain both variable and conserved sequences (Figures 1.1 and 1.2).

1.7.3 Mechanisms of pilin antigenic variation

Pilin antigenic variation occurs by homologous DNA recombination reactions between pilin loci (Koomey *et al.*, 1987). The two pilin loci interact, with one gene (the silent locus) donating DNA sequences to the other gene (the expression locus). A cartoon of this gene conversion is represented in Figure 1.2. Usually only part of a silent copy sequence is transferred into

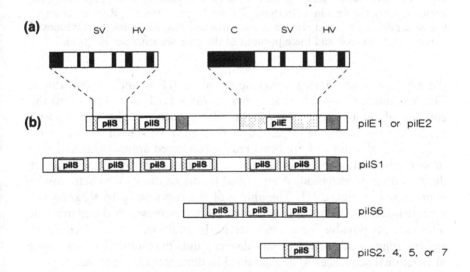

Figure 1.1 Diagram showing pilin gene structure and chromosomal organization of expressed and silent pilin genes (adapted from Haas *et al.*, 1992). (a) Generalized representations of a silent and expressed pilin gene showing the locations of the constant (C), semivariable (SV) and hypervariable (HV) regions of the pilin protein. Also shown are the patchwork of constant sequences (black boxes) and variable sequences (open boxes); (b) Chromosomal organization of pilin loci. Each dotted box represents a silent (pilS) or expressed (pilE) gene. The conserved Sma/Cla repeat present in each locus is shown by the boxes with diagonal lines (Meyer *et al.*, 1984).

9

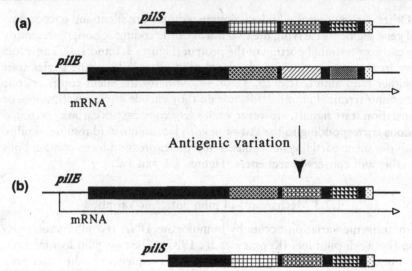

Figure 1.2 Diagram showing a simplified mini-cassette model of antigenic variation. (a) A silent (*pilS*) and expressed (*pilE*) gene copy prior to antigenic variation. Patterned boxes represent different variable amino acid sequences that form mini-cassettes. Black boxes represent conserved sequences. This illustration only depicts three of the six mini-cassettes (Haas and Meyer, 1986); (b) Representation of the same *pilE* and *pilS* loci after antigenic variation. Two mini-cassettes have moved during recombination and these portions of the gene sequence are identical.

the expression site during antigenic variation (Haas and Meyer, 1986). This recombination of short stretches of variant sequence has led to the concept of mini-cassettes of variant pilin sequences. The mini-cassette model derives from the observation that each stretch of variant pilin sequence in all pilin loci are bordered by conserved amino acids and that these mini-cassettes of variant amino acid sequence always move together during antigenic variation. A simplified model of the mini-cassette model is presented in Figure 1.2. The ability of the gonococcus to separate out variant information into mini-cassettes allows increases in the number of pilin variants possible for a single strain. In addition, the arrangement of variant amino acid sequences into discrete units may allow different types of functional sequences to be contained in different mini-cassettes.

A result of antigenic variation is the formation of pilin proteins that are nonfunctional or only partially functional. The production of disabled pilin proteins by antigenic variation results in OFF/ON phase variation. A switch off can result from recombination reactions that produce a nonfunctional pilin. These nonfunctional pilins include: variants that are inefficiently secreted by normal pilin export mechanisms but instead produce small soluble proteins in the medium, S-pilin (Haas *et al.*, 1987), variants that contain small internal deletions within the pilin coding sequence

(Hill *et al.*, 1990), variants encoding two hypervariable regions that produce overlong pilins, L-pilin (Haas *et al.*, 1987) or variants encoding nonsense codons resulting in truncated proteins (Bergström *et al.*, 1986). Each of these phase variants can produce a piliated form when a new functional variant sequence is introduced by recombination into the expression locus.

Interestingly, the first published mechanism of phase variation was not true phase variation since the nonpiliated variants described could not revert (Segal *et al.*, 1985). These nonreverting variants were formed by deletion of the 5'-portions of the expressed gene by recombination between the variable regions of the expressed gene and one of the upstream silent gene copies (Figure 1.1). The deletions removed the promoter region of the gene and because no other promoter sequence was present in the chromosome, these variants could not revert to a piliated form. However, these nonreverting variants probably occurred by the same mechanism that produces nonpiliated variants encoding small internal deletions within the expressed gene (Hill *et al.*, 1990). In both cases, recombination occurred between short repetitive elements within the pilin coding region. The difference between the products is that the nonreverting deletions form between repeats in the expressed gene and an upstream silent gene copy (removing the promoter), while the reverting deletions form between repeats within the expressed gene. One may speculate that the repetitive elements encoded in all pilin copies are used in recombinations leading to antigenic variation as well as deletions. Since most gonococci isolated from patients are piliated, it is unknown if any aspect of phase variation has a function *in vivo*, or if phase variation merely constitutes an inherent sloppiness in antigenic variation.

The major controversy surrounding the mechanistic details of pilin antigenic variation is whether the donor DNA molecules originate in the same or in a different cell. The first models of pilin antigenic variation (Figure 1.3) suggested that recombination occurs intracellularly between a silent and expression locus (Hagblom *et al.*, 1985; Bergström *et al.*, 1986; Segal *et al.*, 1986; Haas and Meyer, 1986). These models resulted from the analysis of *pilE* and *pilS* loci before and after antigenic variation had occurred. The second model proposed that antigenic variation occurs when chromosomal DNA released from a cell by autolysis, is transformed into a sibling cell, with silent sequences recombining with the expression locus (Norlander *et al.*, 1979; Seifert *et al.*, 1988). This model arose from the constitutive ability of the gonococcus to autolyse (Hebeler and Young, 1975) and efficiently transform mutations between strains (Sparling, 1966; Sarubbi and Sparling, 1974). There are reports that support both the intracellular (Swanson *et al.*, 1990) or extracellular models (Gibbs *et al.*, 1989) of antigenic variation but none of the studies published to date is totally convincing. It is possible that both the intracellular and extracellular transfer of DNA can result in antigenic variation but a better understanding of the control of recombination during antigenic variation will be needed before the relative merit of these models can be established.

1.8 PROTEIN II

The second most abundant protein in the gonococcal outer membrane is the opacity protein or Protein II (PII) or OPa. This family of proteins is biochemically defined by a heat modifiable phenotype. These proteins (averaging about 30 Kd molecular weight) migrate differently in SDS polyacrylamide gels, when solubilized at 37°C or 100°C. Every gonococcal strain can express simultaneously from zero to three forms of the PII protein from 12 *opa* gene copies. The PII protein undergoes both phase and antigenic variation and as is the case with the pilus, phase and antigenic variation are produced by the same molecular mechanism.

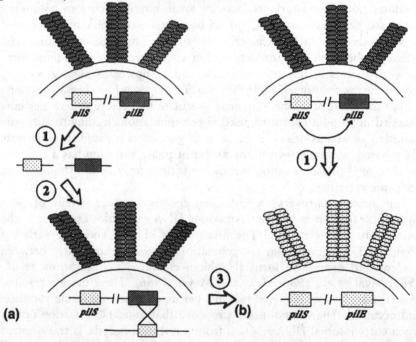

Figure 1.3 Diagram showing two possible mechanisms of pilin antigenic variation. The expressed and silent genes are represented by the dotted or cross-hatched boxes. Pili are shown as idealized structures with the proteins marked similarly to the expressed gene. (a) Antigenic variation by extracellular recombination. (1) A cell in a population autolyses and releases DNA into the medium. (2) A fragment of DNA containing a silent locus is taken up by a sibling cell for transformation. (3) Recombination between the incoming DNA and the expression locus results in a new pilin type; (b) Antigenic variation by intracellular recombination. (1) Nonreciprocal recombination between a silent and expression locus produces a new pilin type.

1.8.1 Functions

The only defined phenotype of PII+ gonococci is to increase adhesion to both eukaryotic cells and other gonococci (Lambden et al., 1979; Virji and Everson, 1981). As mentioned above, expression of most PIIs produces colonies that exhibit a refractile or opaque character when observed through a stereomicroscope. However, not all PII proteins produce an opaque phenotype (Lambden and Heckels, 1979; Swanson, 1982). The leukocyte association protein is defined biochemically and genetically as a PII but produces a translucent colonial phenotype (Rest and Shafer, 1989). The existence of subsets of PII proteins with distinct physical and biological properties has complicated the literature on PII function. It is possible that expression of PIIs also allows the bacterium to respond to other environmental factors but the selection of PIIs in differing environmental conditions has not been shown. Therefore, functionally the PII proteins can be defined as adhesins but other PII functions are likely to emerge from future studies.

1.8.2 Genetic organization

Each PII protein within a strain is expressed from a unique locus (opa). Every opa gene shares a similar overall structure (Figure 1.4). The opa gene structure provides information on how phase and antigenic variation of PIIs occurs (Stern et al., 1986). The promoter regions match the consensus Gram-negative promoter sequence. After a short untranslated region, a ribosome binding site (RBS) precedes an initiating AUG. This is the beginning of the preprotein, since a normal signal sequence directs transport of the protein through the cytoplasmic membrane. Encoded within the signal sequence DNA are a variable number of nucleotide repeats (CTCTT) called the coding repeat (CR). A signal peptidase cleavage site then defines the end of the signal sequence and the first amino acid of the mature protein. When an opa gene is cloned into E. coli, the initiating AUG is always in a different reading frame from the mature protein, even though most of these genes were isolated using monoclonal antibodies directed to the mature protein (section 1.8.3). The mature protein coding sequences are mostly conserved except for two regions called HV1 and HV2. These hypervariable sections of amino acids define the basis for the antigenic difference between PII types.

1.8.3 Mechanisms of PII antigenic variation

The finding that translation initiation signals in every cloned opa gene were out of frame from the protein coding sequence suggested that these genes could not express mature PII. In addition, the different number of CTCTT repeats in each gene indicated that the number of repeats was not fixed.

13

<leader peptide><————————— mature protein —————————————>

Figure 1.4 *Opa* gene structure. The normal Gram-negative promoter sequences are represented by the 'P'. The ribosome binding site (rbs) and initiation codon (AUG) are boxed. The leader peptide is represented by the dotted box showing the variable number of coding repeats $(CTCTT)_N$. The mature protein sequence is represented by the open boxes except for the hypervariable 1 (HV1) and hypervariable 2 (HV2) sequences.

Stern *et al.* (1986) proposed that a change in the number of CTCTT repeats within any gene would align the initiating AUG with the rest of the reading frame and produce the mature protein. When several gonococcal PII phase variants (OFF, ON, OFF, etc.) were screened for the number of CTCTT repeats in their mRNA, the change of PII expression always correlated with a change in CTCTT repeat number (Figure 1.5). Every PII⁺ variant expressed one mRNA species where the initiating AUG was in the same reading frame as the mature protein, while PII⁻ variants had all *opa* loci out of alignment with the initiating AUG. Thus, each *opa* gene is regulated independently at the level of translation by the number of CTCTT repeats. One might expect that if independently regulated *opa* genes would each have the same probability of expressing protein, then strains with multiple PIIs would be common. Although more than one PII protein can be produced in a single culture (Swanson, 1982), this is a rare event. Therefore, either some unknown genetic factor limits the number of PIIs expressed, or the expression of multiple PII types confers a growth disadvantage and is selected against. Finding the genetic factor that regulates PII switching could discriminate between these possibilities.

It was found that homologous recombination could not account for the change in CTCTT repeat numbers during PII phase variation, since a *recA* mutation had no effect on the process (Koomey *et al.*, 1987). Moreover, when PII, alkaline phosphatase fusions were introduced into *E. coli*, the number of CTCTT repeats changed at frequencies similar to those measured in *N. gonorrhoeae* independent of *recA* function (Belland *et al.*, 1989; Murphy *et al.*, 1989). The models proposed from this finding suggest that the change in repeat number occurs during DNA replication, by the formation of an unusual DNA structure (H-DNA). In addition, the production of protein from out of frame genes in *E. coli* was explained by ribosomal frame shifting (Belland *et al.*, 1989). These models conclude that PII phase variation is a random event that occurs continually during growth. This would argue against the existence of a genetic system for regulating PII phase variation and lend support for the idea that multiple PIIs produce a growth disadvantage.

The repertoire of potential PII variants expressed by a single strain is

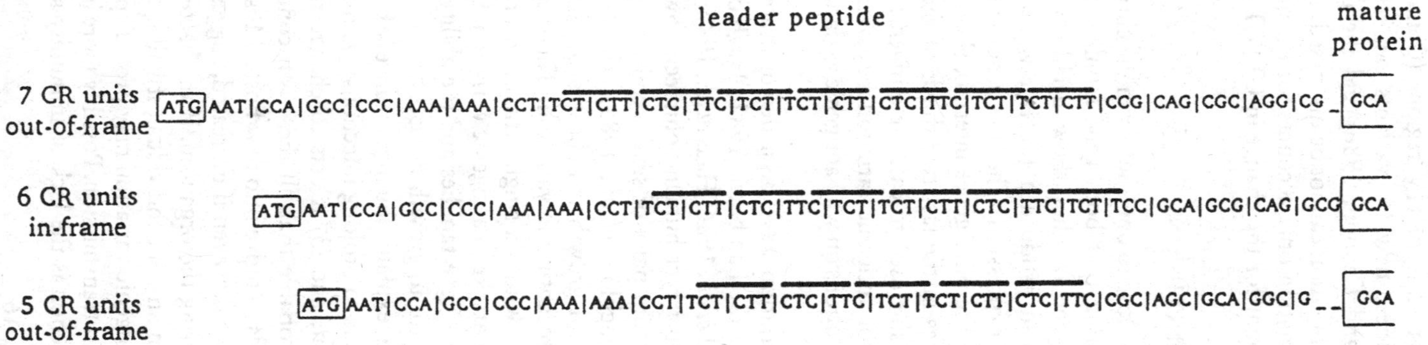

Figure 1.5 Translational control of PII protein expression by the change in coding repeat (CR) unit numbers. The DNA sequence of the leader peptide is shown. The CTCTT repeats are marked by thick lines above the sequence. Only the first codon of the mature protein (GCA) is shown. The reading frame following the initiating ATG is marked by the vertical lines and the missing nucleotides required to produce an in-frame protein are shown by underlined spaces. This figure is adapted from Stern *et al.* (1986).

not fixed. Homologous recombination between PII genes can produce new variants that encode different pairs of HV regions (Connell *et al.*, 1988). As is the case for pilin, these new PII variants can result from DNA transformation (Schwalbe and Cannon, 1986) and the possibility of an intracellular route for recombination cannot be dismissed. It must be stressed that homologous recombination only accounts for the production of new PII genes, and cannot account for the change in CTTCT repeats.

1.9 CONCLUSIONS

Some *opa* and *pil* genes are located close to one another in the chromosome (Stern *et al.*, 1984; 1986) suggesting that both gene families may have coevolved. However, these two gene families show decidedly different characteristics. Every *opa* locus can produce a protein, while only the pilin expression site is capable of expression. The *opa* genes encode two regions of variable sequence while the pilin genes intermix short stretches of conserved and variable sequences in the last two thirds of the protein. Despite the very different mechanisms used to catalyse antigenic variation in these gene families, the reasons for antigenic variation may overlap. This last section will contain some speculations on why antigenic variation occurs in the gonococcus.

Both pilin and PII antigenic variation have been demonstrated during laboratory culture (Virji *et al.*, 1982; Lambden, 1982; Hagblom *et al.*, 1985; Haas and Meyer, 1986; Segal *et al.*, 1986; Haas *et al.*, 1987; Swanson *et al.*, 1990), with bacteria used to infect human volunteers (Swanson *et al.*, 1987) and in strains isolated from pairs of sexual consorts (Tramont *et al.*, 1979; Zak *et al.*, 1984; Hagblom *et al.*, 1985). The importance of antigenic variation during natural growth is indicated by the amount of genetic information dedicated to antigenic variation. There can be up to 19 pilin and 12 *opa* gene copies within a single strain and presumably there are other gene products that partake in antigenic variation. In view of the host restriction of humans, these gene families and the ability to effect antigenic variation must be important to growth *in vivo*.

There are three main ways to explain the importance of antigenic variation: (1) antigenic variation acts to prolong infections; (2) it functions to prevent immunity to reinfection; and (3) it acts to change functional properties of the proteins. It is probable that all these can contribute to pathogenicity. *N. gonorrhoeae* needs to move into new hosts. Most uncomplicated infections will eventually resolve even if untreated, suggesting that effective mechanisms exist for clearing the organism. If the bacterium can extend the average time it spends in any one individual, it will increase the chance of infecting a new host. Antigenic variation may provide prolonged survival to increase the probability of transmission. Possibly more important than prolonging the course of infection is the lack of immunity shown by repeatedly infected individuals. Most sexually transmitted diseases rely on a small highly promiscuous segment of the population as focal centres for transmission. Clearly, if immunity develops in this population the spread of the disease would be curtailed. Pilin antigenic variation may help prevent protective immunity, since vaccination with a single pilus type prevents against challenge with the homologous strain (Brinton *et al.*, 1978) but not against natural infections (Tramont *et al.*, 1985). The contribution of PII antigenic variation to evading the immune response has not been tested.

Finally, a role of antigenic variation in altering the functions of PII and pili cannot be ruled out. Pilin antigenic variants have been demonstrated which gain or lose binding ability to cultured cells as a result of an antigenic switch (Virji *et al.*, 1982). The existence of mini-cassettes of variable pilin sequence provides the potential to have one part of the protein vary to escape an immune response (for instance the hypervariable region) while other parts of the protein vary to change functions (such as adhesion). It is also established that different PII variants occur in different environments during an infection (James and Swanson, 1978; Swanson *et al.*, 1987) and that not all PII proteins possess the same adherence properties. The work of several laboratories in examining the interaction of gonococci with eukaryotic cells should determine if functional properties vary during antigenic variation. Biological systems are often efficient, with one protein encoding multiple activities or a genetic system used in multiple pathways. It is probable that the intricate mechanisms the gonococcus has evolved for changing the primary sequence of surface exposed proteins will be found to contribute to more than one virulence property.

REFERENCES

Ashton, F.E., Pasieka, A.E., Collins, F. *et al.* (1977) Inhibition of attachment of *Neisseria gonorrhoeae* to tissue culture cells by goat milk antigonococcal immunoglobulin G. *Can. J. Microbiol.*, **23**, 975–80.

Barbour, A.G. (1990) Antigenic variation of a relapsing fever *Borrelia* species. *Ann. Rev. Microbiol.*, **44**, 155–71.

Belland, R.J., Morrison, S.G., van der Ley, P. and Swanson, J. (1989) Expression and phase variation of gonococcal P.II genes in *Escherichia coli* involves ribosomal frame shifting and slipped strand mispairing. *Mol. Microbiol.*, **3**, 777–86.

Bergström, S., Robbins, K., Koomey, J.M. and Swanson, J. (1986) Piliation control mechanisms in *Neisseria gonorrhoeae*. *Proc. Natl Acad. Sci. USA*, **83**, 3890–4.

Brinton, C.C., Bryan, J., Dillon, J.-A, *et al.* (1978) Uses of pili in gonorrhoea control: role of bacterial pili in disease, purification and properties of gonococcal pili, and progress in development of a gonococcal pilus vaccine for gonorrhoea, in *Immunobiology of Neisseria gonorrhoeae* (eds G.F. Brooks, E.C. Gotschlich K.K. Holmes, W.D. Sawyer and F.E. Young), American Society for Microbiology, Washington DC, pp. 155–78.

Britigan, B.E. and Cohen, M.S. (1985) Effect of growth in serum on uptake of *Neisseria gonorrhoeae* by human neutrophils. *J. Infect. Dis.*, 152, 330–8.

Brooks, G.F. and Donegan, E.A. (1985) *Gonococcal Infection*, Edward Arnold, Kent.

Bygdeman, S., Danielsson, D. and Sandstrom, E.G. (1983) Gonococcal serogroups in Scandinavia. A study with polyclonal and monoclonal antibodies. *Acta Pathol. Immuno. Scand. Sect. B*, 91, 293–305.

Cannon, J.G. and Sparling, P.F. (1984) The genetics of the gonococcus. *Ann. Rev. Microbiol.*, 38, 111–33.

Casey, S.G. Shafer, W.M. and Spitznagel, J.K. (1986) *Neisseria gonorrhoeae* survives intraleukocytic oxygen-independent antimicrobial capacities of anaerobic and aerobic granulocytes in the presence of pyocin lethal for extracellular gonococci. *Infect. Immun.*, 52, 384–9.

Casey, S.G. Veale, D.R. and Smith, H. (1979) Demonstration of intracellular growth of gonococci in human phagocytes using spectinomycin to kill extracellular organisms. *J. Gen. Microbiol.*, 113, 395–8.

Cheng, D.S.F. (1988) Gonorrhoea, in *Sexually Transmitted Diseases, A Guide for Clinicians* (eds L.C. Parish and F. Gschnait), Springer-Verlag, New York, p. 59.

Connell, T.D., Black, W.J., Kawula, T.H., *et al.* (1988) Recombination among protein II genes of *Neisseria gonorrhoeae* generates new coding sequences and increases structural variability in the protein II family. *Mol. Microbiol.*, 2, 227–30.

Densen, P. and Mandell, G.L. (1978) Gonococcal interactions with polymorphonuclear neutrophils. Importance of the phagosome. *J. Clin. Invest.*, 62, 1161–71.

Donegan, E.A. (1985) Epidemiology of gonococcal infection, in *Gonococcal Infection* (eds G.F. Brooks and E.A. Donegan), Edward Arnold, Kent, chapter 13.

Dorward, D.W., Garon, C.F. and Judd, R.C. (1989) Export and intercellular transfer of DNA via membrane blebs of *Neisseria gonorrhoeae*. *J. Bacteriol.*, 171, 2499–505.

Dorward, D.W. and Garon, C.W. (1989) DNA-binding proteins in cells and membrane blebs of *Neisseria gonorrhoeae*. *J. Bacteriol.*, 171, 4196–201.

Gibbs, C.P., Reimann, B.-Y., Schultz, E. *et al.* (1989) Reassortment of pilin genes in *Neisseria gonorrhoeae* occurs by two distinct mechanisms. *Nature*, 338, 651–2.

Glasgow, A.C., Hughes, K.T. and Simon, M.I. (1989) Bacterial inversion systems, in *Mobile Genetic Elements* (eds D.E. Berg and M.M. Howe), American Society for Microbiology, Washington DC, pp. 637–59.

Gregg, C.R., Melly, M.A., Hellerquist, C.G., *et al.* (1981) Toxic activity of purified lipopolysacharride of *Neisseria gonorrhoeae* for human fallopian tube mucosa. *J. Infect. Dis.*, 143, 432–9.

Haas, R. and Meyer, T.F. (1986) The repertoire of silent pilus genes in *Neisseria gonorrhoeae*: evidence for gene conversion. *Cell*, 44, 107–15.

Haas, R., Schwarz, H. and Meyer, T.F. (1987) Release of soluble pilin antigen coupled with gene conversion in *Neisseria gonorrhoeae*. *Proc. Natl Acad. Sci. USA*, 84, 9079–83.

Haas, R., Veit, S. and Meyer, T.F. (1992) Silent pilin genes of *Neisseria gonorrhoeae* MSII and the occurrence of related hypervariant sequences among gonococcal isolates. *Molec. Microbiol.*, 6, 197–208.

Hagblom, P., Segal, E., Billyard, E. and So, M. (1985) Intragenic recombination leads to pilus antigenic variation. *Nature*, 215, 156–8.

Hebeler, B.H. and Young, F.E. (1975) Autolysis of *Neisseria gonorrhoeae*. *J. Bacteriol.*, 122, 385–92.

Hermodson, M.A., Chen, K.C.S. and Buchanan, T.M. (1978) *Neisseria* pili proteins: amino-terminal amino acid sequences and identification of an unusual amino acid. *Biochemistry*, 17, 442–5.

Hill, S.A. Morrison, S.G. and Swanson, J. (1990) The role of direct oligonucleotide repeats in gonococcal pilin gene variation. *Molec. Microbiol.*, 4, 1341–52.

James, J.F. and Swanson, J. (1978) Studies of gonococcus infection. XIII. Occurrence of color/opacity colonial variants in clinical cultures. *Infect. Immun.*, 19, 332–40.

Johnson, S.C., Chung, R.C.Y., Deal, C.D. *et al.* (1991) Human immunization with Pgh 3–2 gonococcal pilus results in cross-reactive antibody to the cyanogen bromide fragment-2 of pilin. *J. Infect. Dis.*, 163, 128–34.

Joiner, K.A., Scales, R., Warren, K.A., *et al.* (1985) Mechanisms of action of blocking immunoglobulin G for *Neisseria gonorrhoeae*. *J. Clin. Invest.*, 76, 1765–72.

Kellogg, D.S., Peacock, W.L., Deacon, W.E. *et al.* (1963) *Neisseria gonorrhoeae*. I. virulence genetically linked to colonial variation. *J. Bacteriol.*, 85, 1274–9.

Kellogg, D.S., Cohen, I.R. Norins, L.C. *et al.* (1968) *Neisseria gonorrhoeae*. II. Colonial variation and pathogenicity during 35 months *in vitro*. *J. Bacteriol.*, 96, 596–605.

Knapp, J.S., Tam, M.R., Nowinski, R.C. *et al.* (1984) Serological classification of *Neisseria gonorrhoeae* with use of monoclonal antibodies to gonococcal outer membrane protein I. *J. Infect. Dis.*, 150, 44–8.

Knapp, J.S. and Holmes, K.K. (1975) Disseminated gonococcal infections caused by *Neisseria gonorrhoeae* with unique growth requirements. *J. Infect. Dis.* 132, 204–8.

Koomey, M., Gotschlich, E.C., Robbins, K. *et al.* (1987) Effects of *recA* mutations on pilus antigenic variation and phase variations in *Neisseria gonorrhoeae*. *Genetics* 117, 391–8.

Lambden, P.R., Heckels, J.E., James, L.T. and Watt, P.J. (1979) Variations in surface protein composition associated with virulence properties in opacity types of *Neisseria gonorrhoeae*. *J. Gen. Microbiol.*, 114, 305–12.

Lambden, P.R., Robertson, R.N. and Watt, P.J. (1980) Biological properties of two distinct pilus types produced by isogenic variants of *Neisseria gonorrhoeae* P9. *J. Bacteriol.*, 141, 393–6.

Lambden, P.R. (1982) Biochemical comparison of pili from variants of *Neisseria gonorrhoeae* P9. *J. Gen. Microbiol.*, 128, 2105–11.

Lee, B.C. and Schryvers, A.B. (1988) Specificity of the lactoferrin and transferrin receptors in *Neisseria gonorrhoeae*. *Molec. Microbiol.*, 2, 827–9.

Lytton, E.J. and Blake, M.S. (1986) Isolation and partial characterization of the reduction-modifiable protein of *Neisseria gonorrhoeae*. *J. Exp. Med.*, 164, 1749–59.

Marrs, C.F., Schoolnik, G., Koomey, J.M. *et al.* (1986) Cloning and sequencing of the *Moraxella bovis* pilin gene. *J. Bacteriol.*, **163**, 132–9.

McDade, R.L. Jr. and Johnston, K.H. (1980) Characterization of serologically dominant outer membrane proteins of *Neisseria gonorrhoeae*. *J. Bacteriol.*, **141**, 1183–91.

McGee, Z.A., Johnson, A.P. and Taylor-Robinson, D. (1981) Pathogenic mechanisms of *Neisseria gonorrhoeae*: observations on damage to human fallopian tubes in organ culture by gonococci of colony type 1 or type 4. *J. Infect. Dis.*, **143**, 413–22.

McKern, N.M., O'Donnell, I.J., Inglis, A.S. *et al.* (1983) Amino acid sequence of pilin from *Bacteroides nodosus* (strain 198), the causative organism of ovine footrot. *FEBS Letts.*, **164**, 149–53.

Melly, M.A., Gregg, C.R. and McGee, Z.A. (1981) Studies of toxicity of *Neisseria gonorrhoeae* for human fallopian tube mucosa. *J. Infect. Dis.*, **143**, 423–31.

Meyer, T.F., Billyard, E., Haas, R. *et al.* (1984) Pilus gene of *Neisseria gonorrhoeae*: chromosomal organization and DNA sequence. *Proc. Natl Acad. Sci. USA*, **81**, 6110–14.

Meyer, T.F., Gibbs, C.P. and Haas, R. (1990) Variation and control of protein expression in *Neisseria*. *Ann. Rev. Microbiol.*, **44**, 451–7.

Meyer, T.F., Mlawyer, N. and So, M. (1982) Pilus expression in *Neisseria gonorrhoeae* involves chromosomal rearrangement. *Cell*, **30**, 45–52.

Morse, S.A., Johnson, S.R., Biddle, J.W. and Roberts, M.C. (1986) High-level tetracycline resistance in *Neisseria gonorrhoeae* is the result of acquisition of streptococcal *tetM* determinant. *Antimicrob. Agents Chemother.*, **30**, 664–70.

Muir, L.L. Strugnell, A. and Davies, J.K. (1988) Proteins that appear to be associated with pili in *Neisseria gonorrhoeae*. *Infect. Immun.*, **56**, 1743–7.

Murphy, G.L., Connell, T.D., Barritt, D.S. *et al.* (1989) Phase variation of gonococcal protein II: regulation of gene expression by slipped strand mispairing of a repetitive DNA sequence. *Cell*, **56**, 539–47.

Nicolson, I.J., Perry, A.C.F., Virji, M. *et al.* (1987) Localization of antibody-binding sites by sequence analysis of cloned pilin genes from *Neisseria gonorrhoeae*. *J. Gen Microbiol.*, **133**, 825–33.

Norlander, L., Davies, J., Norqvist, A. and Normark, S. (1979) Genetic basis for colonial variation in *Neisseria gonorrhoeae*, *J. Bacteriol.*, **138**, 762–9.

Norrod, E.P. and Williams, R.P. (1978) Growth of *Neisseria gonorrhoeae* in media deficient in iron without detection of siderophore. *Curr. Microbiol.*, **1**, 281–4.

Plaut, A.G., Gilbert, J.V. Artenstein, M.S. and Capra, J.D. (1975) *Neisseria gonorrhoeae* and *Neisseria menigitidis*: extracellular enzyme cleaves human immunoglobulin A. *Science*, **190**, 1103–5.

Rest, R.F. and Shafer, W.M. (1989) Interactions of *Neisseria gonorrhoeae* with human neutrophils. *Clin. Microbiol. Revs.*, **2**(suppl.), S83–S91.

Rest, R.F., Fischer, S.H., Ingram, Z.Z. and Jones, J.F. (1982) Interactions of *Neisseria gonorrhoeae* with human neutrophils: effects of serum and gonococcal opacity on killing and chemiluminescence. *Infect. Immun.*, **36**, 737–44.

Sarubbi, F.A. Jr and Sparling, P.F. (1974) Transfer of antibiotic resistance in mixed cultures of *Neisseria gonorrhoeae*. *J. Infect. Dis.*, **130**, 660–3.

Sastry, P.A., Pearlstone, J.R., Smillie, L.B. and Paranchhych, W. (1983) Amino acid sequence of pilin isolated from *Pseudomonas aeruginosa* PAK. *FEBS Lett.*, 151, 253–5.

Schoolnik, G.K., Buchanan, T.M. and Holmes, K.K. (1976) Gonococci causing disseminated gonococcal infections are resistant to the bactericidal action of normal serum. *J. Clin. Invest.*, 58, 11163–73.

Schwalbe, R.S. and Cannon, J.G. (1986) Genetic transformation of genes for protein II in *Neisseria gonorrhoeae.*, *J. Bacteriol.*, 167, 186–90.

Segal, E., Billyard, E., So, M. *et al.* (1985) Role of chromosomal rearrangement in *N. gonorrhoeae* pilus phase variation. *Cell*, 40, 293–300.

Segal, E., Hagblom, P., Seifert, H.S. and So, M. (1986) Antigenic variation of gonococcal pilus involves assembly of separated silent gene segments. *Proc. Natl Acad. Sci. USA*, 83, 2177–81.

Seifert, H.S. and So, M. (1988) Genetic mechanisms of bacterial antigenic variation. *Microbiol. Revs.*, 52, 327–36.

Seifert, H.S., Ajioka, R.A., Marchal, C. *et al.* (1988) DNA transformation leads to pilin antigenic variation in *Neisseria gonorrhoeae. Nature*, 336, 392–5.

Sparling, P.F. (1966) Genetic transformation of *Neisseria gonorrhoeae* to streptomycin resistance. *J. Bacteriol.*, 92, 1364–70.

Stephans, D.S. (1989) Gonococcal and meningococcal pathogenesis as defined by human cell, cell culture, and organ culture assays. *Clin. Microbiol. Revs.* 2(suppl.) S104–S111.

Stern, A., Nickel, P., Meyer, T.F. and So, M. (1984) Opacity determinants of *Neisseria gonorrhoeae*: gene expression and chromosomal linkage to the pilus gene. *Cell*, 37, 447–56.

Stern, A., Brown, M., Nickel, P. and Meyer, T.F. (1986) Opacity genes in *Neisseria gonorrhoeae*: control of phase and antigenic variation. *Cell*, 47, 61–71.

Swanson, J. (1982) Colony opacity and protein II compositions of gonococci. *Infect. Immun.*, 37, 359–68.

Swanson, J. and Koomey, J.M. (1989) Mechanisms for variation of pili and outer membrane protein II in *Neisseria gonorrhoeae*, in *Mobile DNA* (eds De Berg and M.M. Howe) American Society for Microbiology, Washington DC., Chapter 34.

Swanson, J., Kraus, S.J. and Gotschlich, E.C. (1971) Studies on gonococcal infection I. Pili and zones of adhesion: their relation to gonococcal growth patterns. *J. Exp. Med.*, 134, 886–906.

Swanson, J., Robbins, K., Barrera, O. *et al.* (1987) Gonococcal pilin variants in experimental gonorrhoea. *J. Exp. Med.*, 165, 1344–57.

Swanson, J.S., Morrison, S., Barrera, O. and Hill, S. (1990) Piliation changes in transformation-defective gonococci. *J. Exp. Med.*, 171, 2131–9.

Thomas, D.W., Hill, J.C. and Tyeryar E.J. Jr. (1973) Interactions of gonococci with phagocytic leukocytes from mice and men. *Infect. Immun.*, 8, 98–104.

Tramont, E. (1977) Inhibition of adherence of *Neisseria gonorrhoeae* by human genital secretions. *J. Clin. Invest.*, 59, 117–24.

Tramont, E.C., Boslego, J., Chung, R. *et al.* (1985) Parenteral gonococcal pilus vaccine, in *The Pathogenic Neisseria* (eds G.K. Schoolnik, G.F. Brooks, S. Falkow *et al.*), American Society for Microbiology, Washington DC, pp. 316–22.

Tramont, E.C., Hodge, W.C., Gilbreath, M.J. and Ciak, J. (1979) Differences in attachment antigens of gonococci in reinfection. *J. Lab. Clin. Med.*, 93, 730–5.

Van der Ploeg, L.H.T. (1987) Control of variant surface antigen switching in trypanosomes. *Cell*, 51, 159–61.

Virji, M. and Everson, J.S. (1981) Comparative virulence of opacity variants of *Neisseria gonorrhoeae* P9. *Infect. Immun.*, 31, 965–70.

Virji, M., Everson, J.S. and Lambden, P.R. (1982) Effect of anti-pilus antisera on virulence of variants of *Neisseria gonorrhoeae* for cultured epithelial cells. *J. Gen. Microbiol.*, 128, 1095–100.

Virji, M. and Heckels, J.E. (1984) The role of common and type-specific pilus antigenic domains in adhesion and virulence of gonococci for human epithelial cells. *J. Gen. Microbiol.*, 130, 1089–95.

West, S.E. and Sparling, P.F. (1985) Response of *Neisseria gonorrhoeae* to iron limitation: Alterations in expression of membrane proteins without apparent siderophore production. *J. Bacteriol.*, 47, 388–94.

Wetzler, L.M., Gotschlich, E.C., Blake, M.S. and Koomey, J.M. (1989) The construction and characterization of *Neisseria gonorrhoeae* lacking protein III in its outer membrane. *J. Exp. Med.*, 169, 2199–209.

Zak, K., Diaz, J.-L., Jackson, D. and Heckels, J.E. (1984) Antigenic variation during infection with *Neisseria gonorrhoeae*: detection of antibodies to surface proteins in sera of patients with gonorrhoea. *J. Infect. Dis.*, 149, 166–74.

2

Molecular biology of chlamydiae

MARJORIE A. MONNICKENDAM

2.1 CHLAMYDIAE AND HUMAN DISEASES

Chlamydiae are obligate, intracellular bacteria which cause disease in many species of mammals and birds. There are three species of chlamydiae, *Chlamydia trachomatis*, *Chlamydia pneumoniae* and *Chlamydia psittaci*. Chlamydiae which cause disease in humans are listed in Table 2.1. *C. trachomatis* is a common human mucosal pathogen which causes three forms of disease; trachoma, genital infection and lymphogranuloma venereum (LGV). The only non-human strain is the murine pathogen, mouse pneumonitis agent. *C. trachomatis* is divided into 15 serovars, A, B, Ba, C, D, E, F, G, H, I, J, K, L1, L2 and L3. These serovars are grouped into two sub-species; the B complex which comprises serovars B, Ba, D, E, F, G, L1 and L2, and the C complex which comprises serovars A, C, H, I, J, K and L3. *C. pneumoniae* includes the human pathogens IOL 207 and TWAR. The first isolates came from the eyes of children with trachoma in Iran and Taiwan; more recent isolates have been obtained from patients with respiratory infection. Many people have serum antibodies to *C. pneumoniae*, indicating that infection is common. Most strains of *C. psittaci* are pathogens of non-primate mammals and birds; only a few of these strains are human pathogens.

C. trachomatis serovars D to K are common causes of genital infections throughout the world. There were approximately 146 600 cases of non-specific genital infection in the United Kingdom in 1987 (Adler, 1990), indicating that there were approximately 73 300 cases of chlamydial genital infection if it is assumed that *C. trachomatis* causes at least 50% of non-specific genital infection. There has recently been a decline in the incidence of chlamydial genital infection. In 1984, it was reported that there were approximately 15 500 cases of non-specific genital infection and therefore 77 500 cases of chlamydial genital infection (Monnickendam, 1988). The reduction has been particularly marked in women. The percentage of women

Molecular and Cell Biology of Sexually Transmitted Diseases
Edited by D. Wright and L. Archard
Published in 1992 by Chapman and Hall, London ISBN 0 412 36510 3

attending a clinic in Bournemouth, on the south coast of England, from whom *C. trachomatis* was isolated, fell from 14.6% in 1984 to 3.2% in the first six months of 1989 (Sivakumar and Basu Roy, 1989). In Newcastle, in the north-east of England, similar numbers of women were screened for chlamydial genital infection in 1985 and 1988 but the percentage of women from whom *C. trachomatis* could be isolated fell from 17.4% in 1985 to 7.1% in 1988 (Shanmugaratnam and Pattman, 1989). These reductions presumably reflect changes in sexual practices to prevent the spread of human immunodeficiency virus.

2.2 CHLAMYDIAL LIFE CYCLE

Infection is initiated when the small infectious, but metabolically inert elementary body (EB), which is 200 to 300 nm in diameter, enters a cell. Inside the cell, the EB enlarges to become the metabolically active but non-infectious reticulate body (RB), which is about 10 000 nm in diameter. The RB divides by binary fission and later condenses into EBs within the characteristic intracellular chlamydial inclusion body. The growth of chlamydiae within cells is enhanced by treating the cell monolayer with a cytostatic agent before infection and centrifuging the EBs onto the monolayer. The host cell nucleus is not required for chlamydial growth and the mouse pneumonitis strain of *C. trachomatis* was observed to infect and grow in enucleated mouse fibroblasts (Perara *et al.*, 1990).

Table 2.1 Chlamydiae and human diseases

Species, biovars, serovars and strains	Tissue	Disease
C. trachomatis trachoma biovar,	Conjunctiva	Trachoma
C. trachomatis trachoma biovar, serovars D, E, F, G, H, I, J, K	Conjunctiva Respiratory tract Genital tract	Ophthalmia neonatorum; inclusion conjunctivitis; cervicitis; salpingitis; urethritis; nasopharyngeal infection; pneumonia; epididymitis;
C. trachomatis lymphogranuloma venereum biovar, serovars L_1, L_2, L_3	Lymphoid tissue	Lymphogranuloma venereum
C. pneumonia	Conjunctiva Respiratory tract	Ocular infection; acute respiratory infection
C. psittaci strain (psittacosis agent)	Respiratory tract	Pneumonia
C. psittaci strain (feline keratoconjunctivitis agent)	Conjunctiva	Ocular infection
C. psittaci strain (ovine abortion)	Placenta and foetus	Abortion

Modified from Monnickendam 1988.

For many years, the study of chlamydial antigens was handicapped by the difficulty and expense of producing sufficient quantities of chlamydial antigens and by the presence of contaminating host-cell antigens. Most laboratories used strains of the LGV biovar because they are easier to grow and are more stable than trachoma biovar strains. EB antigens were studied because EBs are easier to isolate and handle than RBs. The study of chlamydiae has been transformed in recent years by the application of modern techniques such as monoclonal antibody production and immunoblotting, gene cloning, and the use of transcription and translation systems. This chapter will survey the advances in our understanding of chlamydiae which have resulted from the application of these techniques and relate these advances to our knowledge of the origins of chlamydiae, their classification, immunology and pathogenicity.

2.3 CHLAMYDIAL CHROMOSOME

2.3.1 Size

Chlamydial DNA comprises a chromosome which is a closed circle of double-stranded DNA. Almost all strains of C. trachomatis and many strains of C. psittaci also contain plasmids. Estimates of the size of chlamydial chromosomal DNA have ranged from 600 to 1440 kilobase pairs (kb). Large DNA molecules are difficult to handle in solution and are easily broken during procedures such as pipetting and mixing. Conventional electrophoresis in agarose gels does not separate large pieces of DNA (>50 kb) effectively. The problems of handling large molecules in liquids can be avoided by embedding EBs in agarose gel and then digesting them to release chlamydial DNA in situ. Pulsed-field gel electrophoresis separates large pieces of DNA by size. These techniques have been applied to chlamydial chromosomal DNA obtained from four strains of chlamydiae; C. trachomatis serovar L2 and three strains of C. psittaci causing ovine abortion (Frutos et al., 1989). Their genomes were very similar with an average size of 1450 kb.

2.3.2 Homology

The DNA molecules of the three species of chlamydiae differ greatly. DNA from three strains of C. trachomatis was compared with DNA from two strains of C. psittaci (Kingsbury and Weiss, 1968). The G+C ratios of the strains of C. trachomatis varied from 45.0% to 45.5%, while the G+C ratios of the strains of C. psittaci were 41.5 and 42.0%. DNA hybridization experiments showed that there was good homology between the three strains of C. trachomatis and between the two strains of C. psittaci.

25

However, the thermal stability of hybrids composed of DNA from the two species was poor and the estimated degree of relatedness between the DNAs of the two species was 10%. The G+C ratios of DNA from three strains of *C. pneumoniae* were intermediate between the ratios of the other two species (Cox *et al.*, 1988). There was 94% or greater homology between DNA from the three strains of *C. pneumoniae*, and 20 to 100% homology between DNA from different strains of *C. psittaci*.

2.3.3 Restriction endonuclease studies

DNA from EBs of *C. trachomatis* serovars B, C, L1, L2 and L3 was digested with a variety of restriction endonucleases, revealing some small differences between the serovars in both chromosomal and plasmid DNA. *C. trachomatis* DNA could be distinguished clearly from DNA from *C. psittaci* strain VR 601 (Peterson and de la Maza, 1983). Further studies of DNA from all 15 *C. trachomatis* serovars digested with restriction endonucleases showed that they could be divided into three groups corresponding to the three forms of disease caused by *C. trachomatis*, i.e. trachoma, genital infection and LGV (Peterson and de la Maza, 1988). The three groups were characterized by the presence or absence of three DNA fragments with molecular weights of 9×10^6; 5.7×10^6 or 4×10^6 following digestion with the restriction enzyme Eco R1. The trachoma and LGV biovars could be distinguished following digestion with the restriction endonuclease Pst 1. Other fragments differentiated between different serovars in the same group or between different strains in the same serovar. There were very few differences between LGV biovar strains. The clear separation between DNA from trachoma biovar strains and LGV biovar strains contrasts with the complex, cross-reactive reactions of serological typing. The authors commented that we are initiating our understanding at the molecular level of the well-known difference in biological properties and pathogenicity of these two biovars.

DNA from ocular and respiratory isolates of *C. pneumoniae* was compared with DNA from *C. trachomatis* and *C. psittaci* (Campbell *et al.*, 1987). When whole chromosomal DNA was used in dot blots, there was no significant homology between *C. pneumoniae* and the other two species. Restriction endonuclease studies revealed that *C. pneumoniae* DNA gave a characteristic pattern of fragments which could be clearly distinguished from the other two species. DNA from different isolates of *C. pneumoniae* revealed some small differences; while eight enzymes produced identical patterns in all isolates, two further enzymes produced an extra band in two isolates.

The differences between DNA from different strains of *C. trachomatis* or different strains of *C. pneumoniae* are small. There are also marked similarities in *C. psittaci* DNA from strains causing the same disease. Eight ovine abortion strains collected from England, Scotland and Wales were identical (McClenaghan *et al.*, 1984). In contrast, there are considerable differences in DNA from strains of *C. psittaci* causing different diseases in different species such as ovine abortion and avian diseases. Strains of *C. psittaci* can be divided into several groups or biovars depending on the different patterns of fragments generated by restriction endonuclease digestion (Fukushi and Hirai, 1989; Anderson and Tappe, 1989).

2.4 CHROMOSOMAL GENES AND PRODUCTS

2.4.1 DNA binding proteins

There are marked differences in the organization of chlamydial EB and RB DNA. The EB has a distinctive, electron-dense, eccentrically located nucleoid with radiating fibres. The RB has a fibrous, loosely organized nucleoid, similar to that seen in other prokaryotes. Nucleoproteins probably play a pivotal role in the ultrastructural changes in the chlamydial nucleoid and the regulation of transcription and chlamydial development, and have been studied in *C. trachomatis* serovars B and L2 and *C. psittaci* strain 6BC (Wagar and Stephens, 1988). DNA-binding proteins were obtained from lysates of EBs, RBs and from chromosomal preparations. Some were specific to EBs, others were specific to RBs or were present in both EBs and RBs. Some bound to double-stranded and others bound to single-stranded chlamydial DNA.

2.4.2 Ribosomal genes

Studies of chlamydial ribosomal RNA genes have encompassed their sequences, location on the chromosome and their promoter regions. The 16s ribosomal RNA gene was present on a 10 kb fragment from an avian psittacosis strain of *C. psittaci* (Weisburg *et al.*, 1986). There was more than 95% identity between the sequences of the 16s ribosomal RNA genes of *C. psittaci* and *C. trachomatis*. The *C. psittaci* 16s ribosomal RNA gene was also compared with the genes of *Escherichia coli* and other bacteria. The chlamydial gene was more closely related to eubacterial gene sequences than to archaebacterial ones. Chlamydiae have also been placed in the eubacteria because of the nature of the cell envelope and because they possess a lipopolysaccharide (LPS). The 16s ribosomal RNA gene sequence clearly demonstrates that chlamydiae are not closely related to Rickettsiae, another important group of obligate intracellular pathogens.

Chlamydiae have two chromosomal genes coding for ribosomal RNA (Palmer and Falkow, 1986a). DNA from EBs was digested with the restriction endonuclease Eco R1 and probed with L2 16s ribosomal RNA. All of the human pathogens (*C. trachomatis* strains B, C, D, E, F, L1,

L2) had one gene on a 9 kb fragment and the other on a 7 kb fragment of DNA. The ribosomal RNA genes of the mouse pneumonitis strain of *C. trachomatis* were located on slightly smaller fragments. The ribosomal RNA genes of guinea-pig inclusion conjunctivitis (GPIC) agent, a strain of *C. psittaci*, were located on considerably smaller fragments than the *C. trachomatis* ribosomal genes. The ribosomal genes were probed further with a 31-base oligonucleotide with inverse complementarity to the predicted unique sequence of L2 16s ribosomal RNA. This probe hybridized with DNA from all the human pathogens but not the two animal pathogens.

The ribosomal RNA genes of the mouse pneumonitis strain of *C. trachomatis* have been sequenced (Engel and Ganem, 1987). They were cloned using a cDNA probe specific to ribosomal RNA sequences. Two ribosomal RNA operons were detected, each of which encoded for 16s and 23s ribosomal RNA. The first 58 bp of the mature 16s ribosomal RNA gene were identical to those of *C. psittaci* and had 70% homology with *E. coli*. The region upstream of the 16s ribosomal RNA gene was sequenced and scanned for possible promoter sequences. Two potential promoters were identified at about 100 and 200 nucleotides upstream and two ribosomal RNA precursor molecules of different sizes were detected in cells seven hours after infection. The smaller precursor molecule was more abundant than the larger precursor molecule throughout the growth cycle. The two promoter sequences had little homology to each other, to the *C. trachomatis* MOMP gene promoter sequences or to *E. coli* promoter sequences.

In a search for chlamydial promoter elements which could function in *E. coli*, a 9.8 kb fragment of DNA from the mouse pneumonitis strain of *C. trachomatis* was cloned. It coded for a protein with a molecular weight of 70 000. The DNA sequence resembled the sequence of *E. coli* ribosomal protein S1 (Sardinia *et al.*, 1989). The predicted amino acid sequence was compared with that for the *E. coli* protein; 43% of the amino acids were identical and 37% of amino acid differences were conservative changes. Messenger RNA was first detected nine hours after infection and was maximal 12 to 18 hours after infection. The DNA sequence around the start of the open reading frame (ORF) was searched for sequences homologous to the putative chlamydial promoter regions for 16s ribosomal RNA and MOMP genes but little homology was found.

2.4.3 *Tuf* genes

The protein elongation factor of *E. coli* is divided into two components, one which is heat-stable and the other which is not. The unstable factor (EF-Tu) which is required for the correct binding of amino-acyl t-RNAs to the receptor site of the ribosome, participates in RNA synthesis and is present in bacterial membranes. EF-Tu comprises up to 5% of the total cellular protein of *E. coli*, and is encoded by two genes, *tufA* and *tufB*

(Jones *et al.*, 1980). *Tuf* genes are highly conserved amongst the eubacteria. Gram-negative bacteria have two *tuf* genes with the possible exception of Cyanobacter, while Gram-positive bacteria have a single *tuf* gene except for some *Clostridia* species which have two *tuf* genes. DNA from EBs of the mouse pneumonitis strain of *C. trachomatis* was digested with restriction endonucleases and probes for the C-terminal and N-terminal sequences of the *E. coli tuf* A gene were used to locate chlamydial *tuf* genes (Goldstein *et al.*, 1989). Only a single fragment of the DNA hybridized with the two probes. This was unexpected because chlamydiae are considered to be Gram-negative bacteria. The chlamydial genome is smaller than genomes of free-living bacteria and it therefore appears that chlamydiae evolved from an ancestor with two *tuf* genes and have lost one of the two *tuf* genes during evolution. Alternatively but less likely, the chlamydial ancestor may have had only one *tuf* gene.

2.5 CHLAMYDIAL ANTIGENS

The chlamydial EB surface comprises a complex molecular mosaic (Allan, 1986). Chlamydial lipopolysaccharide (LPS) and outer membrane proteins (OMPs) have been subject to intense scrutiny. Chlamydial heat shock proteins (HSPs) and proteins which bind to eukaryotic cell surfaces have recently been described and cloned. Immunoblotting studies have revealed the existence of other chlamydial antigens which have been defined on the basis of their size but have not yet been characterized.

2.5.1 Lipopolysaccharide

The first chlamydial EB product to be studied was LPS. The complement-fixation test for the diagnosis of psittacosis was developed 50 years ago. The antigen was heat-stable, and chlamydia genus-specific.

Chlamydial LPS has epitopes which it shares with LPS from other species of bacteria and it also has a chlamydia genus-specific epitope. Antibodies raised against chlamydial LPS react with the deep core (Re) part of enterobacterial LPS which is produced by R mutants of *Salmonella minnesota* and *Salmonella typhimurium* (Nurminem *et al.*, 1983). They also react with LPS in the extracellular slime of *Acinetobacter calcoaceticus* (Nurminem *et al.*, 1984), indicating that these LPS molecules contain common antigenic structures. Chemical analysis of LPS from EBs of *C. trachomatis* serovar L2 and an ovine abortion strain of *C. psittaci* have shown that they contain structures characteristic of LPS, which are D-glucosamine, long-chain 3-hydroxy fatty acids, 3-deoxy-D-*manno*-octulonic acid (KDO) and phosphate (Nurminem *et al.*, 1985; Brade *et al.*, 1986).

Chlamydial LPS also has a genus-specific epitope. This was first demonstrated by a monoclonal antibody raised against *C. trachomatis* serovar

29

L2, which only recognized chlamydial LPS (Caldwell and Hitchcock, 1984). This monoclonal antibody reacted with EBs and RBs of all serovars of *C. trachomatis* and all eight strains of *C. psittaci* tested.

A 6.5 kb DNA fragment from *C. trachomatis* serovar L2 was cloned in *E. coli*. Three forms of LPS were produced, one of which was only expressed by *E. coli* containing the chlamydial DNA fragment (Nano and Caldwell, 1985). This LPS was recognized by the monoclonal antibody which recognized the chlamydia-specific epitope. The chlamydia-specific LPS migrated much faster than the two common forms, indicating that it was smaller than them. It was proposed that the cloned fragment of chlamydial DNA codes for a glycosyl transferase which determines the chlamydia-specific epitope. This enzyme adds a sugar during the synthesis of the common form of LPS, which in turn inhibits further synthesis and produced the smaller chlamydia genus-specific form. The 6.5 kb fragment can also transform R mutants of *S. typhimurium* and *S. minnesota*, which then produces chlamydia genus-specific LPS in quantities sufficient for immunological and chemical analysis (Brade *et al.*, 1987a). The chlamydia genus-specific epitope produced by these transformed bacteria has been identified as an 8-O-subsituted KDO residue (Brade *et al.*, 1987b).

The monoclonal antibody which only recognizes the chlamydia genus-specific epitope of LPS was used to study the synthesis of chlamydial LPS in cultured cells which were infected with *C. trachomatis* serovars F and L2 (Karimi *et al.*, 1989). The method of fixation used before staining determined the pattern of staining. The monclonal antibody reacted with the plasma membrane of infected cells, bleb-like structures on cell surfaces and proximal processes of neighbouring uninfected cells, or it stained the intracellular chlamydial inclusion body. In contrast, a monoclonal antibody directed against the major outer membrane protein (MOMP) failed to stain the surfaces of infected cells.

2.5.2 Outer membrane proteins

Chlamydiae differ from other Gram-negative bacteria because they lack an outer rigid peptidoglycan layer. The structural rigidity of the outer membrane is provided by disulphide-mediated cross-linking of proteins which are cysteine-rich (Newhall and Jones, 1983). The cysteine-rich outer membrane proteins are synthesized late in the growth cycle during the conversion of RBs to EBs and their synthesis is blocked by hydroxyurea and ampicillin (Sardinia *et al.*, 1988). There are three principal protein constituents of the outer membrane. The major outer membrane protein (MOMP) has a molecular weight of about 40 000. The second most common protein is outer membrane protein 2 (OMP2), which has a molecular weight of 60 000 and the third protein has a molecular weight of 15 000.

MOMP has been found in all chlamydiae which have been examined. It contains genus-, species-, sub-species- and serovar-specific antigenic determinants (Stephens *et al.*, 1985; Batteiger *et al.*, 1986). Purified MOMP acts as a porin in liposome swelling assays. The rate of diffusion of sugars across the liposome membrane was clearly dependent on their size, and the MOMP pores had an estimated radius of 0.65 to 0.90 nm (Bavoil *et al.*, 1984).

MOMP genes from several serovars have been cloned and sequenced, their amino acid sequences have been deduced and their molecular weights have been calculated (Table 2.2). They are highly conserved in size and in structure. The first chlamydial MOMP gene to be cloned and sequenced was obtained from *C. trachomatis* serovar L2 (Stephens *et al.*, 1986). A series of fragments spanning the MOMP gene were cloned and sequenced. They included an ORF of 1182 bp, encoding 394 amino acids. The deduced amino acid sequence was compared with the chemical determination of the first 22 amino acids in purified L2 MOMP. The predicted sequence of amino acids 23 to 44 was identical to the chemical determination of the first 22 amino acids. It was therefore concluded that the first 22 amino acids in the predicted sequence were a leader or signal sequence for translocation of MOMP to the outer membrane and that the mature protein consisted of 372 amino acids. The calculated weight of the mature protein was 40 282. There were several possible promoter regions at the 5' end of the gene, none of which showed good homology with *E. coli* promoter sequences.

MOMP genes from other serovars of *C. trachomatis* and from strains of *C. psittaci* have also been cloned and sequenced. The nucleotide sequences and predicted amino acid sequences for serovars B, C and L2 were compared (Stephens *et al.*, 1987). Serovars B and L2 which are closely related had the same numbers of nucleotides with some clustered substitutions, while serovars L2 and C which are more distantly related had differing numbers of nucleotides due to deletions or insertions. Long stretches of the three MOMP genes were relatively invariant; variations were largely confined to four variable sequences or domains known as VDI, VDII, VDIII and VDIV.

The transcription of the *C. trachomatis* MOMP gene appears to be controlled by two tandem promoters (Stephens *et al.*, 1988a), producing two messenger RNA transcripts of the MOMP gene of different sizes; 1550 nucleotides and 1400 nucleotides. The smaller messenger RNA appears early in the chlamydial growth cycle and can be detected four hours after infection, while the larger messenger RNA is delayed until 12 hours after infection. The nucleotide sequences of the two messenger RNAs were determined and compared with the DNA sequence. They showed that the two messenger RNAs are the products of a single gene with two promoter sequences.

The locations of specific epitopes have been mapped to VDI, VDII and VDIV. There are conflicting results. MOMP genes from *C. trachomatis* serovars A, B, C and L2 were cloned and sequenced, the major epitopes were mapped with monoclonal antibodies (Baehr *et al.*, 1988). Monoclonal antibodies which were serovar-specific recognized epitopes in VDI and VDII,

Table 2.2 Chlamydial MOMP genes

Species	Serovar or strain	Amino acids (total)	ORF (mature protein)	Mass (mature protein)	Reference
Chlamydia	L2	394	372	40 282	Stephens *et al.*, 1986
trachomatis	B	394	372	40 282	Stephens *et al.*, 1987
	C	397	375	40 607	Stephens *et al.*, 1987
	L1	393	371	40 300	Pickett *et al.*, 1987
	A		374		Baehr *et al.*, 1988
	H	397	375	40 672	Hamilton and Malinowski, 1989
Chlamydia pneumoniae				39 500	Campbell *et al.*, 1990
Chlamydia psittaci	Unclear (ovine abortion/avian?)	402	380		Pickett *et al.*, 1988a
	Ovine abortion	389	367		Herring *et al.*, 1989
	GPIC agent	389	367	39 677	Zhong *et al.*, 1989
	Meningo-pneumonitis strain Cal 10	402	380	41 000	Zhong *et al.*, 1989

while monoclonal antibodies which were sub-species and species-specific epitopes recognized epitopes in VDIV. In another study, the linear distribution of the antigenic domains of MOMP genes from *C. trachomatis* serovars B, C and L2 were mapped and compared (Stephens *et al.*, 1988b). The serovar-specific domain mapped to VDII and was contained within a sequence of 14 amino acids. The sub-species- and species-specific domains mapped to VDIV; they were located in a sequence of 16 amino acids and the two domains were overlapping. In contrast, the serovar-specific epitope of *C. trachomatis* serovar L3 was encoded by a fragment of the gene which codes for amino acids 59 to 114 and includes VD1 (Carlson *et al.*, 1989). Under stringent conditions, this fragment hybridized exclusively with DNA from serovar C and other members of the C complex or sub-species, which confirms that the serological classification of *C. trachomatis* reflects the antigenic structure of MOMP.

VDs of all 15 serovars of *C. trachomatis* have been sequenced and their predicted amino acid sequences were compared (Yuan *et al.*, 1989). The serovars could be divided into three groups by the amino acid homologies of their VDs. These were group 1 (serovars B, Ba, D, E, L1 and L2), group 2 (serovars F and G) and group 3 (serovars A, C, H, I, J, K and L3). This grouping is similar to the serological classification of *C. trachomatis*.

More detailed studies of the antigenic determinants of *C. trachomatis* have used solid-phase short synthetic peptides and have compared monoclonal antibodies raised against EBs and synthetic peptides (Conlan *et al.*, 1988). MOMP gene sequences from several serovars were compared in order to select peptides. The serovar-specific epitope of L1 was defined as a sequence of four amino acids in VDII; the sub-species and species specificity were defined as two different sequences of four amino acids in VDIV and the genus-specific epitope was defined as a sequence of three amino acids in a conserved region. These epitopes of MOMP appear to be determined by primary structure alone. However, secondary and tertiary structure are also important. Monoclonal antibodies were produced against sequences of 16 amino acids containing the defined epitopes (Conlan *et al.*, 1989). Epitope mapping with these monoclonal antibodies showed they had different binding specificities compared with the monoclonal antibodies raised against MOMP. These monoclonal antibodies also failed to bind to whole EBs or MOMP, suggesting that the epitopes which were defined as sequences of three or four amino acids were also determined by interactions with other parts of MOMP. An alternative approach to the study of antigenic determinants was therefore sought. Fragments of the L2 MOMP gene representing 1/4, 1/2 or 3/4 of the gene were cloned and the incomplete MOMP as produced as insoluble inclusions within the host *E. coli* (Pickett *et al.*, 1988). These MOMP fragments possessed genus-, species-, sub-species and serovar-specific epitopes (Conlan *et al.*, 1990). Monoclonal antibodies raised against these MOMP fragments also reacted with EBs and whole MOMP, showing that the secondary and tertiary structure components of the antigenic determinants were conserved in these fragments and that they could therefore be used in subunit vaccines (section 2.11).

The whole MOMP gene of serovar B was surveyed by epitope mapping with hexapeptides and the molecule was divided into five classes of antigens depending on their reactivity with antibodies (Zhong *et al.*, 1990). The majority of the hexapeptides (215 out of 367) were not immunoreactive. The 152 reactive hexapeptides were divided into immunodominant and immunorecessive sequences. Immunodominant sequences were defined as high frequency/high titre or low frequency/high titre. Immunorecessive regions were defined as high frequency/low titre or low frequency/low titre. The VDIV region contained six epitopes; one serovar-specific, four sub-species-specific and one species-specific epitope.

The MOMP gene has been highly conserved over long periods of time. MOMP genes from two serovar B strains were compared (Hayes *et al.*, 1990). One was isolated from a patient with trachoma in The Gambia in 1985, the other was obtained from a patient with trachoma in Taiwan many years earlier. There were only 12 nucleotide differences, which coded for five amino acid substitutions. One change was in VD1, one in VDII and the other three were in conserved regions and did not affect the binding of serovar- and sub-species-specific antibodies.

C. trachomatis MOMP has been studied intensively because it has potential as a chlamydial vaccine: polyclonal antibodies raised against MOMP from C. *trachomatis* serovar L2 neutralize the infectivity of L2 EBs for HeLa cells (Caldwell and Perry, 1982). The antibodies did not affect the attachment and penetration of EBs into cells, suggesting that they inhibited developmental processes within infected cells. VDI, VDII and VDIV can be cleaved by trypsin and attachment of EBs to HeLa cells is inhibited by trypsin cleavage of VDII and VDIV (Su *et al.*, 1988). These observations suggest that the conformation of MOMP determined these domains functions as a chlamydial ligand. Monoclonal antibodies to VDII and VDIV antigenic determinants of serovar B neutralize the infectivity of chlamydial EBs by interfering with attachment to cells (Su *et al.*, 1990). The antibodies bind to the Ebs and inhibit electrostatic interactions between EBs and cells. However, the binding of EBs to cells is not solely dependent on this electrostatic interaction since binding is unaffected by synthetic peptides corresponding to the epitopes recognized by the blocking monoclonal antibodies. A second component is located in an invariant inaccessible hydrophobic region of MOMP. This component undergoes a conformational change when EBs are heated to 56°C and they are no longer able to bind to cells.

Antibodies against MOMP can also neutralize infectivity *in vivo*. Three monoclonal antibodies which reacted with C. *trachomatis* serovar B MOMP neutralized the infectivity of serovar B EBs for HeLa cells and they inhibited the mouse toxicity test and infection of monkey eyes (Zhang *et al.*, 1987a). Two of these monoclonal antibodies reacted with serovars B and Ba; the third reacted with serovars B, Ba and D. However, monkeys immunized with MOMP were not protected from subsequent ocular infection (Taylor *et al.*, 1988). Oral immunization with MOMP and cholera toxin as adjuvant was followed by the production of antibodies which reacted with MOMP but the course of subsequent ocular infection was unaltered.

MOMPs are also present in C. *psittaci* and C. *pneumoniae*. MOMP genes from two ovine abortion strains of C. *psittaci* have been cloned and sequenced (Pickett *et al.*, 1988; Herring *et al.*, 1989). The first of these was compared with MOMP genes from C. *trachomatis* serovars L1, L2 and B. There was 34% nucleotide homology and 65% amino acid homology with 236 conserved and 72 functionally related amino acids (Pickett *et al.*, 1988a). There were considerable differences between the deduced amino acid sequences of the MOMP genes from the two strains. It was concluded that one strain had become contaminated with an avian strain which had subsequently overgrown the original ovine strain. MOMP genes from two more strains of C. *psittaci*, guinea-pig inclusion conjunctivitis (GPIC) agent strain 1 and meningopneumonitis agent strain Cal 10 have been cloned and sequenced and their deduced amino acid sequences compared (Zhang *et al.*, 1989). They have predominantly conserved regions and four variable domains in the same locations as the C. *trachomatis* variable domains.

All seven isolates of *C. pneumoniae* had a 39 500 protein in the outer membrane (Campbell *et al.*, 1990). This was identified as the species-specific MOMP, which was less immunogenic and less antigenically diverse then MOMP from the other two species.

OMP2 is the second most abundant outer membrane protein, with a molecular weight of 60 000. The OMP2 gene from *C. trachomatis* serovar L2 has been cloned and sequenced (Allen and Stephens, 1989). It contains an ORF of 1641 bp which codes for a protein of 547 amino acids containing 24 cysteine residues and a predicted molecular weight of 58 792. Antibodies raised against the recombinant protein were stain lysed L2 EBs and they reacted with two bands with molecular weights of 56 000 and 58 000. This molecular weight polymorphism results from post-translational cleavage of the precursor molecule at one of two sites, yielding mature proteins consisting of 507 or 525 amino acids.

A 2 kb DNA fragment from *C. trachomatis* serovar L1 has also been cloned and sequenced (Clarke *et al.*, 1988). It comprised two ORFs coding for cysteine-rich proteins. The first ORF encodes for a protein comprising 518 amino acids with a leader sequence of 11 amino acids and a mature protein of 508 amino acids with a predicted molecular weight of 54 525. The second ORF comprises 450 nucleotides which code for a protein of 150 amino acids with a predicted molecular weight of 15 818. A 755 bp stretch of the gene encoding the 60 000 protein was used to probe ten different serovars of *C. trachomatis*. It hybridized with a 4 kb fragment produced by digestion with the restriction endonuclease Sal 1 in all ten serovars and with a larger fragment from *C. psittaci* meningopneumonitis strain Cal 10. These observations showed that there was considerable homology between different serovars and species. Monoclonal antibodies have been raised against the 15 000 outer membrane protein from *C. trachomatis* serovars B and 12, and they show that this protein possesses biovar and species-specific epitopes (Zhang *et al.*, 1987b). The antibodies do not react with intact EBs, indicating that these epitopes are not accessible on the chlamydial cell surface.

2.5.3 Heat shock proteins

Heat shock proteins (HSPs) were first observed in larvae of *Drosophila busckii* which had subjected to a sudden increase in temperature. Other forms of stress induce the synthesis of these proteins so that they are sometimes known as stress proteins. HSPs are present in normal cells of every species and are highly conserved in function and structure. They are major antigens of many pathogens and may also have a role in the chelation of iron. HSPs are also involved in the pathogenesis of autoimmune diseases (Kaufmann, 1990). They are present and active in normal cells, where they are involved in important physiological functions. They are divided by their molecular weights into four groups; HSP90, HSP70, HSP60 and ubiquitin. HSP90 binds to steroid receptors and prevents the receptors from binding to DNA unless steroids are present. HSP70 and HSP60 are involved in the folding and unfolding as well as the translocation of polypeptides and the assembly and disassembly of oligomeric protein complexes. Ubiquitin participates in protein degradation. Both HSP70 and HSP60 share more than 50% homology across a wide range of organisms, from bacteria to yeasts and from plant to animals.

HSP60 and HSP70 proteins have been detected in chlamydial EBs. The genes have been cloned, sequenced and compared with HSP genes from other bacteria.

A fragment of DNA from *C. trachomatis* serovar D, which includes a gene coding for an HSP70, has been cloned and sequenced (Danilition *et al.*, 1990). The gene comprises an ORF of 1956 nucleotides coding for a protein of 652 amino acids with a calculated molecular weight of 70 588. The deduced amino acid sequence of the *C. trachomatis* HSP70 has approximately 70% similarity to DnaK proteins from *E. coli* and *Bacillus megaterium*. Polyclonal antibodies raised against the recombinant chlamydial HSP70 react with serovar L2 and C EBs in a micro-immunofluorescence test and an enzyme-linked immunosorbent assay and reduce the infectivity of EBs for cells *in vitro*. They also react with a 75 000 antigen from both serovars in an immunoblot assay.

The gene encoding the 75 000 protein from *C. trachomatis* serovar L2 has also been cloned and sequenced (Birkelund *et al.*, 1990). It has an ORF of 1980 bp, encoding a polypeptide of 660 amino acids with a calculated molecular weight of 70 875. The DNA sequence had 98% homology with *C. trachomatis* serovar D HSP70 gene, the only major difference was a 27 bp deletion in serovar D. In contrast, there was extensive heterology in the putative promoter regions. The deduced amino acid sequence of serovar L2 HSP70 showed 94% homology with serovar D HSP70, 57% homology with *E. coli* DnaK and 42% homology with human HSP70. A 7.2 kb fragment of DNA from the GPIC agent strain of *C. psittaci* was cloned and sequenced (Morrison *et al.*, 1989). The cloned sequence included two ORFs. One of these comprised 306 nucleotides, encoding a protein of 102 amino acids with a calculated molecular weight of 11 202; the other comprised 1632 nucleotides, encoding a protein of 544 amino acids with a calculated molecular weight of 58 088. The two genes were transcribed as a single molecule of messenger RNA and there was a heat shock promoter sequence upstream. The predicted amino acid sequence of the larger gene was compared with HSP60s from other bacteria. It had 61% identity with the HtpB protein of *Coxiella burnetti*, 60% identity with the GroEL protein of *E. coli*, 58% identity with the 65000 protein of *Mycobacterium tuberculosis* and 53% identity with the mature HSP60 of *Saccharomyces cerevisiae*.

The chlamydial HSP60 evokes an ocular delayed-type hypersensitivity response in animal models of trachoma. The effects of native chlamydial HSP60 and recombinant HSP70 were compared in the monkey model of trachoma (Taylor et al., 1990). Animals were sensitized by ocular infection with C. trachomatis serovar C. They were subsequently challenged with chlamydial HSP60 in one eye and HSP70 in the other eye. The HSP60 elicited an ocular delayed-type hypersensitivity response and the HSP70 had no effect.

Blinding trachoma is caused by multiple episodes of ocular chlamydial infection, with intervening episodes of bacterial conjunctivitis. Bacteria have very similar HSP60s and may also have common epitopes. Episodes of bacterial infection may increase the severity of trachoma by eliciting ocular hypersensitivity reactions to these shared epitopes.

2.5.4 Cell-binding proteins

The binding of chlamydial EBs to the eukaryotic cell surface is the first step in initiating the intracellular growth cycle. C. trachomatis serovars L2 and J both have two proteins with molecular weights of 18 000 and 31 000, which bind to HeLa cell membranes. They are present in EBs but not in RBs. Antibodies raised against these proteins and then absorbed with MOMP inhibit the association of EBs with cells and neutralize their infectivity (Wenman and Meuser, 1986).

A 4.7 kb DNA fragment from C. trachomatis serovar L2, which codes for the 18 000 protein, has been cloned and sequenced (Kaul et al., 1987). The gene codes for a protein of 162 amino acids with a calculated molecular weight of 18 314. The recombinant protein binds to HeLa cell membranes and antibodies raised against it neutralize the infectivity of chlamydial EBs for HeLa cells. Antibodies raised against the native protein prepared from EBs react weakly with the recombinant protein. This difference in reactivity suggests that there may be post-translational modifications of the native protein, or mechanisms for the assembly of the protein which are absent in E. coli. The authors have called the 18 000 protein Clanectin.

2.6 PLASMID

2.6.1 Size and distribution

The C. trachomatis plasmid was detected initially in three strains from serovars B, C and L2 (Lovett et al., 1980; cited in Hyypia et al., 1984). In a later report, plasmids were detected in strains from serovars B, C, D, L1, L2 and L3. The molecular weight of the plasmid was 4.5×10^6, which is equivalent to 7.4 kb (Peterson and de la Maza, 1983). The L2 plasmid was cloned and then used to probe strains of the 15 serovars of *C. trachomatis* which are human pathogens. Every serovar contained DNA homologous to the cloned plasmid (Hyypia *et al.*, 1984). There were ten copies of the plasmid present for each copy of the chlamydial chromosome when *C. trachomatis* serovar L2 was grown in L929 cells. The cloned L2 plasmid was again used to probe for homologous DNA, which again detected in all 15 serovars of *C. trachomatis* and in every strain isolated from over 200 patients (Palmer and Falkow, 1986b). In contrast, the mouse pneumonitis strain of *C. trachomatis* and the GPIC agent strain of *C. psittaci* lacked DNA homologous to the L2 plasmid. These findings led to the development of plasmid DNA probes for the routine diagnosis of *C. trachomatis* (section 2.7). It has been suggested that the plasmid is essential for the survival of *C. trachomatis*. However, a recent report suggests that it may not be essential: a strain of serovar L2 which lacks the plasmid and does not contain plasmid DNA integrated into the chlamydial chromosome was isolated from a patient with proctocolitis (Peterson *et al.*, 1990).

   C. pneumoniae does not have a plasmid (Campbell *et al.*, 1987). Strains of *C. psittaci* showed considerable diversity. Most strains have plasmids which range from 6.2 kb to 7.9 kb, but some strains lack plasmids, e.g. ovine abortion, ovine arthritis and ocular isolates from koala bears (Joseph *et al.*, 1986; Girjes *et al.*, 1988; McClenaghan *et al.*, 1988; Hugall *et al.*, 1989; Lusher *et al.*, 1989).

2.6.2 Restriction endonuclease studies

Studies of *C. trachomatis* plasmid DNA, digested with restriction endonucleases, have demonstrated that plasmids from different serovars are very similar. There was a single site for the enzyme Bam H1, three sites for Eco R1, and varying numbers of sites for Hind 111 (Palmer and Falkow, 1986b). In contrast, restriction endonuclease digestion of plasmid DNA from different strains of *C. psittaci* showed considerable diversity (Joseph *et al.*, 1986; Hugall *et al.*, 1989; Lusher *et al.*, 1988).

2.6.3 Sequence and genes

The DNA sequences of plasmids from serovars B, L1 and L2 have been determined and are very similar (Sriprakash and MacAvoy, 1987a; Hatt *et al.*, 1988; Comanducci *et al.*, 1988).

   The serovar B plasmid was the first to be completely sequenced (Sriprakash and MacAvoy, 1987a). It comprises 7496 bp and it has a G+C ratio of 36% which is distributed evenly throughout the plasmid. There are nine ORFs with the potential to code for proteins with a mass greater than 10 000. There were no recognizable transcription termination signals after several ORFs, suggesting that there is polycistronic transcription. Three plasmid-specific

RNA transcripts were detected in infected cells from 6 to 48 hours after infection. There are four 22 bp tandem repeat sequences between two ORFs, which can fold into a stem-loop structure resembling structures at the origins of replication of several *E. coli* plasmids. The ORF following this structure codes for a negatively charged protein which may control replication. Another ORF had 28% sequence identity with *E. coli* dnaB protein which is essential for the initiation of DNA replication, but it was not complementary in a dnaB mutant of *E. coli* (Sriprakash and MacAvoy, 1987b).

The L1 plasmid had 16 ORFs of at least 100 bp, 11 of which were on one strand and five on the other (Hatt *et al.*, 1988). The putative products ranged from 3500 to 41 000, and one resembled *E. coli* dnaB protein. The plasmid also had four tandem repeats of a 22 bp sequence and a fragment which acted as a promoter for the *E. coli lac Z* gene.

The L2 plasmid had eight major ORFs (Comanducci *et al.*, 1988). The molecular weights of seven putative translation products agreed well with the peptides which were produced in transcription and translation systems (Palmer and Falkow, 1986b). However, the eighth ORF (ORF 7) could not be expressed from the cloned plasmid because it contained the reaction site for the restriction endonuclease Bam H1 which was used to linearize the plasmid before ligation to the vector and cloning. This plasmid also contained a 22 bp sequence tandemly repeated four times which was identical to the sequence in the serovar B plasmid. One ORF coded for a polypeptide which included segments identical to *E. coli* dnaB protein. The first 388 amino acids of this polypeptide were virtually identical to the serovar B protein, with only three amino acid substitutions, two of which were conservative.

Plasmids from *C. trachomatis* serovar B and several strains of *C. psittaci* were compared (Hugall *et al.*, 1989). The plasmids cross-hybridized in only short regions under conditions of high stringency and the homologous regions contained the 22 bp tandem repeat sequence.

2.6.4 Products

The cloning and sequencing of *C. trachomatis* plasmids has been followed by studies of plasmid RNA and proteins. The *C. trachomatis* serovar L2 plasmid has been the most intensively investigated. In the first of these studies, the cloned plasmid was transcribed and translated *in vivo* in *E. coli* minicells and *in vitro* in *E. coli* S30 extract (Palmer and Falkow, 1986b). The *in vivo* translation products comprised eight proteins with molecular weights ranging from 12 000 to 48 000. They failed to react with human convalescent serum or rabbit chlamydial antiserum in an immunoprecipitation assay. The *in vitro* translation products comprised nine proteins with molecular weights ranging from 11 000 to 48 000, and were similar in size to the *in vivo* products. The production of these proteins during the chlamydial growth-cycle could not be studied because they could not be identified in the presence of other proteins. The production of plasmid RNA was studied during the growth of *C. trachomatis* serovar L2 in HeLa cells using Northern blotting and probing with the cloned plasmid. No plasmid RNA was detected in EBs, but was present at 24 and 48 hours after infection and disappeared by 60 hours, towards the end of the chlamydial growth cycle. Five plasmid RNA molecules were detected, with sizes ranging from 480 to 5000 nucleotides. The three smaller molecules probably coded for single proteins, the two larger ones were polycistronic mRNAs.

In a second study, plasmids from *C. trachomatis* serovar L2 and *C. psittaci* meningopneumonitis strain Cal 10 were cloned, transcribed and translated *in vivo* in *E. coli* maxicells and minicells and also *in vitro* (Joseph *et al.*, 1986). The L2 plasmid produced two polypeptides with molecular weights of 38 000 and 48 000 and the Cal 10 plasmid produced three polypeptides with molecular weights of 52 000, 38 000 and 22 000. None of these polypeptides were recognized by antisera which reacted with EBs in immunoprecipitation or immunoblotting assays.

In a third study, the L2 plasmid was cloned, transcribed and translated *in vitro* (Kahane and Sarov, 1987). Five proteins were produced, with molecular weights ranging from 20 000 to 56 000. Four of these were products of the *C. trachomatis* plasmid DNA.

The L1 plasmid was cloned, transcribed and translated in an *E. coli in vitro* system (Clarke and Hatt, 1986). Four proteins were produced, ranging from 15 000 to 46 000, totalling 127 000. The plasmid has the potential to code for a total of over 200 000, and the translation products may represent specific degradation products of plasmid proteins or the chlamydial plasmid may not function in the *E. coli* system

2.7 DNA PROBES FOR THE DETECTION OF CHLAMYDIAE

The cloning of chlamydial DNA was followed by the development of probes for detecting chlamydial DNA in specimens from patients with suspected chlamydial infections. The ideal test for a routine diagnostic laboratory should be sensitive and specific, quick and easy to perform, and be suitable for automation. Many routine laboratories are neither licensed nor equipped to handle radioactive substances and therefore the ideal test should not require the use of radioactively-labelled probes. At present, none of the tests which use DNA probes fulfils all the criteria.

Tests which use DNA probes to detect chlamydiae can be carried out in a number of ways. The major sources of variation are: the treatment of specimen before addition of the probe, the type of probe, the method of labelling the probe and hence the method of detecting hybridization between the probe and the specimen. DNA probes have been tested on specimens collected from the eyes of patients with trachoma and genitourinary tracts of patients with sexually transmitted infections. The probes can detect chlamydial nucleic acids in chlamydial inclusion bodies produced after inoculation of specimens into tissue culture cells. Probes have been used to detect chlamydial nucleic acids *in situ* in smears and in nucleic acids extracted from specimens. the sensitivity of detection can be increased by using the polymerase chain reaction (PCR) which generates extra copies of chlamydial DNA.

The first probes consisted of total chlamydial DNA. They were succeeded by plasmid probes and more recently by probes of cDNA complementary to ribosomal RNA. The first probes were radioactively labelled and hybrids were detected by autoradiography or gamma and liquid scintillation counters. Subsequently, probes have been labelled with biotin and hybrids detected by the reaction of biotin with streptavidin and, recently, probes have been labelled with acrinidium N-hydroxysuccinimide esters and hybrids detected by chemiluminescence. A number of studies have compared DNA probes with established diagnostic methods to detect chlamydial DNA in clinical specimens. Several of these studies are listed in Table 2.3 and are discussed below.

2.8 CHLAMYDIAL NUCLEIC ACIDS IN TISSUE CULTURE

A biotinylated probe was used to detect C. *trachomatis* DNA in McCoy cells which had been inoculated with urethral and endocervical specimens and

Table 2.3 Detection of chlamydial nucleic acids in clinical specimens

Probe	Detection of hybrids	Specimens	Reference
Chromosome	Autoradiography	Cervix	Hyypia et al., 1985
Chromosome	Autoradiography	Genitourinary tract	Palva, 1985
Plasmid	Autoradiography	Cervix	Horn et al., 1986
Plasmid	Autoradiography	Eye	Schachter et al., 1988
Plasmid	Autoradiography	Eye	Dean et al., 1989a
Plasmid	Autoradiogrpahy	Eye	Dean et al., 1989b
Plasmid and MOMP	Autoradiography and liquid scintillation counter	Female genital tract	Pao et al., 1987
cDNA	Gamma counter	Cervix, Urethra	Naher et al., 1989
cDNA	Chemiluminescence	Cervix	Peterson et al., 1989
cDNA	Chemiluminescence	Cervix, Urethra	Gratton et al., 1990
cDNA	Chemiluminescence	Cervix	Woods et al., 1990

incubated for up to 72 hours (Naher *et al.*, 1988). It was compared with a monoclonal antibody directed against chlamydial LPS which was used to stain inoculated cells. The DNA probe was highly specific, but it was not as sensitive as the monoclonal antibody for detecting inclusions. The DNA probe also had the disadvantages of being laborious and expensive.

2.9 CHLAMYDIAL NUCLEIC ACIDS IN SPECIMENS

Chlamydial chromosomal DNA probes were the first to be developed and tested on clinical specimens. Specimens were collected from the cervix and urethra of patients with sexually transmitted infections. Samples were spotted onto nitrocellulose filters and were probed with chromosomal DNA from *C. trachomatis* serovar L1 (Hyypia *et al.*, 1985). The probe was radioactively labelled and hybrids were detected by autoradiography. The probe gave a positive reaction with 24 out of 30 (80%) of specimens from *C. trachomatis* that had been isolated in tissue culture, and 7 out of 30 (23%) from which *C. trachomatis* had not been isolated. It was concluded that either a more sensitive probe or better methods of collecting specimens were required.

A cloned 10 kb fragment of serovar L2 chromosomal DNA was used to probe for chlamydial DNA in clinical specimens (Palva, 1985). DNA was extracted from the specimens and partially purified in order to prevent background or non-specific hybridization. The extracted DNA was spotted onto nitrocellulose filters. The probe was radioactively labelled and hybrids were detected by autoradiography. The probe gave a positive reaction with 69 out of 107 (64%) of specimens from which *C. trachomatis* had been isolated, most of which contained large numbers of chlamydial inclusion-forming units. The probe frequently failed to detect chlamydial DNA in specimens which contained low numbers of chlamydial inclusion-forming units. This probably reflected the loss of chlamydial DNA in the extraction and purification procedures which preceded hybridization. However, the probe also gave a positive reaction with 30 out of 109 (28%) of specimens from which *C. trachomatis* had not been isolated. It was concluded that some represented true positives, specimens which were toxic to cells so that they did not produce intracellular inclusions in tissue culture.

C. trachomatis plasmid DNA has been used to detect chlamydial DNA in ocular and genital specimens *in situ* and in DNA extracted from specimens. The chlamydial plasmid was used for *in situ* hybridization in cervical smears (Horn *et al.*, 1986). Forty-six smears from 31 patients were tested and the results were compared with the isolation of *C. trachomatis* in tissue culture. There were 39 concordant results and seven discrepant results. Nineteen specimens gave positive results in both tests, two were only positive in tissue culture and five were only positive in *in situ* hybridization. The *in situ* hybridization technique detects low numbers of chlamydia while cell morphology is maintained. It may therefore be useful for screening asymptomatic patients and populations with low prevalences of chlamydial infections.

Probes for *C. trachomatis* plasmid and MOMP DNA were used to test specimens from 172 women at high or low risk of chlamydial genital infection (Pao *et al.*, 1987). The two probes gave very similar results. They were sensitive and specific, detecting 1pg chlamydial DNA. They were more sensitive than the 'Chlamydiazyme' antigen detection system, an enzyme immunoassay which detects chlamydial LPS, but not as sensitive as culture.

C. trachomatis plasmid probes have also been tested on ocular specimens collected from patients with trachoma in Egypt and Nepal. In Egypt, specimens were collected from children (Schachter *et al.*, 1988). DNA was extracted from the specimens and applied to nylon filters before hybridization. This technique was compared with tissue culture and two established immunological methods for the direct detection of chlamydial antigens in specimens. These two methods were Microtrak, which is a fluorescent labelled monoclonal antibody which reacts with *C. trachomatis* MOMP and is used to detect EBs in smears and Chlamydiazyme, an enzyme immunoassay which reacts with chlamydial LPS. The DNA probe detected 38 out of 45 (84%) positive specimens and was the most sensitive method. Several of these positive specimens were also positive in the immunological assays but were negative in tissue culture system. It was concluded that these were true positives and therefore that some children intermittently produced chlamydial antigens.

Two methods of generating and radioactively labelling the *C. trachomatis* plasmid DNA probe have been tested on ocular specimens collected from patients with trachoma in Nepal. A conventionally-generated and labelled probe was used in the first study and a probe which had been generated by the polymerase chain reaction (PCR) and labelled during amplification was used in the second (Dean *et al.*, 1989a; Dean *et al.*, 1989b). The probes were compared with the immunological assay 'Microtrak' which detects EBs in smears and with a double passage tissue culture technique. The probe generated by PCR was more sensitive and specific than the conventionally generated probe. Both probes were considerably more sensitive that the immunological assay, and slightly less sensitive than tissue culture.

A cDNA probe complementary to chlamydial ribosomal RNA has been developed and used in a number of studies to detect *C. trachomatis* in specimens from patients with sexually transmitted infections. Specimens are collected using swabs which were then placed in transport medium provided by the manufacturer and contained substances which release ribosomal RNA from micro-organisms. In the first study, the probe was radioactively labelled with ^{125}I, hybrids were separated because they bound to magnetic particles and the radioactivity was counted in a gamma counter (Naher *et al.*, 1989). The probe gave a positive reaction in 31 out of 37 (84%) specimens from which *C. trachomatis* was isolated in tissue culture and also in a further 11 specimens from which *C. trachomatis* was not grown. Tests with a wide range of micro-organisms showed that they could all evoke a non-specific false-positive reaction when present in large numbers.

A recent modification has been introduced in which this cDNA probe was labelled with acrinidium N-hydroxysuccinimide ester which is covalently linked to the probe. The ester reacts with hydrogen peroxide to produce light (chemiluminescence) which is measured in a luminometer. The hybridized probe is protected from hydrolysis and is chemiluminescent. The unhybridized probe is not protected and it is no longer chemiluminescent after hydrolysis. This method has been used in several recent studies (Peterson *et al.*, 1989; Gratton *et al.*, 1990; Woods *et al.*, 1990). The probe was tested on tissue culture cells which had been inoculated with strains of all 15 serovars which are human pathogens. The lower limit of sensitivity of the probe varied widely; ranging from 99 chlamydial inclusion forming units for serovar C to 2930 chlamydial inclusion forming units for serovar L1 (Peterson *et al.*, 1989). The probe was used to test cervical specimens and compared with tissue culture and two immunological assays; a direct fluorescent antibody staining EBs and RBs and Chlamydiazyme. The DNA probe was the least sensitive method, particularly when low numbers of chlamydial inclusion-forming units were present and it was also the least specific method, with the highest number of false-positive results. The two other studies also found that this method was not as sensitive or as specific as established methods (Gratton *et al.*, 1990; Woods *et al.*, 1990).

2.10 AMPLIFIED CHLAMYDIAL NUCLEIC ACIDS

PCR is a rapid method of producing multiple copies of a piece of DNA, known as the target DNA, *in vitro*. It requires two oligonucleotides which act as primers for DNA polymerase. One primer is complementary to the 5′ end of one strand of the target DNA, and the second primer is complementary to the 5′ end of the target DNA on the other strand. The target DNA is denatured and the primers anneal to the strands. In the presence of DNA polymerase and nucleotides, a copy of each strand is then synthesized and the whole cycle is repeated many times to produce many copies of both strands of the target DNA. The advantages of PCR are that it permits the geometric amplification of a specific sequence and hence makes it easier to detect very small amounts of target DNA. Specimens do not need complex purification procedures to remove other DNA before amplification because primers and conditions of hybridization can be selected so that only the target DNA is amplified. However, great care must be taken to ensure that reagents and equipment are not contaminated with previous samples. For diagnostic purposes, the identity of the PCR product must be confirmed. This is usually done by Southern blotting to check that the correct sized product hybridizes radioactively-labelled chlamydial DNA.

The application of PCR to the diagnosis of infection with *C. trachomatis* is at an early stage. Primers have been developed and used to amplify the plasmid and the 16s ribosomal RNA gene. Amplification of the *C. trachomatis* plasmid has been used to detect chlamydial DNA in specimens collected from the genital tracts of patients with chlamydial genital infection (Griffais and Thibon, 1989) and in synovial fluids from patients with sexually-acquired reactive arthritis (Wordsworth *et al.*, 1990). PCR has also been used to generate radioactively-labelled plasmid probes to detect *C. trachomatis* in ocular specimens from patients with trachoma living in Nepal (Dean *et al.*, 1989b).

A study of 200 genital specimens was carried out to compare PCR of the *C. trachomatis* plasmid with isolation of *C. trachomatis* in tissue culture (Griffais and Thibon, 1989). A 201 bp fragment of plasmid DNA was amplified and the product was identified by agarose gel electrophoresis, followed by blotting and hybridization with radioactively-labelled RNA specific to the target sequence. A total of eight specimens out of 200 (4%) were positive by PCR, six were positive by tissue culture (3%) while a further four were doubtful and immunofluorescent staining of these four specimens showed that two were positive, giving a total of eight positives by tissue culture and immunological staining. It was concluded that PCR was both highly sensitive and highly specific.

PCR was used to probe for chlamydial DNA in synovial biopsies and fluids obtained from the inflamed joints of patients with sexually-acquired reactive arthritis (Wordsworth *et al.*, 1990). PCR produced a 590 bp segment of the *C. trachomatis* plasmid and a minimum of ten copies of the plasmid could be detected, which is equivalent to one EB, assuming that there are ten copies of the plasmid to each copy of the chlamydial chromosome. Immunofluorescent staining with 'Microtrak', which reacts with *C. trachomatis* MOMP, showed that EBs were present in specimens from seven out of 34 patients. However, no plasmid DNA was produced by PCR in any specimens. The reasons for this discrepancy between the two tests are not understood.

Amplification of the 16S ribosomal RNA gene with two sets of primers was carried out using McCoy cells infected with several serovars of *C. trachomatis* and *C. psittaci* (Pollard *et al.*, 1989). This test may to be more sensitive and more specific than tests with the cDNA probe which are unsatisfactory for the direct detection of *C. trachomatis* in clinical specimens. The test can also be used to differentiate between *C. trachomatis* and *C. psittaci*. The products were identified by electrophoresis in agarose gels. When *C. trachomatis* DNA was amplified, the product was a 240 bp fragment, when *C. psittaci* DNA was amplified, there were two products of 240 bp and 119 bp. The test was also very sensitive; it could detect one or two molecules of chlamydial DNA in 10^5 McCoy cells. Other cell lines and other bacteria were used to check the specificity of the test. No PCR products were formed. Nucleic acids were extracted from specimens before PCR, and after extraction, the test could be completed in a day. This test was not carried out on clinical specimens and it was not considered to be suitable for use in a routine diagnostic laboratory because of the time taken to carry out the test. The test is being modified so that DNA is not extracted before PCR and the test can be completed in five hours.

2.11 THE STORY TO DATE

The application of molecular biology to the study of chlamydiae has greatly increased our understanding of their origins, classification, immunology and pathogenesis. Studies of chlamydial ribosomal RNA genes, ribosomal protein S1, LPS, HSP70 and HSP60 all support the hypothesis that chlamydiae are Gram-negative bacteria. The observation that chlamydiae only have one *tuf* gene is the only conflicting finding.

The degree of homology between the genomes of the three species of chlamydiae is very low (<10%). In contrast, when the sequences of genes from the different species are compared, they show a high degree of homology (>70%). There are evolutionary pressures on structural genes to conserve function. The high degree of homology between genes of the three species is therefore not surprising. The presence or absence of the chlamydial plasmid is the only major difference which has been described. Virtually all strains of C. *trachomatis* have a plasmid which is highly conserved in size and sequence. Many strains of C. *psittaci* have plasmids, which vary considerably in size and sequence and have little homology with the C. *trachomatis* plasmid. Plasmids are not present in some strains of C. *psittaci* and have not been found in any isolates of C. *pneumoniae*. The chlamydial plasmid DNA is approximately less than 1% the chlamydial chromosome DNA, and assuming that there are ten copies per chromosome, the plasmid comprises less than 10% of the total DNA. The plasmid DNA therefore can only account for a small portion of the non-homologous DNA.

The serological classification of C. *trachomatis* is based upon antigenic determinants located in the outer membrane proteins, and the techniques of molecular biology have been used to map the genetic differences which give rise to particular epitopes and hence to serological differences. Studies using restriction endonucleases have shown that particular fragments are associated either with the trachoma or LGV biovar, or with members of the trachoma biovar causing ocular or genital infection. However, the genetic basis of the differences in pathogenicity remains unclear.

Molecular biology has identified three possible candidates for a chlamydial vaccine because antibodies raised against them neutralize the infectivity of *C. trachomatis* EBs *in vitro*. They are MOMP, HSP70 and the 18 000 cell-binding protein. Antibodies against MOMP also neutralize the infectivity of EBs *in vivo*. MOMP was used to immunize monkeys against subsequent ocular infection, but the results were disappointing, with no evidence of any protective effect. Molecular biology has also identified HSP60 as a chlamydial antigen or allergen. This elicits the ocular delayed-type hypersensitivity reaction which is associated with repeated infection and ocular damage. HSP60 may also cause delayed-type hypersensitivity responses in the genital tract, leading to tissue damage and resultant infertility.

Molecular biology has already made a major impact on our understanding of chlamydiae and will continue to do so. This chapter attempts to put the contribution of molecular biology into the context of our knowledge of chlamydiae from other disciplines. Much remains to be done, one is reminded of the first century sage who wrote: 'You are not required to complete the task, nor are you at liberty to abstain from it' (Tarfon cited Blackman, 1983).

ACKNOWLEDGEMENTS

I thank Dr Pamela Greenwell, Polytechnic of Central London; Dr A. Keat, Westminster Hospital and Professor S. Darougar, Institute of Ophthalmology, with whom I have discussed various aspects of this chapter.

REFERENCES

Adler, M. (1990) Epidemiology of sexually transmitted diseases, in *Sexually Transmitted Diseases. A textbook of genito-urinary medicine* (eds G.W. Csonka and J.K. Oates), Bailière-Tindall, London, Philadelphia, Toronto, Sydney, Tokyo, pp. 6–16.

Allan, I. (1986) Chlamydial antigenic structures and genetics, in *Chlamydial Infections* (eds D. Oriel, G. Ridgway, J. Schachter, D. Taylor-Robinson and M. Ward), Cambridge University Press, Cambridge, London, New York, New Rochelle, Melbourne, Sydney, pp. 73–80.

Allen, J.E. and Stephens, R.S. (1989) Identification by sequence analysis of two-site posttranslational processing of the cysteine-rich outer membrane protein 2 of *Chlamydia trachomatis* serovar L2. *J. Bacteriol.*, 171, 285–91.

Anderson, A.A. and Tappe, J.P. (1989) Genetic, immunologic, and pathologic characterization of avian chlamydial strains. *J. Am. Vet. Med. Assoc.*, 195, 1512–16.

Baehr, W., Zhang, Y.-X., Joseph, T. *et al.* (1988) Mapping antigenic domains expressed by *Chlamydia trachomatis* major outer membrane protein genes. *Proc. Nat. Acad. of Sci.*, 85, 4000–4.

Batteiger, B.E., Newhall, W.J., Terho, P. *et al.* (1986) Antigenic analysis of the major outer membrane protein of *Chlamydia trachomatis* with murine monoclonal antibodies. *Infect. and Immun.*, 53, 530–3.

Bavoil, P., Ohlin, A. and Schachter, J. (1984) Role of disulfide bonding in outer membrane structure and permeability in *Chlamydia trachomatis*. *Infect. and Immun.*, **44**, 479–85.

Birkelund, S., Lundemose, A.G. and Christiansen, G. (1990) The 75-kilodalton cytoplasmic *Chlamydia trachomatis* L2 polypeptide is a DnaK-like protein. *Infect. and Immun.*, **58**, 2098–104.

Blackman, P. (1985) citing Tarfon in, *Translations of the Mishna*, **4**, Judaica Press Ltd, Gateshead, p. 505.

Brade, H., Brade, L. and Nano, F.E. (1987a) Chemical and serological investigations on the genus-specific lipopolysaccharide epitope of *Chlamydia*. *Proc. Natl Acad. Sci. USA*, **84**, 2508–12.

Brade, L., Nano, F.E., Schlecht, S. *et al.* (1987b) Antigenic and immunogenic properties of recombinants from *Salmonella typhimurium* and *Salmonella minnesota* rough mutants expressing in their lipopolysaccharide a genus-specific chlamydial epitope. *Infect. and Immun.*, **55**, 482–6.

Brade L., Schramek, S., Schade U. and Brade H. (1986) Chemical, biological and immunochemical properties of the *Chlamydia psittaci* lipopolysaccharide. *Infect. and Immun.*, **54**, 568–74.

Caldwell, H.D. and Perry, L.J. (1982) Neutralization of *Chlamydia trachomatis* infectivity with antibodies to the major outer membrane protein. *Infect. and Immun.*, **38**, 745–54.

Caldwell, H.D. and Hitchcock P.J. (1984) Monoclonal antibody against a genus-specific antigen of *Chlamydia* species: location of the epitope on chlamydial lipopolysaccharide. *Infect. and Immun.*, **44**, 306–14.

Campbell, L.A., Kuo, C.-C. and Grayston, J.T. (1987) Characterisation of the new Chlamydia agent, TWAR, as a unique organism by restriction endonuclease analysis and DNA–DNA hybridisation. *J. Clin. Microbiol.*, **25**, 1911–16.

Campbell, L.A., Kuo, C.-C. and Grayston, J.T. (1990) Structural and antigenic analysis of *Chlamydia pneumoniae*. *Infect. and Immun.*, **58**, 93–7.

Carlson, E.J., Peterson, E.M. and de la Maza, L.M. (1989) Cloning and characterization of a *Chlamydia trachomatis* L3 DNA fragment that codes for an antigenic region of the major outer membrane protein and specifically hybridizes to the C- and C-related-complex serovars. *Infect. and Immun.*, **57**, 487–94.

Clarke, I.N. and Hatt, C. (1986) *In vitro* transcription/translation of cloned plasmid DNA from *Chlamydia trachomatis* serovar L2, in *Chlamydial Infections* (eds D. Oriel, G. Ridgway, J. Schachter *et al.*), Cambridge University Press, Cambridge, London, New York, New Rochelle, Melbourne, Sydney, pp. 85–8.

Clarke, I.N., Ward, M.E. and Lambden, P.R. (1988) Molecular cloning and sequence analysis of a developmentally regulated cysteine-rich outer membrane protein from *Chlamydia trachomatis*. *Gene*, **71**, 307–14.

Comanducci, M., Ricci, S. and Ratti, G. (1988) The structure of a plasmid of *Chlamydia trachomatis* believed to be required for growth within mammalian cells. *Mol. Microbiol.*, **2**, 531–8.

Conlan, J.W., Clarke, I.N. and Ward, M.E. (1988) Epitope mapping with solid-phase peptides: identification of type-, sub-species-, species-, and genus-reactive antibody binding domains on the major outer membrane protein of *Chlamydia trachomatis*. *Mol. Microbiol.*, **2**, 673–9.

Conlan, J.W., Ferris, S., Clarke, I.N and Ward, M.E. (1990) Isolation of recombinant fragments of the major outer membrane protein of *Chlamydia trachomatis*: their potential as subunit vaccines. *J. Gen. Microbiol.*, 136, 2013–20.

Conlan, J.W., Kajbaf, M., Clarke, I.N. *et al.* (1989) The major outer membrane protein of *Chlamydia trachomatis*: critical binding site and conformation determine the specificity of antibody to viable chlamydia. *Mol. Microbiol.*, 3, 311–18.

Cox, R.L., Kuo, C.-C., Grayston, J.T. and Campbell, L.A. (1988) Deoxyribonucleic acid relatedness of *Chlamydia* sp. strain TWAR to *Chlamydia trachomatis* and *Chlamydia psittaci*. *Int. J. Syst. Bacteriol.*, 38, 265–8.

Danilition, S.L., Maclean, I.W., Peeling R. *et al.* (1990) The 75-kilodalton protein of *Chlamydia trachomatis*: a member of the heat shock protein 70 family? *Infect. and Immun.*, 58, 189–96.

Dean, D., Palmer, L., Pant, C.R. *et al.* (1989a) Use of a *Chlamydia trachomatis* DNA probe for detection of ocular chlamydiae. *J. Clin. Microbiol.*, 27, 1062–7.

Dean, D., Pant, C.R. and O'Hanley, P. (1989b) Improved sensitivity of a modified polymerase chain reaction amplified DNA probe in comparison with serial tissue culture for detection of *Chlamydia trachomatis* in conjunctival specimens from Nepal. *Diag. Microbiol. Infect. Dis.*, 12, 133–7.

Engel, J.N. and Ganem, D. (1987) Chlamydial rRNA operons: gene organization and identification of putative tandem promoters. *J. Bacteriol.*, 169, 5678–85.

Frutos, R., Pages, M., Bellis, M. *et al.* (1989) Pulsed-field gel electrophoresis determination of the genome size of obligate intracellular bacteria belonging to the genera *Chlamydia, Ricketsiella,* and *Porochlamydia*. *J. Bacteriol.*, 171, 4511–13.

Fukushi, H. and Hirai, K. (1989) Genetic diversity of avian and mammalian *Chlamydia psittaci* strains and relation to host origin. *J. Bacteriol.*, 171, 2850–5.

Girjes, A.A., Hugall, A.F., Timms, P. and Lavin, M.F. (1988) Two distinct forms of *Chlamydia psittaci* associated with disease and infertility in *Phascolarctos cinereus* (koala). *Infect. and Immun.*, 56, 1897–1900.

Goldstein, B.P., Zaffaroni G., Tiboni, O. *et al.* (1989) Determination of the number of *tuf* genes in *Chlamydia trachomatis* and *Neisseria gonorrhoeae*. *FEMS Microbiol. Letts.*, 60, 305–10.

Gratton, C.A., Lim-Fong, R., Prasad, E. and Kibsey, P.C. (1990) Comparison of a DNA probe with culture for detecting *Chlamydia trachomatis* directly from genital specimens. *Mol. and Cell. Probes*, 4, 25–31.

Griffais, R. and Thibon, M. (1989) Detection of *Chlamydia trachomatis* by the polymerase chain reaction. *Res. Microbiol.*, 140, 139–41.

Hamilton, P.T. and Malinowski, D.P. (1989) Nucleotide sequence of the major outer membrane protein gene from *Chlamydia trachomatis* serovar H. *Nucl. Acids Res.*, 17, 8366.

Hatt, C., Ward, M.E. and Clarke, I.N. (1988) Analysis of the entire nucleotide sequence of the cryptic plasmid of *Chlamydia trachomatis* serovar L1. Evidence for involvement in DNA replication. *Nucl. Acids Res.*, 16, 4053–67.

Hayes, L.J., Pickett, M.A., Conlan, J.W. *et al.* (1990) The major outer membrane proteins of *Chlamydia trachomati* serovars A and B: intra serovar amino acid changes do not alter specificities of serovar- and C subspecies-reactive domains. *J. Gen. Microbiol.*, **136**, 1559–66.

Herring, A.J., Tan, T.W., Baxter, S. *et al.* (1989) Sequence analysis of the major outer membrane protein gene of an ovine abortion strain of *Chlamydia psittaci. FEMS Microbiol. Letts.*, **65**, 153–8.

Horn, J.E., Hammer, M.L., Falkow, S. and Quin, T.C. (1986) Detection of *Chlamydia trachomatis* in tissue culture and cervical scrapings by in situ hybridisation. *J. Infect. Dis.*, **153**, 1155–9.

Hugall, A., Timms, P., Girjes, A.A. and Lavin, M.F. (1989) Conserved DNA sequences in chlamydial plasmids. *Plasmid*, **22**, 91–8.

Hyypia, T., Larsen, S.H., Stahlberg, T. and Terho, P. (1984) Analysis and detection of chlamydial DNA. *J. Gen. Microbiol.*, **130**, 3159–64.

Hyypia, T., Jalava, A., Larsen, S.H. *et al.* (1985) Detection of *Chlamydia trachomatis* in clinical specimens by nucleic acid spot hybridisation. *J. Gen. Microbiol.*, **131**, 975–8.

Jones, M.D., Peterson, T.B., Nielsen, K.M. *et al.* (1980) The complete amino acid sequence of elongation factor Tu from *Escherichia coli. Europ. J. Biochem.*, **108**, 507–26.

Joseph, T., Nano, F.E., Garon, C.F. and Caldwell, H.D. (1986) Molecular characterisation of *Chlamydia trachomatis* and *Chlamydia psittaci* plasmids. *Infect. and Immun.*, **51**, 699–703.

Kahane, S. and Sarov, I. (1987) Cloning of a chlamydial plasmid: its use as a probe and in vitro analysis of encoded polypeptides. *Current Microbiol.*, **14**, 255–8.

Karimi, S.T., Schloemer, R.H. and Wilde, C.E. (1989) Accumulation of chlamydial lipopolysaccharide antigen in the plasma membranes of infected cells. *Infect. and Immun.*, **57**, 1780–5.

Kaufmann, S.H.E. (1990) Heat shock proteins and the immune response. *Immunol. Today*, **11**, 129–36.

Kaul, R., Roy, K.L. and Wenman, W.N. (1987) Cloning, expression and primary structure of a *Chlamydia trachomatis* binding protein. *J. Bacteriol.*, **169**, 5152–6.

Kingsbury, D.T. and Weiss, E. (1968) Lack of deoxyribonucleic acid homology between species of the genus *Chlamydia. J. Bacteriol.*, **96**, 1421–3.

Lovett, M., Kuo, C.-C., Holmes, K.K. and Falkow, S. (1980) Plasmids of the genus Chlamydia. *Current Chemotherapy and Infectious Disease*, **2**, 1250–2. American Society for Microbiology, Washington.

Lusher, M., Storey, G.C. and Richmond, S.J. (1988) Plasmid diversity within the genus *Chlamydia. J. Gen. Microbiol.*, **135**, 1145–51.

McClenaghan, M., Herring, A.J. and Aitken, I.D. (1984) Comparison of *Chlamydia psittaci* isolates by DNA restriction endonuclease analysis. *Infect. and Immun.*, **45**, 384–9.

McClenaghan, M., Honeycombe, J.R., Bevan, B.J. and Herring A.J. (1988) Distribution of plasmid sequences in avian and mammalian strains of *Chlamydia psittaci. J. Gen. Microbiol.*, **134**, 559–64.

Monnickendam, M.A. (1988) Chlamydial genital infections, in *Immunology of Sexually Transmitted Diseases* (ed. D.J.M. Wright), Kluwer Academic Publishers; Dordrecht, Boston, London, pp. 117–61.

Morrison, R.P., Belland, R.J., Lyng, K. and Caldwell, H.D. (1989) Chlamydial disease pathogenesis. The 57-KD chlamydial hypersensitivity antigen is a stress response protein. *J. Exp. Med.*, 170, 1271–83.

Naher, H., Niebauer, B., Hartmann, M. *et al.* (1989) Evaluation of a radioactive rRNA: cDNA hybridisation assay for the direct detection of *Chlamydia trachomatis* in urogenital specimens. *Genitourinary Med.*, 65, 319–22.

Naher, H., Petzoldt, D. and Sethi, K.K. (1988) Evaluation of non-radioactive in situ hybridisation method to detect *Chlamydia trachomatis* in cell culture. *Genitourinary Med.*, 64, 162–4.

Nano, F.E. and Caldwell, H.D. (1985) Expression of the chlamydial genus-specific lipopolysaccharide epitope in *Escherichia coli. Science*, 230, 1279–81.

Newhall, W.J. and Jones B.R. (1983) Disulfide-linked oligomers of the major outer membrane protein of Chlamydiae. *J. Bacteriol.*, 154, 998–1001.

Nurminen, M., Leionen, M., Saikku, P. and Makela, P.H. (1983) The genus-specific antigen of *Chlamydia*: resemblance to the lipopolysaccharide of enteric bacteria. *Science*, 228, 742–4.

Nurminen, M., Rietschel, E.T. and Brade, H. (1985) Chemical characterisation of *Chlamydia trachomatis* lipopolysaccharide. *Infect. and Immun.*, 48, 573–5.

Nurminen, M., Wahlstrom, E., Kleemolu, M. *et al.* (1984) Immunologically related ketodeoxy-octonate-containing structures in *Chlamydia trachomatis*, Re mutants of *Salmonella* species, and *Acinetobacter calcoaceticus* var. *anitratus. Infect. and Immun.*, 44, 609–13.

Palmer, L. and Falkow, S. (1986a) 16s ribosomal RNA genes of *Chlamydia trachomatis*, in *Chlamydial Infections* (eds D. Oriel, G. Ridgway, J. Schachter, D. Taylor-Robinson and M. Ward), Cambridge University Press, Cambridge, London, New York, New Rochelle, Melbourne, Sydney, pp. 89–92.

Palmer, L. and Falkow, S. (1986b) A common plasmid of *Chlamydia trachomatis. Plasmid*, 16, 52–62.

Palva, A. (1985) Nucleic acid spot hybridisation for detection of *Chlamydia trachomatis. FEMS Microbiol. Letts*, 28, 85–91.

Pao, C.C., Lin, S.-S., Yang, T.-E. *et al.* (1987) Deoxyribonucleic acid hybridization analysis for the detection of urogenital *Chlamydia trachomatis* infection in women. *Am. J. Obstet. Gynecol.*, 156, 195–9.

Perara, E., Yen, T.S. and Ganem, D. (1990) Growth of *Chlamydia trachomatis* in enucleated cells. *Infect. and Immun.*, 53, 3816–18.

Peterson E.M. and de la Maza, L.M. (1983) Characterisation of *Chlamydia* DNA by restriction endonuclease cleavage. *Infect. and Immun.*, 41, 604–8.

Peterson E.M. and de la Maza, L.M. (1988) Restriction endonuclease analysis of DNA from *Chlamydia trachomatis* biovars. *J. Clin. Microbiol.*, 26, 625–9.

Peterson, E.M., Markoff, B.A., Schachter, J. and de la Maza, L.M. (1990) Absence of the 7.5 kb plasmid in a *Chlamydia trachomatis* isolate, in *Chlamydial Infections* (eds W.R. Bowie, H.D. Caldwell, B.R. Jones, P.A. Mardh, G.L. Ridgway, J. Schachter, W.E. Stamme and M.E. Ward), Cambridge University Press, Cambridge, New York, Port Chester, Melbourne, Sydney, pp. 144–8.

Peterson, E.M., Oda, R., Alexander, R. *et al.* (1989) Molecular techniques for the detection of *Chlamydia trachomatis. J. Clin. Microbiol.*, 27, 2359–63.

Pickett, M.A., Ward, M.E. and Clarke, I.N. (1987) Complete nucleotide sequence of the major outer membrane protein gene from *Chlamydia trachomatis* serovar L1. *FEMS Microbiol. Letts*, 42, 185–90.

Pickett, M.A., Everson, J.S. and Clarke, I.N. (1988a) *Chlamydia psittaci* ewe abortion agent; complete nucleotide sequence of the major outer membrane protein gene. *FEMS Microbiol. Letts*, 55, 229–34.

Pickett, M.A., Ward, M.E. and Clarke, I.N. (1988b) High-level expression and epitope localisation of the outer membrane protein of *Chlamydia trachomatis* serovar L1. *Mol. Microbiol.*, 2, 681–5.

Pollard, D.R., Tyler, S.D., Ng, C.-W. and Rozee, K.R. (1989) A polymerase chain reaction (PCR) protocol for the specific detection of *Chlamydia* spp. *Mol. and Cell. Probes*, 3, 383–9.

Sardinia, L.M., Engel, J.N. and Ganem, D. (1989) Chlamydial gene encoding a 70-kilodalton antigen in *Escherichia coli*; analysis of expression signals and identification of the gene product. *J. Bacteriol.*, 171, 335–51.

Sardinia, L.M., Segal, E. and Ganem, D. (1988) Developmental regulation of the cysteine-rich outer membrane protein of murine *Chlamydia trachomatis*. *J. Gen. Microbiol.*, 134, 997–1004.

Schachter, J., Moncada, J., Dawson, C.R. *et al.* (1988) Nonculture methods for diagnosing chlamydial infections in patients with trachoma: a clue to the pathogenesis of the disease? *J. Infect. Dis.*, 158, 1347–52.

Shanmugaratnam, K. and Pattman, R.S. (1989) Declining incidence of *Chlamydia trachomatis* in women attending a provincial genitourinary medicine clinic. *Genitourinary Med.*, 65, 400.

Sivakumar, K. and Basu Roy, R. (1989) Falling prevalence of *Chlamydia trachomatis* infection among female patients attending the Department of Genitourinary Medicine, Bournemouth. *Genitourinary Med.*, 65, 400.

Sriprakash, K.S. and MacAvoy, E.S. (1987a) Characterisation and sequence of a plasmid from the trachoma biovar of *Chlamydia trachomatis*. *Plasmid*, 18, 205–14.

Sriprakash, K.S. and MacAvoy, E.S. (1987b) A gene for dnaB like protein in chlamydial plasmid. *Nuc. Acids Res.*, 15, 10 596.

Stephens, R.S., Kuo, C.-C., Newport, G. and Agabian, N. (1985) Molecular cloning and expression of *Chlamydia trachomatis* major outer membrane protein antigens in *Escherichia coli*. *Infect. and Immun.*, 47, 713–18.

Stephens, R.S., Mullenbach, G., Sanchez-Pescador, R. and Agabian, N. (1986) Sequence analysis of the major outer membrane protein gene from *Chlamydia trachomatis* serovar L2. *J. Bacteriol.*, 168, 1277–82.

Stephens, R.S., Sanchez-Pescador, R., Wagar, E.A. *et al.* (1987) Diversity of *Chlamydia trachomatis* major outer membrane protein genes. *J. Bacteriol.*, 169, 3879–85.

Stephens, R.S., Wagar, E.A. and Edman, U. (1988a) Developmental regulation of tandem promoters for the major outer membrane protein gene of *Chlamydia trachomatis*. *J. Bacteriol.*, 170, 744–50.

Stephens, R.S. Wagar, EA. and Schoolnik, G.K. (1988b) High resolution mapping of serovar-specific and common antigenic determinants of the major outer membrane protein of *Chlamydia trachomatis*. *J. Exp. Med.*, 167, 817–31.

Su, H., Zhang, Y.-X., Barrera, O. *et al.* (1988) Differential effect of trypsin on infectivity of *Chlamydia trachomatis*: loss of infectivity requires cleavage of major outer membrane protein variable domains II and IV. *Infect. and Immun.*, 56, 2094–100.

Su, H., Watkins, N.G., Zhang, Y.-X. and Caldwell, H.D. (1990) *Chlamydia trachomatis* – host cell interactions; role of the chlamydial major outer membrane protein as an adhesin. *Infect. and Immun.*, 58, 1017–25.

Taylor, H.R., Maclean, I.W., Brunham, R.C. *et al.* (1990) Chlamydial heat-shock proteins and trachoma. *Infect. and Immun.*, 58, 3061–3.

Taylor, H.R., Whittum-Hudson, J., Schachter, J. *et al.* (1988) Oral immunization with chlamydial major outer membrane protein (MOMP). *Invest. Ophthalm. Vis. Sci.*, 29, 1847–53.

Wagar, E.A. and Stephens, R.S. (1988) Developmental-form-specific DNA-binding proteins in *Chlamydia* spp. *Infect. and Immun.*, 56, 1678–84.

Weisburg, W.G., Hatch, T.P. and Woese, C.R. (1986) Eubacterial origin of chlamydiae. *J. Bacteriol.*, 167, 570–4.

Wenman, W.M. and Meuser, R.U. (1986) *Chlamydia trachomatis* elementary bodies possess proteins which bind to eucaryotic cells. *J. Bacteriol.*, 165, 602–7.

Woods, G.L., Young, A., Scott, J.C. *et al.* (1990) Evaluation of a non-isotopic probe for detection of *Chlamydia trachomatis* in endocervical specimens. *J. Clin. Microbiol.*, 28, 370–2.

Wordsworth B.P., Hughes, R.A., Allan, I. *et al.* (1990) Chlamydial DNA is absent from the joints of patients with sexually acquired reactive arthritis. *Brit. J. Rheumatol.*, 29, 208–10.

Yuan Y., Zhang, Y.-X., Watkins, N.G. and Caldwell, H.D. (1989) Nucleotide and deduced amino acid sequences for the variable domains of the outer membrane proteins of the 15 *Chlamydia trachomatis* serovars. *Infect. and Immun.*, 57, 1040–9.

Zhang, Y.-X., Morrison, S.G., Caldwell, H.D. and Baehr, W. (1989) Cloning and sequence analysis of the major outer membrane protein genes of two *Chlamydia psittaci* strains. *Infect. and Immun.*, 57, 1621–5.

Zhang, Y.-X., Stewart, S., Joseph, T. *et al.* (1987a) Protective monoclonal antibodies recognise epitopes located on the major outer membrane protein of *Chlamydia trachomatis*. *J. Immunol.*, 138, 575–81.

Zhang, Y.-X., Watkins, N.G., Stewart, S. and Caldwell, H.D. (1987b) The low-molecular-weight mass, cysteine-rich outer membrane protein of *Chlamydia trachomatis* possesses both biovar and species-species-specific epitopes. *Infect. and Immun.*, 55, 2570–3.

Zhong, G., Reid, R.E. and Brunham, R.C. (1990) Mapping antigenic sites on the major outer membrane protein of *Chlamydia trachomatis* with synthetic peptides. *Infect. and Immun.*, 58, 1450–5.

Sexually transmitted mycoplasmas in humans

ALAIN BLANCHARD, LYN D. OLSON AND MICHAEL F. BARILE

3.1 CLASSIFICATION

The current classification of the Kingdom, *Prokaryotae*, includes four divisions, three of which encompass the Gram-negative bacteria, the Gram-positive bacteria and the archaeobacteria. The remaining division, *Tenericutes*, (soft skin) covers the prokaryotes which lack a cell wall. It contains only one class, *Mollicutes*, three orders, four families and six genera (Subcomm. Tax. *Mollicutes*, 1979; Freundt and Edward, 1979; Razin and Freundt, 1984; Freundt and Razin, 1984; Murray, 1984) (Table 3.1). Mollicute(s) is used as the trivial term to denote all organisms belonging to the class, and the trivial names; mycoplasmas, ureaplasmas, acholeplasmas, etc., denote members of the corresponding genus.

Of the six genera, the *Mycoplasma* and *Ureaplasma* species colonize and infect humans and animals and some produce a variety of debilitating diseases. Of the 80 distinct species of *Mycoplasma*, ten colonize the respiratory, urogenital, and/or gastrointestinal tract of humans and one of the five *Ureaplasma* species colonizes the respiratory and urogenital mucosal tissues of man. *Acholeplasma laidlawii* has also colonized human tissues on rare occasions.

3.2 SEXUALLY TRANSMITTED HUMAN SPECIES OF MYCOPLASMAS

Of the 12 *Mollicute* species that colonize humans, four have been associated with various urogenital diseases. They are *Mycoplasma genitalium*, *M. fermentans*, *M. hominis* and *Ureaplasma urealyticum*.

Colonization of infants with urogenital mycoplasmas is believed to occur during passage through the birth canal (Taylor-Robinson *et al.*, 1984) or *in*

Molecular and Cell Biology of Sexually Transmitted Diseases
Edited by D. Wright and L. Archard
Published in 1992 by Chapman and Hall, London ISBN 0 412 36510 3

utero (Cassell *et al.*, 1988), while adult colonization is primarily the result of sexual transmission (Hill *et al.*, 1985).

3.3 DISEASES ASSOCIATED WITH *M. GENITALIUM*

This recently described species (Tully *et al.*, 1981; Tully *et al.*, 1983) has

Table 3.1 Classification of the *Mollicutes*

Division *Tenericutes* (wall-less prokaryotes)
Class *Mollicutes*
  Order I *Mycoplasmatales*
    Sterol required for growth
    Genome size $5 \times 10^8 - 1 \times 10^9$ daltons
    Reduced nicotinamide oxidase in cytoplasm
    Family I *Mycoplasmataceae*
      Genome size about 5×10^8 daltons
      G+C content of DNA from 22% to 41 mol %
    Genus I *Mycoplasma* (77 species)
      Some ferment glucose and/or hydrolyse arginine
      Habitat mainly human and animals.
      Human pathogens *M. pneumoniae*
                           M. genitalium;
                           M. hominis
    Genus II *Ureaplasma* (3 species)
      Hydrolyse urea
      Habitat in human and animals
      Human pathogen U. *urealyticum*
    Family II *Spiroplasmataceae*
      Genome size about 1×10^9 daltons
      Helical organisms during some phase of growth
    Genus I *Spiroplasma* (10 species)
      Ferment glucose and most hydrolyse arginine
      Habitat in arthropods and plants
  Order II *Acholeplasmatales*
    Sterol not required for growth
    Genome size about 1×10^9 daltons
    Reduced nicotinamide adenine dinucleotide oxidase in cell membrane
    Family I *Acholeplasmataceae*
      Genus I *Acholeplasma* (10 species)
      G+C of DNA from 27 to 36 mol %
      Habitat mainly in animals and some in plants and insects
  Order III *Anaeroplasmatales*
    Family I *Anaeroplasmataceae*
      Obligate anaerobe (oxygen sensitive)
      Habitat in bovine and ovine rumen
      Genus I *Anaeroplasma*
    Genus II *Asteroleplasma*

been implicated in non-gonococcal urethritis (NGU) and other urogenital diseases. The organism is extremely fastidious and very difficult to isolate which has hampered investigations. In fact, the two original isolates were recovered after three months of incubation in broth culture (Tully *et al.*, 1981)

Only a few urethral strains have been isolated, primarily from homosexual patients with non-gonococcal urethritis. It was therefore believed to reside primarily in the urogenital tract. Hooten *et al.* (1988) detected M. *genitalium* more frequently in homosexual than in heterosexual men, and Taylor-Robinson *et al.*, (1985a) have proposed that this agent may actually reside in the gastrointestinal tract. Recently, however, four strains were also isolated in mixed cultures with M. *pneumoniae* from the sputum of patients with pneumonia (Baseman *et al.*, 1988), indicating that this pathogen can also cause respiratory disease and suggesting that urogenital pathogens can gain access to the respiratory tract and cause pulmonary disease. It might be considered that perhaps, under certain circumstances, the reverse involving M. *pneumoniae* infection may be possible. On one occasion, M. *pneumoniae* was observed to cause a tubo-ovarian abscess (Thomas *et al.*, 1975).

Experimentally induced urogenital infections with M. *genitalium* in male and female chimpanzees, monkeys, tamarins and marmosets resulted in colonization of the mycoplasma, overt signs of clinical disease and pronounced serum antibody responses (Taylor-Robinson *et al.*, 1985b; Taylor-Robinson *et al.*, 1987; Tully *et al.*, 1986). In addition to NGU, M. *genitalium* has also been associated with various other urogenital diseases, such as pelvic inflammatory disease (Møller *et al.*, 1984). This diagnosis was based primarily on serologic data. About 40% of the patients examined in this study had four-fold or greater antibody titres one month following onset. In monkeys experimentally infected via the oviduct, M. *genitalium* was shown capable of producing salpingitis (Møller *et al.*, 1985). Thus, it would appear that M. *genitalium* is a potentially important human pathogen. However, the significance of its role as a major cause of human disease remains to be determined.

3.4 DISEASES ASSOCIATED WITH M. FERMENTANS

This organism was first isolated from the urogenital tract of two patients with genital infections and was then partially characterized by Ruiter and Wentholt (1953). Similar observations were made shortly thereafter by Nicol and Edward (1953).

Recovery of M. *fermentans* from the human urogenital tract is relatively low, about 1%, and it is even less frequently isolated from the human oral cavity (Mardh and Westrom, 1970; Lelijveld *et al.*, 1981). Because of the low incidence, M. *fermentans* has not been considered until recently to be a pathogen.

One strain of *M. fermentans* was recovered from the cervix of a *Cercophithecus* African green monkey in a survey of 21 tissues examined (Del Giudice and Carski, 1969). When *M. fermentans* was inoculated directly into the uterine tubes of female grivet monkeys, it produced an acute, self-limiting salpingitis and parametritis with a significant rise in serum antibody titre (Møller *et al.*, 1980).

M. fermentans has also been isolated from patients with various clinical diseases. Murphy *et al.* (1967) used cell culture procedures to isolate this agent from the bone marrow tissues of several patients with acute and chronic leukemia. Subsequently, it was shown to produce a leukemoid disease in mice (Plata *et al.*, 1973). Gabridge *et al.*, (1972) reported that they had produced a fatal necrotizing disease in mice which was similar to endotoxic shock by the intraperitoneal inoculation of large inocula (10^{10}) of *M. fermentans*. In this experiment, the level of acid phosphatase remained normal but the level of beta-glucuronidase was increased markedly in the serum of the animals (Gabridge *et al.*, 1975). However, *M. fermentans* does not contain classical lipopolysaccharides and does not induce the release of bacteria through the intestinal wall. Vulfovich and coworkers (1981) demonstrated that infection of rats with *M. fermentans* resulted in the prolonged persistence, up to a year, of the agent in body tissues, mainly in lymphoid tissues, bone marrow, articular cartilage and blood.

M. fermentans was also reported to play a role in the pathogenesis of rheumatoid arthritis (Williams *et al.*, 1970; Williams and Bruckner, 1971; see review by Cole *et al.*, 1985) Williams and colleagues isolated the agent from the joint fluids of patients using a special sucrose density gradient procedure of the synovial specimen prior to culture, but these findings could not be confirmed by others using similar procedures. An experimentally induced polyarthritis was produced in rabbits by intraperitoneal injection of the organism by Kagan *et al.* (1982). In addition to joint disease, these investigators observed the development of *M. fermentans*-specific IgG and rheumatoid factor, deposition of immune complexes and the presence of phagocytes in the synovia and synovial fluids.

Recently, Lo *et al.* detected what appeared to be a novel virus in the tissues of both AIDS (1989a) and non-AIDS patients (1989c) and named it 'virus-like-infectious-agent' (VLIA). The six non-AIDS patients examined were diagnosed initially as having a mycoplasma pneumonia, but the illness rapidly developed into a fulminating necrotizing disease which is quite uncharacteristic of a mycoplasma disease. Subsequently, each patient died abruptly within a seven week period from onset. The VLIA DNA was detected in various organs, including the liver, spleen, kidney, lymph nodes and brain of these patients. The VLIA agent was subsequently shown to be a mycoplasma (Lo *et al.*, 1989d).

The recent demonstration that certain mycoplasmas can survive phagocytosis by neutrophils in the absence of antibody (Webster *et al.*, 1988) suggests a possible mechanism for the dissemination of the pathogen from the original site of colonization to tissues throughout the body. Based on their antigenic and genomic analyses, Lo and colleagues believed the organism to be a new species and named it *incognitus* (1989d). Extensive DNA, protein and serological analysis, however, demonstrated clearly that the VLIA was a unique strain of *M. fermentans* (Saillard *et al.*, 1990). Thus, the findings by Lo and colleagues have suggested a possible role for *M. fermentans* in the cause of human disease.

In other studies, Lemaître *et al.* (1990) have shown that the *in vitro* cytopathogenicity of human immunodeficiency viruses may be due in some cases to the presence of a mycoplasma contaminant in the cell cultures and this contaminant was recently identified as *M. fermentans* (L. Montagnier, personal communication). The significance of these findings as it relates to human HIV disease remains to be determined.

3.5 DISEASES ASSOCIATED WITH *M. HOMINIS*

Genital colonization with *M. hominis* was shown to be directly related to frequency of sexual encounters and number of sexual partners (McCormack *et al.*, 1972). Although *M. hominis* can be isolated from the cervix of up to 20% of healthy women, as many as 70% of patients with clinical symptoms involving the upper or lower genital tract are found to be colonized (Krohn *et al.*, 1989).

M. hominis has been associated with a number of urogenital diseases including pyelonephritis (Thomsen, 1978), pelvic inflammatory disease (Taylor-Robinson and McCormack, 1980; Mardh *et al.*, 1983;), vaginosis (Krohn *et al.*, 1989), and post-abortal and post-partum fever (Mardh *et al.*, 1983; Platt *et al.*, 1980). *M. hominis* is also involved in non-genitourinary infections (for reviews see Mardh *et. al.*, 1983; Madoff and Hooper, 1988) and can also cause septic arthritis (McDonald *et al.*, 1983; Sneller *et al.*, 1986; O'Meara and McQueen, 1988), septicaemia, wound infections (Bøe *et al.*, 1983; Dan and Robertson, 1988), septic thrombophlebitis (Martinez *et al.*, 1989), chronic central nervous system infections in newborn infants (Siber *et al.*, 1977), meningitis and pneumonia in preterm babies (Kirk and Kovar, 1987; Waites *et al.*, 1988) and exudative pharyngitis in adults (Limb, 1989).

3.6 DISEASES ASSOCIATED WITH *U. UREALYTICUM*

There are 14 distinct serological groups (serovars) of the human pathogen, *Ureaplasma urealyticum*. These serovars form two genomic clusters based on DNA–DNA homology, DNA restriction patterns, PAGE patterns, and their sensitivity to magnesium salts (Barile, 1986; Blanchard, 1990).

Although this agent was reputed to play a role in the pathogenesis of infertility (Gnarpe and Friberg, 1973), these findings have not been confirmed and its role remains to be determined (Taylor-Robinson and McCormack, 1980; Hill et al., 1985). Ureaplasmas can attach to and alter the motility of spermatozoa (Fowlkes et al., 1975), and they were shown to induce degeneration of Sertoli, Leydig and germ cells in experimentally infected rats (Audring et al., 1988)

Between 50 and 75% of sexually active individuals are colonized with U. urealyticum in the lower urogenital tract and colonization appears to increase with the number of sexual partners (Taylor-Robinson and McCormack, 1980). Because this organism is commonly found in asymptomatic individuals, it has been very difficult to establish its role in the pathogenesis of urogenital disease.

Although a role for Ureaplasma urealyticum in the cause of nongonococcal urethritis (NGU) was proposed well over 30 years ago, it still remains controversial. In 1980, Taylor-Robinson and McCormack summarized the available data obtained on over 700 NGU patients and found no convincing evidence that U. urealyticum was involved, based on incidence and/or numbers of organisms in patients with or without chlamydial-induced urethritis. However, the intraurethral inoculation of two adult male volunteers provided compelling evidence that this agent is capable of causing NGU (Taylor-Robinson et al., 1977). In experimentally infected chimpanzees, ureaplasmas grown in broth failed to stimulate a urethral cellular response, but inoculation of exudates (unpropagated in culture) containing ureaplasma produced an urethral exudate similar to that induced in patients with chlamydial infections (Taylor-Robinson et al., 1977). In another study, large numbers of ureaplasmas were isolated from the male urethra of a patient with hypogammaglobulinemia, and because no other micro-organism was recovered, U. urealyticum was reputed to be the cause of urethritis in this man (Webster et al., 1982).

Experimental and clinical studies have suggested that this pathogen has an aetiological role in the development of urinary infectious stones (Grenabo et al., 1988). They isolated the organism from the voided urine of 30% of patients with infectious stones as compared with 13% of patients with metabolic stones.

U. urealyticum has been isolated also from the purulent synovial fluids of septic arthritis patients with hypogammaglobulinemia in Canada, England, and the United States (Barile et al., 1987). Some cases of septic arthritis responded to appropriate antibiotic therapy (Kraus et al., 1988; Jorup-Rönström et al., 1989). In other cases, the repeated and prolonged episodes of septic arthritis have been associated with antibiotic-resistant ureaplasma, and responded only to treatment with hyperimmune serum (Webster et al., 1988).

U. urealyticum has also been associated with severe complications during pregnancy and it has been reputed to cause chorioamnionitis (Shurin *et al.*, 1975; Gibbs *et al.*, 1986; Cassell *et al.*, 1986), low birth weight (Foy *et al.*, 1970), and perinatal mortality and morbidity (Kundsin *et al.*, 1984; Quinn *et al.*, 1985) and neonatal meningitis (Garland and Morton, 1987). In a prospective study, *U. urealyticum* and/or *M. hominis* were found to be common pathogens in neonates with meningitis (Waites *et al.*, 1988). The organism was isolated from cerebrospinal fluids (CSF) of 10 of the 100 premature newborns examined in a high risk population. Six of these infants had severe intraventricular haemorrhage and three developed hydrocephalus. In four of the babies, multiple isolations were made over several weeks. Half of the infants with positive CSF cultures were delivered by caesarean procedure, suggesting a possible *in utero* route of infection. *In utero* transmission was also reported by Cassell *et al.* (1988) in very low birth weight infants with chronic lung disease from which *U. urealyticum* was isolated. In this study, *U. urealyticum* was isolated from 34 out of 200 low birth weight infants (< 2500g) who had evidence of respiratory disease. In 29 of them, the organism was recovered in pure culture from endotracheal aspirates and, in some cases, from the blood. Fifteen of the infants died. The findings provide compelling evidence that *U. urealyticum* can cause lung disease and death in preterm infants.

3.7 CELL BIOLOGY OF MOLLICUTES

Mollicutes are small, filterable, pleomorphic prokaryotes totally devoid of cell walls and incapable of synthesizing peptidoglycans or their precursors. They are therefore resistant to penicillin and other beta-lactam antibiotics and to lysozyme, but are sensitive to lysis by osmotic shock and detergents (Brunner and Laber, 1985). Mollicutes have a typical trilayered plasma membrane comprised of phospholipids, glycolipids, lipoglycans, sterols and proteins but are devoid of quinones and cytochromes.

In electron micrographs, the morphology of the mollicute cell is seen to consist of three basic elements and include the cell membrane, ribosomes and a network of nucleic acids. Mollicutes have no inner membranes, fimbriae, pili, nor flagellar motility. Most pathogenic species examined possess a tip structure which plays an important role in attachment to the membranes of infected tissues (Barile, 1965; Biberfeld and Biberfeld, 1970; Collier, 1979).

Mollicutes produce microscopic colonies (50 to 300 micrometres in diameter) and grow submerged beneath the surface of the agar medium, frequently producing a characteristic fried egg appearance. On primary isolation, however, many mycoplasma pathogens produce a granular colony without defined borders.

Most species are sensitive to tetracycline, chloramphenicol, and other antibiotics which inhibit protein synthesis (Brunner and Laber, 1985). Macrolides have been shown to be effective against ureaplasmas (Palu *et al.*, 1989; Taylor-Robinson and Furr, 1986; Davis and Hanna, 1981), but not *Mycoplasma hominis* (Bygdeman and Mardh, 1983), so that erythromycin sensitivity has been used to differentiate between these two urogenital species.

About 40% of the tetracycline-resistant strains in the London area were found to be resistant to erythromycin in contrast to about 10% of the tetracycline-sensitive strains (Taylor-Robinson and Furr, 1986). Development of resistance to tetracycline, erythromycin, rosaramicin and spectinomycin by a *Ureaplasma* strain, serovar II, was followed after successive treatment with these antibiotics, but multiple antibiotic resistance of this kind is thought to be rare. All tetracycline-resistant clinical isolates examined each contained DNA sequences homologous to the streptococcal determinant *tet*M (Roberts and Kenny, 1986; Brunet *et al.*, 1989; Robertson *et al.*, 1988) suggesting the spread of this extrachromosomal element to the genus *Ureaplasma* (section 3.9). A mechanism for *U. urealyticum* resistance to macrolides was proposed by Palu *et al.* (1989) who reported a six-fold reduction in intracellular macrolide influx and accumulation and a reduction in the antibiotic binding to ribosomes of the resistant strains as compared with susceptible strains.

The increasing resistance of genital mollicutes to tetracycline poses a problem to clinicians because this antibiotic is one of the few antimicrobial agents active against these micro-organisms. Some of the newer quinolones, such as the 6-fluoroquinolones, could alleviate this problem because these drugs have been shown to have promise in treating mycoplasmal infections (Kenny *et al.*, 1989).

Antibiotic sensitivities and chemotherapeutic approaches to treatment of mycoplasmal infections and prevention against disease by immunization are reviewed elsewhere (Brunner and Laber, 1985; Barile, 1985; Barile *et al.*, 1985).

3.8 DISTINGUISHING PROPERTIES OF THE GENERA AND SPECIES

Mollicutes can be separated into two metabolic groups, the glycolytic fermenters and the non-fermenters. The fermenters can be further separated based on serum requirement, cell morphology, and optimal growth environment. For example, *M. genitalium* ferments glucose while *M. hominis* which is incapable of fermentation utilizes arginine to generate ATP via the arginine dihydrolase pathway. Interestingly, very few species, such as *M. fermentans* possess both the glycolytic and the arginine dihydrolase pathway. In contrast, the genus, *Ureaplasma*, requires urea for energy and growth (Ford and MacDonald, 1967; also see Razin and Freundt, 1984). Because each genus has different nutritional, energy and carbon requirements, different specialized media formulations may be necessary for their isolation and growth. Members of the *Mycoplasma* and *Ureaplasma* genera are facultative aerobes, require serum and are distinguished by different requirements for an energy and carbon source.

There are ten human species of *Mycoplasma* and 14 human serovars of *Ureaplasma urealyticum*. Speciation and diagnostic identification of pathogenic and non-pathogenic mollicutes have been based on antigenic properties using immunofluorescence or growth (disc) inhibition procedure with species-specific antiserum. The antibody response to mycoplasmal infections and the antigenic relatedness among strains within a species are determined by the metabolism or tetrazolium inhibition test, ELISA, complement fixation, microimmuno-fluorescence, or agglutination procedures

Table 3.2 Association of mollicutes with human disease

Disease	M. pneumoniae	M. genitalium	M. hominis	U. urealyticum	Comments
Atypical pneumonia	++	++/+	–	-	New association with M. genitalium
Pharyngitis	+	ND	–	-	Data still conflicting
Septic arthritis cases in hypogammaglobuli- naemic patients	+	ND	++	++	Several well defined cases in M. hominis and U. urealyticum
Inflammatory polyarthritis	++	ND	+	+	
Nongonococcal urethritis	–	++	–	+	Data still unclear
Pyelonephritis	–	ND	++	–	
Infertility	–	-	-	+	Data still conflicting
Pelvic inflammatory disease	–	++	++	–	Strong association with M. genitalium and M. hominis
Chorioamnionitis	–	–	-	++	
Post abortal/ partum fever	–	ND	++	–	
Low birth weight Stillbirth	–	ND	–	++	Usually associated with chorioamnionitis
Chronic lung disease in new born infants	–	–	++	++	Incidence higher with U. urealyticum than with M. hominis
Central nervous system infection in new born infants	–	–	++	++	Incidence higher with U. urealyticum than with M. hominis

++, strong association; +, weak association; –, no association, ND, not determined.

(Razin and Tully, 1983; Tully and Razin, 1983). Genomic relatedness among genera, species, or strains has been determined by DNA–DNA hybridization (Chandler *et al.*, 1982; Barile *et al.*, 1983; Harasawa *et al.*, 1990), G+C content of the genome, DNA restriction patterns, protein electrophoretic patterns (SDS-PAGE) (Freundt and Razin, 1984), two-dimensional gel electrophoresis and silver staining (Swenson *et al.*, 1983).

3.9 MOLECULAR BIOLOGY OF MYCOPLASMAS

The mollicute genome is circular double-stranded DNA with a characteristically low guanine plus cytosine content (G+C) which ranges from 23 to 41 mol%. The *Mycoplasma* and *Ureaplasma* genera have a genome of approximately 5×10^8 daltons (Razin *et al.*, 1983a; Carle and Bove, 1983; Wenzel and Herrmann, 1988). In comparison, the genome size of rickettsia is approximately 1000 MDa (Myers *et al.*, 1980) and the chlamydiae genome is approximately 660 MDa (Sarov and Becker, 1969). Thus, the *Ureaplasma* and *Mycoplasma* species are the smallest of the self-replicating, free-living prokaryotes. The human ureaplasma strains examined showed diverse genome sizes, as determined by pulsed-field electrophoresis (Robertson *et al.*, 1990). The size of mycoplasma genomes calculated by pulsed field electrophoresis are larger than the values obtained using DNA renaturation, electron microscopy or restriction fragment analyses (Razin *et al.*, 1983a) or two-dimensional denaturing gradient gel electrophoresis (Poddar and Maniloff, 1989). However, physical mapping (Wenzel and Herrmann, 1988) and cloning (Wenzel and Herrmann, 1989) of the complete *Mycoplasma pneumoniae* genome, as well as the *Mycoplasma genitalium* genome (Coleman *et al.*, 1990), supports the somewhat smaller estimated genome size.

The minimal content of genomic material would indicate that the number of genes are probably fewer than 500 (Muto, 1987). In comparison, the *E. coli* genome encodes approximately 2500 genes. The mollicutes have only one or two copies of the rRNA operon as compared with the other prokaryotes which may have as many as ten copies (Amikam *et al.*, 1984; Sawada *et al.*, 1981). They also possess a reduced number of tRNAs (Razin, 1985; Andachi *et al.*, 1989). Mollicutes lack many enzyme systems and metabolic pathways, including those responsible for cell wall synthesis, and as might be expected of obligate parasites, require complex and enriched media for growth. In fact, many pathogens grow best in cell culture systems rather than artificial media.

Although few isoaccepting tRNAs are present, analysis of the complete nucleotide sequence of the P1 adhesin gene of the human respiratory pathogen, *M. pneumoniae*, demonstrated that in spite of the low G+C content of the genome, codon usage is not unusually restricted in that virtually all possible codons are found within the gene and surrounding reading frames (Su *et al.*, 1987; Inamine *et al.*, 1988). Several investigators have shown that some tRNAs from the animal species, *M. mycoides* subsp. *mycoides*, exhibit an extreme form of wobble, virtually ignoring the third base in the codons within the same family boxes (Samuelsson *et al.*, 1987; Guindy *et al.*, 1989). If this is true of all mycoplasmas, it represents a rather efficient use of DNA which may have developed during evolution.

Another molecular characteristic of the *Mycoplasma* species, including such human pathogens as *Mycoplasma genitalium* and *Ureaplasma urealyticum*, is the use the opal stop codon, UGA, as a codon for tryptophan (Yamao *et al.*, 1985; Inamine *et al.*, 1989; Inamine *et al.*, 1990; Blanchard, 1990). The difficulties that this presents for getting expression of cloned mollicute genes into hosts such as *E. coli* which use the more common UGA signal of transcriptional termination are obvious (Schaper *et al.*, 1987). Successful cloning of *Mycoplasma pneumoniae* DNA and the expression adhesin P1-epitopes in *Escherichia coli* was reported by Frydenberg *et al.*, (1987). However, some of the clones that were expected to express the C-terminal of the P1 protein did not express the same epitopes. It was proposed, as one of the possibilities, that some of these clones may code for an internal part of the P1 protein that is prematurely terminated in *E. coli* by the UGA codon. Recently, Renbaum *et al.* (1990) described a procedure which allowed for the controlled expression of genes in an opal codon suppressor strain of *Escherichia coli*. However, this system depends on the effectiveness of the suppressor and has limitations. Another strategy proposed by P.-C. Hu (Chapel Hill, personal communication) is to site specifically mutagenize UGA codons to UGG, which would allow expression of large amounts of proteins in *E. coli* systems.

Research performed thus far on transcription, translation and regulatory signals used by mollicutes suggests similarities to other prokaryotes. Promoters and terminators are structurally similar to regulatory signals in other bacteria (Razin, 1985; Taschke and Herrmann, 1986) and Shine-Dalgarno sequences are present (Bove *et al.*, 1989). However, the *tuf* gene of *M. genitalium* has been reported to have an atypical translation initiation sequence (Loechel *et al.*, 1989).

Mycoplasma and *Ureaplasma* species possess only a single DNA polymerase, although members of the other genera appear to have the usual complement of three polymerases (Maurel *et al.*, 1989). Furthermore, the DNA-dependent RNA polymerases of mycoplasmas have the same enzyme subunit structure observed in eubacteria (Gadeau *et al.*, 1986) but unlike eubacterial enzymes, mollicute RNA polymerases are rifampin resistant (Razin, 1985; Gadeau *et al.*, 1986).

A number of bacteriophage infect mycoplasmas (Maniloff, 1988) and plasmids have also been detected in various *Mycoplasma* species. The presence of plasmids in *Mycoplasma* species is based mainly on the demonstration of extrachromosomal bands in CsCl density gradients (Zouzias *et al.*, 1973) or by detection of DNA distinct from chromosomal DNA in agarose gels following electrophoresis (Harasawa and Barile, 1983). Extrachromosomal covalently-closed, circular double stranded DNA from *Mycoplasma hominis* were also observed by electron microscopy (Zouzias *et al.*, 1973). Mycoplasma plasmids are considered cryptic, having no established functions. However, plasmids of mollicutes which infect plants and insects (*Spiroplasma* species) were also considered cryptic initially (Mouches *et al.*, 1984), but recent reports indicate that they can be involved in the transfer from cell to cell of erythromycin resistance (Barroso and Labarere, 1989; Salvado *et al.*, 1989). The complete nucleotide sequence of a plasmid from *M. mycoides* subsp. *mycoides* has been published and an open reading frame was shown to exhibit homology to that of the *repF* gene product of staphylococcal plasmid pE194 (Bergemann *et al.*, 1989). One cannot rule out the possibility that these mobile extrachromosomal elements are nevertheless replicative forms of a virus.

The lack of a defined medium (medium contains 5–20% serum, 10% yeast extract), and thus the inability to detect auxotrophs, the UGA codon usage and other unique properties of the mollicutes have presented difficult problems to resolve and have tended to inhibit advances into genetics. Nevertheless, the *tetM* resistance determinant was shown by Roberts *et al.* (1985) not only to be present in clinical samples of *M. hominis* but also to be transferable by conjugation in mixed cultures (Roberts and Kenny, 1986, 1987). Furthermore, Dybvig and coworkers (1987, 1988) as well as Roberts and Kenny (1987) have shown that the TN916 transposon can be transferred by conjugation and can be integrated into the chromosome of at least four different species (see also Mahairas and Minion, 1989a, b). These data indicate that genetic manipulations to study mollicutes will be more fruitful in the near future.

There is less than 10% DNA homology among the different *Mycoplasma* (Chandler *et al.*, 1982; Barile *et al.*, 1983) and *Ureaplasma* species (Harasawa *et al.*, 1990). Genomic heterogeneity also occurs among strains within a species and among strains isolated from different or similar tissues or host animals producing similar or unrelated diseases (Barile *et al.*, 1983). An example of this heterogeneity is the range of 40 to 100% DNA homology among the 14 serologically distinct serovars of *U. urealyticum* (Christiansen *et al.*, 1981). Strains of *M. hominis* vary from 50–100% DNA homology (Barile *et al.*, 1983).

Razin and coworkers (1983a) have used both restriction patterns and probes of conserved DNA regions, the rRNA operon and the *tuf* gene (Razin *et al.*, 1983b; Razin *et al.*, 1984; Yogev *et al.*, 1988; Loechel *et al.*, 1989) to demonstrate genetic polymorphisms among strains within the same species. The hybridization patterns of the probes to chromosomal DNAs were used also to group genomically related clusters of strains within species. However, the genetic heterogeneity among strains of a single species is such that a probe containing a different conserved region of the genome, the *atp* genes (Christiansen *et al.*, 1987a) may produce different groupings based on these hybridization patterns.

The study of the molecular biology of all genera among the mollicutes is much more advanced than has been covered by this overview which has concentrated on the sexually transmitted human pathogens. A considerable amount of work has been published on agents that infect plants and animals and the reader is referred to reviews of the subject by Razin (1985, 1990).

3.10 NUCLEIC ACID AND IMMUNOLOGICAL PROBES FOR DETECTING PATHOGENIC MYCOPLASMAS

A number of investigators have used immunological probes to detect mycoplasmas in contaminated cell cultures (see McGarrity and Kotani, 1985, 1986). Various nucleic acid probes have also been used successfully for this purpose (Razin *et al.*, 1984; Johansson and Bolske, 1989; Johannsson *et al.*, 1990).

Synthetic oligonucleotide probes complementary to rRNA were used for group-specific and species-specific detection of mycoplasmas (Gobel *et al.*, 1987a; Gobel *et al.*, 1987b). DNA probes were also used for the detection of mycoplasmas in genital specimens (Roberts *et al.*, 1987a).

Numerous cross-reactive antigens exist among *M. genitalium*, the human respiratory pathogen, *M. pneumoniae* (Taylor-Robinson *et al.*, 1983; Hu, *et al.*, 1984; Hu *et al.*, 1987; Lind *et al.*, 1984; Bredt *et al.*, 1987; Dallo *et al.*, 1989a,b; Sasaki *et al.*, 1989), and *Acholeplasma laidlawii* (Cimolai *et al.*, 1987). Thus, the serodiagnoses of diseases associated with *M. genitalium* and *M. pneumoniae* and adventitious colonization by the acholeplasma must be interpreted with caution. Because *M. genitalium* and *M. pneumoniae*, are genomically distinct (Yogev and Razin, 1986; Lind *et al.*, 1984), investigators have developed DNA probes (Hyman *et al.*, 1987; Bernet *et al.*, 1989) and monoclonal antibodies (Chandler *et al.*, 1989) for identification of these human pathogenic agents. Hooten *et al.* (1988) used whole genomic DNA probes to evaluate the presence of *M. genitalium* in human urethral infections and reported that this pathogen probably accounts for some NGU cases, especially those that are persistent or recurring. Risi *et al.* (1987) cloned *M. genitalium* DNA by transfection into *Escherichia coli* and used the cDNA to probe clinical specimens to determine the ecologic niche and spectrum of disease caused by *M. genitalium*. Hyman *et al.*, (1987) also developed DNA probes for detection and identification of both *Mycoplasma pneumoniae* and *Mycoplasma genitalium*. In other studies, various DNA probes were compared with culture procedures for the rapid diagnosis of *M. pneumoniae* with varying results (Tilton *et al.*, 1988; Harris *et al.*, 1988; Hata *et al.*, 1990). In addition, *Mycoplasma pneumoniae*

DNA was detected within diseased gingiva by *in situ* hybridization using a biotin-labelled probe (Saglie *et al.*, 1988).

In situ hybridization, polymerase chain reaction (PCR) and ultramicroscopic study were used for identification of *Mycoplasma incognitus* in tissues of infected patients with AIDS (Lo *et al.*, 1989e). PCR was also used by Jensen *et al.* (1989) to detect *M. pneumoniae* in simulated clinical samples. Bernet *et al.* (1989) examined throat swabs obtained from hamsters that were experimentally infected with *M. pneumoniae* and claimed that PCR was more sensitive and reliable than conventional culture techniques for the detection of *M. pneumoniae*. In experiments on artificially seeded human bronchoalveolar lavages, they showed that PCR could detect 10^2 to 10^3 organisms. However, sensitive, well-standardized culture medium should be able to detect less than 10^1 organisms.

3.11 PATHOGENIC, DISEASE-RELATED AND IMMUNOGENIC COMPONENTS OF MYCOPLASMAS

The toxic and virulence components of a pathogen are generally the most effective protective antigens because the host must defend itself from such noxious, tissue damaging substances to avoid or prevent serious disease. Specialized compounds produced by pathogenic mycoplasmas that affect eukaryotic cells and assist in the establishment of colonization include neurotoxins, ciliotoxic activities, peroxides, polysaccharides and endotoxinlike factors, among others. The pathogenic factors of the mycoplasmas are reviewed elsewhere (Gabridge *et al.*, 1985).

3.12 PATHOGENIC MECHANISMS OF *M. GENITALIUM*

M. genitalium possesses specialized attachment components at the tip of its pear-shaped cell that mediates attachment to the sialic acid moieties of host membrane glycoproteins as does *M. pneumoniae*. Hu *et al.* (1987) showed that a 140 kDa *M. genitalium* surface protein (MgPa) is antigenically related to the P1 attachment component protein of *M. pneumoniae* and the gene has been sequenced and cloned and showed 45% homology with the P1 adhesin of *M. pneumoniae* (Inamine *et al.*, 1989). *M. genitalium* attaches to and grows as colonies on the inert surfaces of glass or plastic, providing a useful means of propagating this agent and producing large populations of cells for study. This species attaches to cultured cells such as Vero cells (African green monkey kidney cells) and WiDr cells (human adenocarcinoma cells), grows to high titres, and produces severe cytopathic effects. The attachment and cytopathic components of *M. genitalium* are considered important virulence and immunogenic factors (Hu *et al.*, 1987; Morrison-Plummer *et al.*, 1987; Baseman *et al.*, 1988).

Using human oviduct organ cultures, *M. genitalium* attaches to and infects both the ciliary cells and the basal mucosal cells as demonstrated by scanning electron microscopy. Some organisms were observed to penetrate the cell membrane via the specialized tip structure. Attachment was inhibited by either treatment with trypsin or with monoclonal antibodies specific to the MgPa protein (Collier *et al.*, 1988).

3.13 PATHOGENIC MECHANISMS OF *M. FERMENTANS*

Lo *et al.* (1989b) reported that when cell cultures infected with the *M. incognitus (M. fermentans)* were inoculated intraperitoneally into monkeys, each of the five animals inoculated died within nine months. Probes shown to be specific for the *M. incognitus* isolates and for established *M. fermentans* strains were used for PCR and *M. incognitus* DNA was detected in the autopsy tissues of these experimentally infected monkeys. Using the same probes, they also detected specific *M. incognitus* DNA in various biopsy tissues obtained from AIDS and non-AIDS patients. Electron micrographs of infected human and animal tissues suggested that this pathogen was an invasive, intracellular parasite. This is the first suggestion that a mycoplasma might be anything other than an obligate, extracellular parasite. Further research is required to clarify the pathogenic properties of this organism.

3.14 PATHOGENIC MECHANISMS OF *M. HOMINIS*

The pathogenic properties of *M. hominis* have been difficult to define because of the diversity of diseases they cause and the significant antigenic and genomic heterogeneity among the strains isolated from various infected tissues in patients with widely diverse diseases. Strains isolated from different tissue sources, different animal species, and from contaminated cell cultures indicate that there is marked heterogeneity in antigenic properties (Hollingdale and Lemcke, 1970), isozyme expression (O'Brien *et al.*, 1981), polypeptide composition (Christiansen *et al.*, 1987b), DNA homology and DNA restriction patterns (Barile *et al.*, 1983; Christiansen and Andersen, 1988). Genomic clusters can occur among strains isolated from a similar source, such as the blood of post-partum fever patients or isolated from contaminated cell cultures, which may indicate selective pressures provided by the host and/or environment (Barile *et al.*, 1983).

Attempts are being made to examine the pathogenic properties of *M. hominis* by using various animal models (Barile *et al.*, 1983, 1985, 1987, 1990). A potentially important pathogenic property of *M. hominis* is the ability to attach to cell culture substrates. Those strains which attach the best appear to be most pathogenic (Izumikawa *et al.*, 1987). Freshly isolated strains from patients with various clinical diseases were found to attach to tissue cells better than strains maintained in high broth passage in the laboratory (Barile *et al.*, 1983; Izumikawa *et al.*, 1987). For example, strain 1620 isolated from the joint exudates of patients with septic arthritis and used at low broth passage has good cytadsorbing activity and produces a severe arthritis in experimentally infected chimpanzees whereas the high broth passage laboratory, non-cytadsorbing type strain PG21 produces no overt disease (Barile *et al.*, 1987). Thus, attachment appears to be a critical virulence factor.

Another potentially important virulence component as determined in the experimental chimpanzee arthritis model is the ability to induce severe inflammation and to cause the recruitment and accumulation of a large number of leukocytes ($>10^5$ cells/mm^3) in joint exudates (Barile *et al.*, 1987). Sneller *et al.* (1986) reported findings on the repeated isolation over a two year period of M. *hominis* from a prosthetic joint of a patient with rheumatoid arthritis who was suffering from repeated bouts of septic arthritis. An analysis of the strains isolated from the patient over six years, as well as strains isolated from various other tissue sources with a series of monoclonal antibodies specific for surface exposed lipoproteins suggest that M. *hominis* might be capable of antigenic variation of its membrane proteins during infection (Olson *et al.*, 1991). A mechanism of expressing antigenic heterogeneity during infection may provide this pathogen with the means by which it may elude the protective immune systems and persist in the host causing chronic disease. Similar findings have been demonstrated earlier for other animal species of mycoplasmas grown in broth culture (Watson *et al.*, 1988; Boyer and Wise, 1989; Rosengarten and Wise, 1990).

Another important property of M. *hominis* is its ability to metabolize arginine with the resultant release of large quantities of ammonia. Arginine-utilizing M. *hominis* alters the metabolism and function of infected fibroblastic cell cultures, lymphocyte cultures and other tissue cells grown *in vitro* by rapidly depleting this essential amino acid from the medium (Barile and Leventhal, 1968; Simberkoff *et al.*, 1969; Hahn and Kenny, 1974). Attachment to tissues permits the release of enzymatic and/or toxic activity (perhaps including ammonia) directly on the host cell membrane which could adversely affect the cell membrane. There are multiple forms of arginine deiminase, the initial enzyme in the three arginine dihydrolase pathway, and they can vary in type and amount among the various strains. It has been suggested, but not proven, that strains with the highest amount of arginine dihydrolase activity are most pathogenic (Weickmann and Farhney, 1977).

3.15 PATHOGENIC MECHANISMS OF *U. UREALYTICUM*

Ureaplasmas differ from all other genera within the class *Mollicutes* by possessing urease activity and they are considered the only organisms known to depend on urea for growth (Romano *et al.*, 1986). Urea hydrolysis appears to play a major role in the energy metabolism of ureaplasmas by promoting ATP synthesis. As with other prokaryotes that have urease activity, *U.*

urealyticum can induce kidney stones and calculi in experimentally infected animals and in patients resulting in upper urogenital disease. Thus, urease may play a role in the pathogenesis of certain urinary diseases (Friedlander and Braude, 1974; Clerc *et al.*, 1984).

Secretory IgA antibodies are considered a major defence mechanism in the prevention of mucosal disease (Kilian *et al.*, 1988). It is now well-established that a number of bacterial mucosal pathogens possess IgA-specific proteases and these enzymes have been considered potentially important virulence factors which permit the pathogens to gain entry at or cause diseases of mucosal tissue membranes (Kilian *et al.*, 1983). This activity is associated with bacterial mucosal pathogens, but not with nonpathogenic species of the same genus (Robertson *et al.*, 1984; Kapatai-Zoumbos *et al.*, 1985). *Ureaplasma urealyticum*, the human pathogen, also possesses specific protease activity which can degrade human IgA_1 into the Fab and Fc fragments. The 27 human strains of *U. urealyticum* examined by Kapatai-Zoumbos *et al.* (1985), representing nine different serovars, were each capable of cleaving human IgA_1, but not human IgA_2 nor IgA from cattle, dogs, or other animals. On the other hand, ureaplasma strains isolated from dogs can specifically cleave canine IgA_1 but not IgA derived from human and other animal sources. Thus, the IgA_1 specific protease activity is believed to be a virulence factor for *U. urealyticum*. However, no other species examined had IgA protease activity, including 14 human strains of *Mycoplasma hominis*, seven pathogenic stains of *Mycoplasma pneumoniae* or strain G37C of *M. genitalium* (Kapatai-Zoumbos *et al.*, 1985).

Another potentially important virulence component as determined in the experimental chimpanzee arthritis model is the ability of strain 2010B of *Ureaplasma urealyticum* to induce severe arthritis, characterized by inflammation and the recruitment and accumulation of large amounts of synovial fluids containing a large number of leukocytes ($>10^5$ cells/mm^3) at the site of joint infection (Barile *et al.*, 1987). Cytotoxins, such as phospholipases, have been identified as virulence factors in bacterial pneumonia (Dennis, 1983). The association of the activity of bacterial phospholipases and birth prematurity has also been suggested (Bejar *et al.*, 1981), since these enzymes could degrade the membrane phospholipids of the placental tissues. The activities of phospholipases A and C have been reported in *U. urealyticum* by De Silva and Quinn (1986, 1987). However, the role of these enzymes, if any, in the pathogenicity of the ureaplasmas remains undetermined.

REFERENCES

Amikam, D., Glaser, G. and Razin, S. (1984) Mycoplasmas (Mollicutes) have a low number of ribosomal RNA genes. *J. Bacteriol.*, 158; 376–8.

Andachi, Y., Yamao, F., Muto, A. and Osawa, S. (1989) Codon recognition patterns as deduced from sequences of the complete set of transfer RNA species in *Mycoplasma capricolum*: resemblance to mitochondria. *J. Mol. Biol.*, **209**, 37–54.

Audring, H., Klug, H., Sokolowska–Köhler, W. and Engel, S. (1988) *Ureaplasma urealyticum* and male infertility: an animal model. II. Morphologic changes of testicular tissue at light microscopic level and electron microscopic findings. *Andrologia*, **21**, 166–75.

Barile, M.F. (1965) Mycoplasma (PPLO), Leukemia and Autoimmune Disease, in *Methodological Approaches to the Study of Leukemias* (ed. V. Defendi). Wistar Institute Press, Philadelphia, PA, pp. 171–86.

Barile, M.F. and Leventhal, B.G. (1968) Possible mechanism for mycoplasma inhibition of lymphocyte transformation induced by phytohemagglutinin. *Nature*, **219**, 751–2.

Barile, M.F., Grabowski, M.W., Stephens, E.B. *et al.* (1983) *Mycoplasma hominis* – tissue cell interactions: a review with new observations on phenotypic and genotypic properties. *Sexually Trans. Dis.*, 10; 345–54.

Barile, M.F. (1985) Immunization against mycoplasma infections, in *The Mycoplasmas, Vol. IV*, (eds S. Razin and M.F. Barile), Academic Press, Inc., Orlando, Fla., pp. 452–92.

Barile, M.F., Bove, J.M., Bradbury, J.M., *et al.* (1985) Control of mycoplasma diseases of man, animals, plants and insects. *Bulletin de Pasteur*, **83**, 339–73.

Barile, M.F. (1986) DNA homologies and serologic relationships among urea-plasmas from various hosts. *J. Pediat. Inf. Dis.*, **5**, S296–9.

Barile, M.F., Grabowski, M.W., Snoy, P.J. and Chandler, D.K.F. (1987) The superiority of the chimpanzee animal model to study the pathogenicity of known (*M. pneumoniae*) and reputed mycoplasma pathogens. *Israel J. Med. Sci.*, **23**, 556–60.

Barile, M.F., Snoy, P.J., Miller, L.M. *et al.* (1990) Mycoplasma-induced septic arthritis in chimpanzees: A brief review. *Zentrablatt. Bakt. Mikrob. Hyg.* (*Intern. J. Microbiol. and Hyg.*) (In press).

Barroso, G. and Laberere, J. (1988) Chromosomal gene transfer in *Spiroplasma citri*. *Science*, **241**, 959–61.

Baseman, J.B., Dallo, S.F., Tully, J.G. and Rose, D.L. (1988) Isolation and characterization of *Mycoplasma genitalium* strains from the human respiratory tract. *J. Clin. Microbiol.*, **26**, 2266–9.

Bejar, R., Curbelo, V., Dairs, V. and Gluck, L. (1981) Premature labour. Bacterial sources of phospholipase. *Obstet. Gynecol.*, **57**, 479–82.

Bergemann, A.D., Whitley, J.C. and Finch, L.R. (1989) Homology of mycoplasma plasmid pADB201 and staphylococcal plasmid pE194. *J. Bacteriol.*, **171**, 593–5.

Bernet, C., Garret, M., De Barbeyrac, B. *et al.* (1989) Detection of *Mycoplasma pneumoniae* by using the polymerase chain reaction. *J. Clin. Microbiol.*, **27**, 2492–6.

Biberfeld, G. and Biberfeld, P. (1970) Ultrastructural features of *Mycoplasma pneumoniae*. *J. Bacteriol.*, **120**, 855–61.

Blanchard, A. (1990) *Ureaplasma urealyticum* urease genes; use of a UGA tryptophan codon. *Molec. Microbiol.*, **4**, 669–76.

Boe, O., Iverson, O.E. and Mehl, A. (1983) Septicemia due to *M. hominis*. *Scan. J. Infect. Dis.*, **15**, 87–90.

Bove, J.M., Carle, P., Garnier, M. *et al.* (1989) Molecular and cellular biology of mycoplasmas, in *The Mycoplasmas, Vol. 5* (eds R.F. Whitcomb and J.G. Tully), Academic Press, Inc., San Diego, CA. pp. 244–364.

Boyer, M.J. and K.S. Wise. (1989) Lipid-modified surface protein antigens expressing size variation within the species *Mycoplasma hyorhinis. Infect. Immun.*, 57, 245–54.

Bredt, W., Kleinmann, B. and Jacobs, E. (1987) Antibodies in the sera of *Mycoplasma pneumoniae*-infected patients against proteins of *Mycoplasma genitalium* and other mycoplasmas of man. *Zentralbl. Bakteriol. Mikrobiol. Hyg. [A].*, 266, 32–42.

Brunet, B., De Barbeyrac, B., Renaudin, H. and Bebear, C. (1989) Detection of tetracycline-resistant strains of *Ureaplasm urealyticum* by hybridization assays. *Eur. J. Clin. Microbiol. Infect. Dis.*, 8, 636–8.

Brunner, H. and Laber, G. (1985) Chemotherapy of mycoplasma infections, in *The Mycoplasmas, Vol. IV.* (eds S. Razin and M.F. Barile), Academic Press, New York, pp. 403–51.

Bygdeman, S.M. and Mardh, P.A. (1983) Antimicrobial susceptibility and susceptibility testing of *Mycoplasma hominis*: A review. *Sex. Transm. Dis.*, 10, 366–70.

Carle, P. and Bove, J.M. (1983) Genome size determination, in *Methods in Mycoplasmology, Vol. 1* (eds S. Razin and J.G. Tully), Academic Press, Inc., New York, N.Y. pp. 309–12.

Cassell, G.H., Waites, K.B., Gibbs, R.S. and Davis, J.K. (1986) Role of *Ureaplasma urealyticum* in amnionitis. *Pediatr. Infect. Dis.*, 5, S247–52.

Cassell, G.H., Waites, K.B., Crouse, D.T. *et al.* (1988) Association of *Ureaplasma urealyticum* infection of the lower respiratory tract with chronic lung disease and death in very low birth weight infants. *Lancet*, i, 240–5.

Chandler, D.K.F., Razin, S., Stephens, E.B. *et al.* (1982) Genomic and phenotypic analyses of *Mycoplasma pneumoniae* strains. *Infect. Immun.*, 38, 604–9.

Chandler, D.K.F., Olson, L.D., Kenimer, J.G. *et al.* (1989) Biological activities of monoclonal antibodies to *Mycoplasma pneumoniae* membrane glycolipids. *Infect. Immun.*, 57, 1131–6.

Chanock, R.M., Hayflick, L. and Barile, M.F. (1962) Growth on artificial medium of an agent associated with atypical pneumonia and its identification as a PPLO. *Proc. Natl Acad. Sci.*, 48, 41–9.

Christiansen, C., Black, F.T. and Freundt, E.A. (1981) Hybridization experiments with deoxyribonucleic acid from *Ureaplasma urealyticum* serovars I to VIII. *Int. J. Syst. Bacteriol.*, 31, 259–62.

Christiansen, C., Christiansen, G. and Rasmussen, O.F. (1987a) Heterogeneity of *M. hominis* as detected by a probe for *atp* genes. *Isr. J. Med. Sci.*, 23, 591–4.

Christiansen, G., Andersen, H., Birkelund, S. and Freundt, E.A. (1987b) Genomic and gene variation in *Mycoplasma hominis* strains. *Isr. J. Med. Sci.*, 23, 595–602.

Christiansen, G. and Andersen, H. (1988) Heterogeneity among *Mycoplasma hominis* strains as detected by probes containing parts of ribosomal ribonucleic acid genes. *Int. J. Syst. Bacteriol.*, 38, 108–15.

Cimolai, B., Bryan, L. E., To, M. and Woods, D. E. (1987) Immunological cross-reactivity of a *Mycoplasma genitalium* and *Acholeplasma laidlawii*. *J. Clin. Microbiol.*, 25, 2136–9.

Clerc, M., Texier, J. and Bebear, C. (1984) Experimental production of bladder calculi in rats by *Ureaplasma urealyticum*. *Ann. Microbiol. (Inst. Pasteur)*, 135A, 135–40.

Cole, B. C., Washburn, L. R. and Taylor–Robinson, D. (1985) Mycoplasma-induced arthritis, in *The Mycoplasmas, Vol. IV* (eds S. Razin and M.F. Barile), Academic Press, New York. pp. 108–60.

Collier, A.M. (1979) Mycoplasma in organ culture, in *The Mycoplasmas, Vol. II* (eds J.G. Tully and R.F. Whitcomb), Academic Press, New York, pp. 475–94.

Collier, A.M., Carson, J.S., Hu, P.C. *et al.* (1988) Attachment of *Mycoplasma genitalium* to the ciliated epithelium of human fallopian tubes. *7th Intern. Congress Intern. Organ. Mycoplasmology, Compendium of Abstracts*, Baden, Austria, June 2–9, p. 69 (Abstract).

Coleman, S.D., Hu, P.–C., Litaker, W. and Bott, K.F. (1990) A physical map of the *Mycoplasma genitalium* genome. *Molec. Microbiol.*, 4, 683–7.

Dallo S.F, Horton J.R, Su C.J. and Baseman J.B. (1989a) Homologous regions shared by adhesin genes of *Mycoplasma pneumoniae* and *Mycoplasma genitalium*. *Microb. Pathog.*, 6, 69–73.

Dallo, S.F., Chavoya, A., Su, C.L. and Baseman, J.B. (1989b) DNA and protein sequence homologies between the adhesins of *Mycoplasma genitalium* and *Mycoplasma pneumoniae*. *Infect. Immun.*, 57, 1059–65.

Dan, M. and Robertson, J. (1988) *Mycoplasma hominis* septicemia after heart surgery. *Amer. J. Med.*, 84, 976–7.

Davis, J.W. and Hanna, B.A. (1981) Antimicrobial susceptibility of *Ureaplasma urealyticum*. *J. Clin. Microbiol.*, 13, 320–5.

Del Giudice, R.A. and Carski, T.R. (1969) Recovery of human mycoplasmas from simian tissues. *Nature*, 222, 1088–9.

Dennis, E.A. (1983) Phospholipases, in *Enzymes, vol 16*. (Ed. P.D. Boyer.), Academic Press, Inc., New York, NY, pp. 307–53.

De Silva, N.S. and Quinn, P.A. (1986) Endogenous activity of phospholipases A and C in *Ureaplasma urealyticum*. *J. Clin. Microbiol.*, 23, 354–9.

De Silva, N.S. and Quinn, P.A. (1987) Rapid screening assay for phospholipase C activity in mycoplasmas. *J. Clin. Microbiol.*, 25, 729–31.

Dybvig, K. and Cassell, G.H. (1987) Transposition of Gram-positive Tn916 in *Acholeplasma laidlawii* and *Mycoplasma pulmonis*. *Science*, 235, 1392–4.

Dybvig, K. and Alderete, J. (1988) Transformation of *Mycoplasma pulmonis* and *Mycoplasma hyorhinis*: transposition of Tn916 and formation of cointegrate structures. *Plasmid*, 20, 33–41.

Ford, D.K. and MacDonald, J. (1967) Influence of urea on the growth of T-strain mycoplasmas. *J. Bacteriol.*, 93, 1509–12.

Fowlkes, D.L., MacLeod, J. and O'Leary, W.M. (1975) T-mycoplasmas and human infertility: correlation of infection with alterations in seminal parameters. *Fertil. Steril.*, 26, 1212–18.

Foy, H.M., Kenny, G.E., Wentworth, B.B. *et al.* (1970) Isolation of *Mycoplasma hominis*, T-strains, and cytomegalovirus from the cervix of pregnant women. *Am. J. Obstet. Gynecol.*, 106, 635–43.

Freundt, E.A. and Edward, D.G. (1979) Classification and taxonomy, in *The*

Mycoplasmas, Vol. I (eds M.F. Barile and S. Razin), Academic Press, New York, pp. 1–42.

Freundt, E.A. and Razin, S. (1984). Division Tenericutes, Div. Nov., in *Bergey's Manual Of Systematic Bacteriology. Vol. I* (eds N. R. Kreig and J.G. Holt), Williams & Wilkins, Baltimore, pp. 740–75.

Friedlander, A.M. and Braude, A.I. (1974) Production of bladder stones by human T-mycoplasmas. *Nature*, 247, 67–9.

Frydenberg, J., Lind, K. and Hu, P.C. (1987) Cloning of Mycoplasma pneumoniae DNA and expression of P1-epitopes in *Escherichia coli. Isr. J. Med. Sci.*, 23, 759–62.

Gabridge, M.G., Abrams, D.G. and Murphy, W.H. (1972) Lethal toxicity of *Mycoplasma fermentans* for mice. *J. Infect. Dis.*, 125, 153–60.

Gabridge, M.G., Yip, D.-M. and Hedges, K. (1975) Levels of lysosomal enzymes in tissues of mice infected with *Mycoplasma fermentans. Infect. Immun.*, 12, 233–9.

Gabridge, M.G., Chandler, D.F.K. and Daniels, M.J. (1985) Pathogenicity factors in mycoplasmas and spiroplasmas, in *The Mycoplasmas, Vol. IV*, (eds S. Razin and M.F. Barile), Academic Press, New York, pp. 313–52.

Gadeau, A.-P., Mouches, C. and Bove, J.M. (1986) Probable insensitivity of Mollicutes to rifampin and characterization of spiroplasmal DNA-dependent RNA polymerase. *J. Bacteriol.*, 166, 824–8.

Garland, S.M. and Murton, L.J. (1987) Neonatal meningitis caused by *Ureaplasma urealyticum. Pediatr. Infect. Dis. J.*, 6, 868–9.

Gibbs, R.S., Cassell, G.H., Davis, J.K. and St Clair, P.J. (1986) Further studies on genital mycoplasmas in intra-amniotic infection: blood cultures and serologic response. *Am. J. Obstet. Gynecol.*, 154, 717–26.

Gobel, U.B., Geiser, A. and Stanbridge, E.J. (1987a) Oligonucleotide probes complementary to variable regions of ribosomal RNA discriminate between Mycoplasma species. *J. Gen. Microbiol.*, 133, 1969–74.

Gobel, U., Maas, R., Haun, G. et al. (1987b) Synthetic oligonucleotide probes complementary to rRNA for group- and species-specific detection of mycoplasmas. *Isr. J. Med. Sci.*, 23, 742–6

Gnarpe, H. and Friberg, J. (1973) T-mycoplasmas as a cause for reproductive failure. *Nature*, 242, 120–1.

Grenabo, L., Hedelin, H. and Petterson, S. (1988) Urinary infection stones caused by *Ureaplasma urealyticum:* a review. *Scand. J. Infect. Dis.*, S53, 46–9.

Guindy, Y.S., Samuelsson, T. and Johansen, T.-I. (1989) Unconventional codon reading by *Mycoplasma mycoides* tRNAs as revealed by partial sequence analysis. *Biochem. J.*, 258, 869–73.

Hahn, R. G. and Kenny, G. E. (1974) Differences in arginine requirement for growth among arginine-utilizing *Mycoplasma* species. *J. Bacteriol.*, 117, 611–18.

Harasawa, R. and Barile, M.F. (1983) Survey of plasmids in various mycoplasmas. *The Yale Journal of Biology and Medicine*, 56, 783–8.

Harasawa, R., Stephens, E.B., Koshimizu, K. et al. (1990) DNA Relatedness among established *Ureaplasma* species and unidentified feline and canine serogroups. *Intern. J. System. Bacteriol.*, 40, 52–5.

Harris, R., Marmion, B.P., Varkanis, G., et al. (1988). Laboratory diagnosis of *Mycoplasma pneumoniae* infection. 2. Comparison of methods for the direct detection of specific antigen or nucleic acid sequences in respiratory exudates. *Epidemiol. Infect.*, 101, 685–94.

Hata, D., Kuze, F., Mochizuki, Y. *et al.* (1990) Evaluation of DNA probe test for rapid diagnosis of *Mycoplasma pneumoniae* infections; *J. Pediatr.*, 116, 273–6.

Hill, G.B., Eschenbach, D.A. and Holmes, K.K. (1985) Bacteriology of the vagina. *Scand. J. Urol. Nephrol.*, 86 (suppl), 23–9.

Hollingdale, M.R. and Lemcke, R.M. (1970) The antigens of *Mycoplasma hominis*. *J. Hyg. (Camb.)*, 67, 585–602.

Hooton, T.M., Roberts, M.C., Roberts, P.L. *et al.* (1988) Prevalence of *Mycoplasma genitalium* determined by DNA probe in men with urethritis. *Lancet*, i, 266–8.

Hu, P.-C., Clyde Jr, W.A. and Collier, A.M. (1984) Conservation of pathogenic mycoplasma antigens. *Isr. J. Med. Sci.*, 20, 916–19.

Hu, P.-C., Schaper, V., Collier, A.M. *et al.* (1987) A *Mycoplasma genitalium* protein resembling the *Mycoplasma pneumoniae* attachment protein. *Infect. Immun.*, 55, 1126–31.

Hyman, H.C., Yogev, D. and Razin, S. (1987) DNA probes for detection and identification of *Mycoplasma pneumoniae* and *Mycoplasma genitalium*. *J. Clin. Microbiol.*, 25, 726–8.

Inamine, J.M., Denny, T.P., Loechel, S. *et al.* (1988) Nucleotide sequence of the P1-attachment protein gene of *Mycoplasma pneumoniae*. *Gene*, 64, 217–29.

Inamine, J.M., Loechel, S. Collier, A.M. *et al.* (1989) Nucleotide sequence of MgPa (mgp) operon of *Mycoplasma genitalium* and comparison to the P1 (mpp) operon of *Mycoplasma pneumoniae*. *Gene*, 82, 259–67.

Inamine, J.M., Ho, K.C., Loechel, S. and Hu, P.C. (1990) Evidence that UGA is read as a tryptophan codon rather than as a stop codon by *Mycoplasma pneumoniae*, *Mycoplasma genitalium*, and *Mycoplasma gallisepticum*. *J. Bacteriol.*, 172, 504–6.

Izumikawa, K., Chandler, D.K., Grabowski, M.W. and Barile, M.F. (1987) Attachment of *Mycoplasma hominis* to human cell cultures. *Isr. J. Med. Sci.*, 23, 603–7.

Jensen, J.S. Sndergard-Andersen, J., Uldum, S.A. and Lind, K. (1989) Detection of *Mycoplasma pneumoniae* in simulated clinical samples by polymerase chain reaction. Brief report; *APMIS.*, 97, 1046–8.

Johansson, K.E. and Bolske, G. (1989) Evaluation and practical aspects of the use of a commercial DNA probe for detection of mycoplasma infections in cell cultures. *J. Biochem. Biophys. Methods*, 19, 185–99.

Johansson, K.E., Johansson, I. and Gobel, U.B. (1990) Evaluation of different hybridization procedures for the detection of mycoplasma contamination in cell cultures. *Mol. Cell. Probes*, 4, 33–42.

Jorup-Rönström, C., Ahl, T., Hammarström, L. *et al.* (1989) Septic osteomyelitis and polyarthritis with ureaplasma in hypogammaglobulinemia. *Infection*, 17, 301–3.

Kagan, G.Y., Vulfovich, Y.V., Zilfyan, A.V. *et al.* (1982) Experimental polyarthritis induced by *Mycoplasma fermentans* in rabbits. *Zh. Mickrob. E.*, 3, 107.

Kapatai-Zoumbos, K., Chandler, D.K.F. and Barile, M.F. (1985) Survey of immunoglobulin A protease activity among selected species of *Ureaplasma* and *Mycoplasma*: specificity for host immunoglobulin A. *Infect. Immun.*, **47**, 704–9.

Kenny, G.E., Hooton, T.M., Roberts, M.C. *et al.* (1989) Susceptibilities of genital mycoplasmas to the newer quinolones as determined by the agar dilution method. *Antimicrob. Agents Chemotherapy*, **33**, 103–7.

Kilian, M., Thomsen, B., Petersen, T.E. and Bleeg, H.S. (1988) Occurrence and nature of bacterial IgA proteases. *New York Acad. Sci.*, **409**, 612–24.

Kirk, N. and Kovar, I. (1987) *Mycoplasma hominis* meningitis in a preterm infant. *J. Infect.*, **15**, 109–10.

Kraus, V., Baraniuk, V., Hill, G. and Bates, W. (1988) *Ureaplasma urealyticum* septic arthritis in hypogammaglobulinemia. *J. Rheumatol.*, **15**, 369–71.

Krohn, M.A., Hillier, S.L. and Eschenbach, D.A. (1989) Methods for diagnosing bacterial vaginosis among pregnant women. *J. Clin. Microbiol.*, **27**, 1266–71.

Kundsin, R.B., Driscoll. S.G., Monson, R.R. *et al.* (1984) Association of *Ureaplasma urealyticum* in the placenta with perinatal morbidity and mortality. *N. Engl. J. Med.*, **310**, 941–5.

Lelijveld, J. L. M., Leentvaar-Kuijpers, A., Hekker, A. C. *et al.* (1981) Some sexually transmitted diseases in female visitors of an out-patient department for venereal diseases in Utrecht, Netherlands. *Ned. Tijdschr. Geneeskd.*, **125**, 463–6.

Lemaître, M., Guétard, D., Hénin, H. *et al.* (1990) Protective activity of tetracycline analogs against the cytopathic effect of the human immunodeficiency viruses in CEM cells. *Res. Virol. (Inst, Pasteur)*, **141**, 5–16.

Limb, D.I. (1989) Mycoplasmas of the human genital tract. *Med. Lab. Sci.*, **46**, 146–56.

Lin, J.S. (1985) Human mycoplasmal infections: serologic observations. *Rev. Infect. Dis.*, **7**, 216–31.

Lind, K., Lindhardt, B.O., Schutten, H.J. *et al.* (1984) Serological cross-reactions between *Mycoplasma genitalium* and *Mycoplasma pneumoniae*. *J. Clin. Microbiol.*, **20**, 1036–43.

Lo, S.-C., Shih, J.W.-K., Yang, N.-Y. *et al.* (1989a) A novel virus-like infectious agent in patients with AIDS. *Am. J. Trop. Med. Hyg.*, **40**, 213–26.

Lo, S.-C., Wang, R. Y.-H., Newton, P.B. III *et al.* (1989b) Fatal infection of silvered leaf monkeys with a virus-like infectious agent (VLIA) derived from a patient with AIDS. *Am. J. Trop. Med. Hyg.*, **40**, 399–409.

Lo, S.-C., Dawson, M.S., Newton, P.B. III *et al.* (1989c) Association of the virus-like infectious agent originally reported in patients with AIDS with acute fatal disease in previously healthy non-AIDS Patients. *Am. J. Trop. Med. Hyg.*, **41**, 364–76.

Lo, S,-C., Shih, J. W.-K., Newton, P.B. III *et al.* (1989d) Virus-like infectious agent (VLIA) is a novel pathogenic mycoplasma: *Mycoplasma incognitus*. *Am. J. Trop. Med. Hyg.*, **41**, 586–600.

Lo, S.-C., Dawson, M.S., Wong, D.M. *et al.* (1989e) Identification of *Mycoplasma incognitus* infection in patients with AIDS: An immunohistochemical, in situ hybridization and ultrastructural study. *Am. J. Trop. Med. Hyg.*, **41**, 601–16.

Loechel, S., Inamine, J.M. and Hu, P.C. (1989). Nucleotide sequence of the tuf gene from *Mycoplasma genitalium*. *Nucleic Acids Res.*, **17**, 10127.

Madoff, S. and Hooper, D.C. (1988) Nongenitourinary infections caused by *Mycoplasma hominis* in adults. *Rev. Infect. Dis.*, **10**, 602–13.

Mahairas, G.G. and Minion, F.C. (1989a) Transformation of *Mycoplasma pulmonis*: Demonstration of homologous recombination, introduction of cloned genes, and preliminary description of an intergrating shuttle system. *J. Bacteriol.*, 171, 1775–80.

Mahairas, G.G. and Minion, F.C. (1989b) Random insertion of the gentamicin resistance transposon Tn4001 in *Mycoplasma pulmonis*. *Plasmid*, 21, 43–7.

Maniloff, J. (1988) Mycoplasma viruses. *Crit. Rev. Microbiol.*, 15, 339–92.

Mardh P.A. and Westrom L. (1970) T-mycoplasma in the genito-urinary tract of the female. *Acta Pathol Microbiol Scand Sect B Microbiol. Immunol.* 78, 374–6.

Mardh, P.-A., Moller, B. and McCormack, W.M. (1983) International symposium on *Mycoplasma hominis* – a human pathogen. *Sexually Trans. Dis.*, 10, 225–385.

Martinez, O.V., Chan, J., Cleary, T. and Cassell, G.H. (1989) *Mycoplasma hominis* septic thrombophlebitis in a patient with multiple trauma: a case report and literature review. *Diagn. Microbiol. Infect. Dis.*, 12, 193–6.

Maurel, D., Charron, A. and Bebear, C. (1989) Mollicutes DNA polymerases: characterization of a single enzyme from *Mycoplasma mycoides* and *Ureaplasma urealyticum* and three enzymes from *Acholeplasma laidlawii*. *Res. Microbiol. Inst. Pasteur*, 140, 191–250.

McCormack, W.M., Almeida P.C., Bailey P.E. *et al.* (1972) Sexual activity and vaginal colonisation with genital mycoplasmas. *JAMA*, 221, 1375.

McDonald, M.I., Moore, J.O., Harrelson, J.M. *et al.* (1983) Septic arthritis due to *Mycoplasma hominis*. *Arthritis Rheum.*, 26, 1044–7.

McGarrity G.J. and Kotani, H. (1985) Cell culture mycoplasmas, in *The Mycoplasmas, Vol. IV* (eds S. Razin and M.F. Barile), Academic Press, Inc., Orlando, Fla., pp. 353–90.

McGarrity G.J. and Kotani, H. (1986) Detection of cell culture mycoplasmas by a genetic probe. *Exp. Cell. Res.*, 163, 273–8.

Møller, B.R., Freundt, E.A. and Mardh, P.A. (1980) Experimental pelvic inflammatory disease provoked by *Chlamydia trachomatis* and *Mycoplasma hominis* in grivet monkeys. *Am. J. Obstet. Gyn.*, 138, 990–5.

Møller, B.R., Taylor-Robinson, D. and Furr, P.M. (1984) Serological evidence implication *Mycoplasma genitalium* in pelvic inflammatory disease. *Lancet*, 1102–3.

Møller B.R., Taylor-Robinson D., Furr P.M., and Freundt E.A. (1985) Acute upper genital-tract disease in female chimpanzee provoked experimentally by *Mycoplasma genitalium*. *Brit. I. Exp. Pathol.*, 66, 417–26.

Morrison-Plummer, J., Lazzell, A. and Baseman, J.B. (1987) Shared epitopes between *Mycoplasma pneumoniae* major adhesin protein P1 and a 140-kilodalton protein of *Mycoplasma genitalium*. *Infect. Immun.*, 55, 49–56.

Mouches, C., Barroso, G., Gadeau, A. and Bove, J.M. (1984) Characterization of two cryptic plasmids from *Spiroplasma citri* and occurrence of their DNA sequences among various spiroplasmas. *Ann. Microbiol. (Inst Pasteur)*, 135A, 17–24.

Murphy, W. H., Bullis, C., Ertel, E. J. and Zarafonetis, C. J. (1967) Mycoplasma studies of human leukemia. *Ann. N. Y. Acad. Sci.*, 143, 544–56.

Murray, R. G. E. (1984) The higher taxa or a place for everything and kingdom prokaryotae, in *Bergey's Manual Of Systematic Bacteriology Vol. 1* (eds N. R. Kreig and J.G. Holt), Williams & Wilkins, Baltimore, pp. 310–36.

Muto, A. (1987) The genome structure of *Mycoplasma capricolum*. *Isr. J. Med. Sci.*, **23**, 334–41.

Myers, W. F., Baca, O. G. and Wisseman, C. L., Jr. (1980) Genome size of *Rickettsia burnetii*. *J. Bacteriol.*, **146**, 460–1.

Nicol, C. S. and Edward, D. G. (1953) Role of organisms of the pleuropneumonia group in human genital infections. *Brit. J. Vener. Dis.*, **29**, 141–50.

O'Brien, S.J., Simonson, J.M., Grabowski, M.W. and Barile, M.F. (1981) An analysis of multiple isoenzyme expression among twenty-two species of Mycoplasma and Acholeplasma. *J. Bacteriol.*, **146**, 222–32.

Olson, L.D., Shane, S.W., Karpas, A.A. *et al.* (1990) Monoclonal antibodies to pathogenic *Mycoplasma hominis* surface antigens. Manuscript submitted.

O'Meara, P. and McQueen, D.A. (1988) *Mycoplasma hominis* septic arthritis with hypogammaglobulinemia: a case report. *Contemp. Orthopaed.*, **16**, 29–34.

Palu, G., Valisena, S., Barile, M.F. and Meloni, G.A. (1989) Mechanisms of macrolide resistance in *Ureaplasma urealyticum*: a study on collection and clinical strains. *Eur. J. Epidemiol.*, **5**, 146–53.

Plata, E. J., Abell, M. R. and Murphy, W. H. (1973) Induction of leukemoid disease in mice by *Mycoplasma fermentans*. *J. Infect. Dis.*, **128**, 588–97.

Platt, R., Warren, J.W., Edelin, K.C. *et al.* (1980) Infection with *Mycoplasma hominis* in post-partum fever. *Lancet*, **ii**; 1217–21.

Poddar, S.K. and Maniloff, J. (1989) Determination of microbial genome sizes by two-dimensional denaturing gradient gel electrophoresis. *Nucleic Acids Res.*, **17**, 2889–95.

Quinn, P.A., Butany, J., Chipman, M. *et al.* (1985) A prospective study of microbial infection in stillbirths and early neonatal death. *Am. J. Obstet. Gynecol.*, **151**, 238–49.

Razin, S., Barile, M.F., Harasawa, R. *et al.* (1983a) Characterization of the Mycoplasma genome. *Yale J. Biol. Med.* **56**, 357–66.

Razin, S., Harasawa, R. and Barile, M.F. (1983b) Cleavage patterns of the mycoplasma chromosome, obtained by using restriction endonucleases, as indicators of genetic relatedness among strains. *Intl. J. Syst. Bacteriol.*, **33**, 201–6.

Razin, S. and Tully, J.G. (eds) (1983) *Methods in Mycoplasmology*. Academic Press, Inc., New York.

Razin, S., Tully, J.G., Rose, D.L. and Barile, M.F. (1983b) DNA cleavage patterns as indicators of genotypic heterogeneity among strains of Acholeplasma and Mycoplasma species. *J. Gen. Micro.*, **129**, 1935–44.

Razin, S., Amikam, D. and Glaser, G. (1984) Mycoplasma ribosomal RNA genes and their use as probes for detection and identification of Mollicutes. *Isr. J. Med. Sci.*, **20**, 758–61.

Razin, S. and Freundt, E.A. (1984) Biology and pathogenicity of mycoplasmas. *Isr. J. Med.*, **20**, 749–1027.

Razin, S. (1985) Molecular biology and genetics of mycoplasmas (*Mollicutes*). *Microbiol. Rev.*, **49**, 419–55.

Razin, S. (1990) Class: *Mollicutes*, in *The Prokaryotes* (2nd edn), (eds A. Balows, H.G. Truper, M. Dworkin *et al.*) Springer-Verlag. (in press).

Renbaum, P., Abrahamove, D., Fainsod, A. *et al.* (1990) Cloning, characterization, and expression in *Escherichia coli* of the gene coding for the CpG DNA methylase from *Spiroplasma* sp. strain MQ1 (M.Sssl). *Nucl. Acids Res.*, **18**, 1145–52.

Risi, G.F. Jr, Martin, D.H. Silberman, J.A. and Cohen, J.C. (1987) A DNA probe for detecting *Mycoplasma genitalium* in clinical specimens. *Mol. Cell. Probes*, **1**, 327–35.

Roberts, M.C. and Kenny, G.E. (1986) Dissemination of the *tet*M tetracycline resistance determinant to *Ureaplasma urealyticum. Antimicrob. Agents Chemother.*, **29**, 350–2.

Roberts, M.C., Koutski, L.A., Holmes, K.K. *et al.* (1985) Tetracycline-resistant *Mycoplasma hominis* strains contain streptococcal tet M sequences. *Antimicrob. Agts. Chemother.*, **28**, 141–3.

Roberts, M.C., Hooton, M., Stamm, W., Holmes, K.K. and Kenny G.E. (1987) DNA probes for the detection of mycoplasmas in genital specimens. *Isr. J. Med. Sci.*, **23**, 618–20.

Roberts, M.C. and Kenny, G.E. (1987) Conjugal transfer of transposon Tn916 from *Streptococcus faecalis* to *Mycoplasma hominis. J. Bacteriol.*, **169**, 3836–9.

Robertson, J.A., Stemke, G.W., MacLellan, S.G. and Taylor, D.E. (1988) Characterization of tetracycline-resistant strains of *Ureaplasma urealyticum. J. Antimicrob. Chemotherapy*, **21**, 319–32.

Robertson, J.A., Stemler, M.E. and Stemke, G.W. (1984) Immunoglobulin A protease activity of *Ureaplasma urealyticum. J. Clin. Microbiol.*, **19**, 255–8.

Robertson, J.A., Pyle, L.E., Stemke, G.W. and Finch, L.R. (1990) Human ureaplasmas show diverse genome sizes by pulsed-field electrophoresis. *Nucleic Acids Res.*, **18**, 1451–5.

Romano, N., La Licata, R. and Alesi, D.R. (1986) Energy production in *Ureaplasma urealyticum. Pediatr. Infect. Dis.*, **5**, S308–12.

Rosengarten, R. and K.S. Wise. (1990) Phenotypic Switching in Mycoplasmas: Phase Variation of Diverse Surface Lipoproteins. *Science*, **247**, 315–18.

Ruiter, M. and Wentholt, H.M.M. (1953) Isolation of a pleuropneumonia-like organism (G-strain) in a case of fusospirillary vulvovaginitis. *Acta Dermat. Vener.*, **33**, 130.

Saglie, R., Cheng, L. and Sadighi, R. (1988) Detection of *Mycoplasma pneumoniae*-DNA within diseased gingiva by in situ hybridization using a biotin-labeled probe. *J. Periodontol.*, **59**, 121–3.

Saillard, C., Carle, P., Bove, J.M. *et al.* (1990) Genetic and serologic relatedness between *Mycoplasma fermentans* and a mycoplasma recently identified in tissues of AIDS and non-AIDS patient. *Res. Virol. (Inst. Pasteur)*, **141**, 385–95.

Salvado, J.C., Barroso, G. and Labarrere, J. (1989) Involvement of a *Spiroplasma citri* plasmid in the erythromycin-resistance transfer. *Plasmid*, **22**, 151–9.

Samuelsson, T., Guindy, Y.S., Lustig, F. *et al.* (1987) Apparent lack of discrimination in the reading of certain codons in *Mycoplasma mycoides. Proc. Natl. Acad. Sci. USA*, **84**, 3166–70.

Sarov, L. and Becker, Y. (1969) Trachoma agent DNA. *J. Mol. Biol.*, **42**, 581–9.

Sasaki, Y., Shintani, M., Watanabe, H. and Sasaki, T. (1989) *Mycoplasma genitalium* and *Mycoplasma pneumoniae* share a 67 kilodalton protein as a main cross-reactive antigen. *Microbiol. Immunol.*, 33, 1059–62.

Sawada, M., Osawa, S., Kobayashi, H. *et al.* (1981) The number of ribosomal RNA genes in *Mycoplasma capricolum*. *Mol. Gen. Genet.*, 182, 502–4.

Schaper, U., Chapman, J.S. and Hu, P.-C. (1987) Preliminary indication of unusual codon usage in the DNA coding sequence of the attachment protein of *Mycoplasma pneumoniae*. *Isr. J. Med. Sci.*, 23, 361–7.

Shurin, P.A., Alpert, S., Rosner, B. *et al.* (1975) Chorioamnionitis and colonization of the newborn infant with genital mycoplasmas. *N. Engl. J. Med.*, 293, 5–8.

Siber, G.R., Alpert, S., Smith, A. L. *et al.* (1977) Neonatal central nervous system infection due to *Mycoplasma hominis*. *J. Pediatr.*, 90, 625–7.

Simberkoff, M. S., Thorbecke, G. J. and Thomas, L. (1969) Studies of PPLO infection. V. Inhibition of lymphocyte mitosis and antibody formation by mycoplasmal extracts. *J. Ext. Med.*, 129, 1163–81.

Sneller, M. F., Wellborne, M. F., Barile and P. Plotz. (1986) Prosthetic joint infection with *Mycoplasma hominis*. *J. Infect. Dis.*, 153, 174–5.

Su, C.J., Tryon, V.V. and Baseman, J.B. (1987) Cloning and sequence analysis of cytadhesin P1 gene from *Mycoplasma pneumoniae*. *Infect. Immun.*, 55, 3123–9.

Subcommittee on the Taxonomy of Mollicutes (1979) Proposal of minimal standards for descriptions of new species of the class Mollicutes. *Int. J. Syst. Bacteriol.*, 29, 172–80.

Swenson, C.E., VanHamont, J. and Dunbar, B.S. (1983) Specific protein differences among strains of *Ureaplasma urealyticum* as determined by two-dimensional gel electrophoresis and a sensitive silver stain. *Intl. J. Syst. Bact.*, 33, 417–21.

Taschke, C. and Herrmann, R. (1988) Analysis of transcription and processing signals in the 5′ regions of the two *Mycoplasma capricolum* rRNA operons. *Mol. Gen. Genet.*, 212, 522–30.

Taylor-Robinson, D., Csonka, G.W. and Prentice, M.J. (1977) Human intra-urethral inoculation of ureaplasmas. *Q. J. Med.*, 46, 309–26.

Taylor-Robinson, D. and McCormack, W.M. (1980) The genital mycoplasmas. 1. *N. Eng. J. Med.*, 302, 1003–10.

Taylor-Robinson, D., Furr, P.M. and Tully, J.G. (1983) Serological cross-reactions between *Mycoplasma genitalium* and *M. pneumoniae*. *Lancet*, i, 527.

Taylor-Robinson, D., Furr, P.M. and Liberman, M.M. (1984) The occurrence of genital mycoplasmas in babies with and without respiratory distress. *Acta Paed. Scand.*, 73, 383–6.

Taylor-Robinson, D. (1985) Mycoplasmal and mixed infections of the human male urogenital tract and their possible complications, in *The Mycoplasmas*, Vol. *IV* (eds S. Razin and M.F. Barile), Academic Press, Inc., pp. 27–63.

Taylor-Robinson, D., Furr, PM. and Hanna, N.F. (1985a) Microbiological and serological study of non-gonococcal urethritis with special reference to *Mycoplasma genitalium*. *Genitourin Med.*, 61, 319–324.

Taylor-Robinson, D., Tully, J.G., and Barile, M.F. (1985b) Urethral infection in male chimpanzees produced experimentally by *Mycoplasma genitalium*. *Brit. J. Exp. Path.*, 66, 95–101.

Taylor-Robinson, D. and Furr, P.M. (1986) Clinical antibiotic resistance of Ureaplasma urealyticum. *Pediatr. Infec. Dis.*, 5, S335–7.

Taylor-Robinson, D., Furr, P.M., Tully, J.G., Barile, M.F., and Moller, B.R. (1987) Animal models of *Mycoplasma genitalium* urogenital infections. *Isr. J. Med. Sci.*, 23, 561–4.

Thomas, M., Jones, M., Ray, S. and Andrews, B., (1975) *Mycoplasma pneumoniae* in a tubo-ovarian abscess. *Lancet*, ii (7938), 774–5.

Thomsen, C. (1978) Occurrence of mycoplasmas in urinary tracts of patients with chronic pyelonephritis. *J. Clin. Microbiol.*, 8, 84–8.

Tilton, R.C., Dias, F., Kidd, H. and Ryan, R.W. (1988) DNA probe versus culture for detection of *Mycoplasma pneumoniae* in clinical specimens. *Diagn. Microbiol. Infect. Dis.*, 10, 109–12.

Tully, J.G. and Razin, S., eds (1983) *Methods in Myoplasmology, Vol. II.* Academic Press, Inc, New York. pp. 1–440.

Tully, J.G., Taylor-Robinson, D., Cole, R.M., and Rose, D.L. (1981) A newly discovered mycoplasma in the human urogenital tract. *Lancet*, i, 1288–91.

Tully, J.G., Taylor-Robinson, D., Rose, D.L., Cole, R.M. and Bove, J.M. (1983) *Mycoplasma genitalium*, a new species from the human urogenital tract. *Int. J. Syst. Bacteriol.*, 33, 387–96.

Tully, J.G., Taylor-Robinson, D., Rose, D.L. *et al.* (1986) Urogenital challenge of primate species with *Mycoplasma genitalium* and characteristics of infection induced in chimpanzees. *J. Infect. Dis.*, 153, 1046–54.

Vulfovich, Y. V., Zilfyan, A. V., Zheverzheeva, I.V. *et al.* (1981) Prolonged persistence of *Mycoplasma arthriditis* and *Mycoplasma fermentans* in the body of experimentally-infected rats, *Zh Mikrobiol. Epidemiol. Immunobiol.*, 1, 14–17.

Waites, K.B., Crouse, D.T. and Cassell, G.H. (1990) Ureaplasma and mycoplasma CNS infections in newborn babies. *Lancet*, 335, 658–9.

Waites, K.B., Rudd, P.T., Crouse, D.T. *et al.* (1988) Chronic *Ureaplasma urealyticum* and *Mycoplasma hominis* infections of central nervous systems in preterm infants. *Lancet*, i, 17–23.

Waites, K.B., Crouse, D.T., Philips III, J.B. *et al.* (1989) Ureaplasmal pneumonia and sepsis associated with persistent pulmonary hypertension of the newborn. *Pediatrics*, 83, 79–85.

Watson, H.L., McDaniel, L.S., Blalock, D.K. *et al.* (1988) Heterogeneity among strains and a high rate of variation within strains of a major surface antigen of *Mycoplasma pulmonis*. *Infect. Immun.*, 56, 1358–63.

Webster, D.B., Taylor-Robinson, D., Furr, P.M. and Asherson, G.L. (1982) Chronic cystitis and urethritis associated with ureaplasmal and mycoplasmal infection in primary hypogammaglobulinemia. *Br. J. Urol.*, 54, 287–91.

Webster, A.D.B., Furr, P.M., Hughes-Jones, N.C. *et al.* (1988) Critical dependence on antibody for defence against mycoplasmas. *Clin. Exp. Immunol.*, 71, 383–7.

Weickmann J.L. and Farhney, D.E. (1977) Arginine deiminase from *Mycoplasma arthritidis*: Evidence of multiple forms. *J. Biol. Chem.*, 252, 2615–20.

Wenzel, R. and Herrmann, R. (1988) Physical mapping of the *Mycoplasma pneumoniae* genome. *Nuc. Acids Res.*, 16, 8323–36.

Williams, M. H., Brostoff, J. and Roitt, I. M. (1970) Possible role of *Mycoplasma fermentans* in pathogenesis of rheumatoid arthritis. *Lancet*, ii, 277–80.

Williams, M. H. and Bruckner, F. E. (1971) Immunological reactivity of *Mycoplasma fermentans* in patients with rheumatoid arthritis. *Annal. Rheumatic Disease*, 30, 271–3.

Yamao, F., Muto, A., Kawauchi, Y. *et al.* (1985) UGA is read as tryptophan in *Mycoplasma capricolum. Proc. Natl Acad. Sci., USA*, 82, 2306–9.

Yogev, D. and Razin, S. (1986) Common deoxyribonucleic acid sequences in *Mycoplasma genitalium* and *Mycoplasma pneumoniae* genomes. *Intern. J. Syst. Bacteriol.*, 36, 426–30.

Yogev, D., Sela, S., Bercovier, H. and Razin, S. (1988) Elongation factor (Ef-Tu) gene probe detects polymorphism in *Mycoplasma* strains. *FEMS Microbiol. Lett.*, 50, 145–9.

Zouzias, D., Mazaitis, A.J., Simberkoff, M. and Rush, M. (1973) Extrachromosomal DNA of *Mycoplasma hominis. Biochim. Biophys. Acta*, 312, 484–91.

4

Molecular biology of
Treponema pallidum

LEO M. SCHOULS

The spirochaete *Treponema pallidum* subsp. *pallidum*, the causative agent of the venereal disease syphilis, has been the subject of extensive studies ever since its discovery over eighty years ago. The study of this pathogen has been hampered by inability to cultivate the pathogenic treponemes *in vitro*. The last decade a number of investigators have used modern molecular biology techniques to circumvent this problem. This chapter reviews some of the progress made in elucidating the molecular basis of the pathogenesis of *Treponema pallidum* subsp. *pallidum* (*T. pallidum*). It also discusses how recombinant DNA technology has been applied to diagnosis of treponemal infection and potential vaccine production.

4.1 SYPHILIS

Approximately 95% of all syphilis is transmitted by sexual exposure. Other less frequent means of transmission are prenatal infection and blood transfusions. The portal of entry of the highly infectious *T. pallidum*, with an ID_{50} of 57 organisms (Magnusson *et al.*, 1956), is usually mucous membranes or abraded skin. The spirochaetes start to multiply locally and within hours gain access to the circulatory system, disseminating throughout the body. The initial site of infection is infiltrated with mononuclear cells, forming the primary lesion or chancre. On average, the chancre will appear three weeks after infection and persist for 2–6 weeks. The secondary stage follows after the primary chancre has healed; it is characterized by systemic skin lesions and, in 85% of the cases, by generalized lymphadenopathy. In 15–40% of the cases the central nervous system is infected, although this does not necessarily lead to neurological symptoms. Secondary syphilis may

Molecular and Cell Biology of Sexually Transmitted Diseases
Edited by D. Wright and L. Archard
Published in 1992 by Chapman and Hall, London ISBN 0 412 36510 3

last for several weeks or months and is followed by a latency period, in which about one-fourth of the patients suffer relapses. The patient then becomes asymptomatic and can only be diagnosed by serological testing (latent stage). In the pre-antibiotic era, 30% of the untreated syphilis patients developed manifestations of late or tertiary syphilis. Half of these individuals develop late benign or gummatous syphilis. The gumma, closely resembling the granulomatous infiltration observed in tuberculosis, usually occurs in the skin, mucous membranes or bones. These lesions do not contain detectable treponemes by darkfield microscopy at the time of diagnosis. Serial sections will reveal the occasional spirochaete if silver stain examination is used. It is the spirochaetes that are responsible for the intense immune response associated with the gumma. The other half of the patients with tertiary disease have cardiovascular or central nervous system involvement.

The incidence of syphilis declined rapidly after the introduction of penicillin therapy. Furthermore, improved hygiene, better physical conditions and the constant exposure to antibiotics used for infections other than syphilis, have probably contributed to this decline. The incidence of syphilis in most industrialized countries is still very low. In Great Britain the number of primary and secondary cases of syphilis in 1988 was about two per 100 000 inhabitants. However, an alarmingly rapid increase in syphilis cases has been reported in the United States; 42 600 new cases of infectious syphilis (17 per 100 000 inhabitants) were reported in 1989, the highest number of cases in 40 years. The trend in 1990 indicates a further increase. A striking example of the return of syphilis as a major public health problem is the increase in congenital syphilis in New York City. It is estimated that the number of cases in 1989 is about 900 as opposed to the 57 cases in 1986 (CDC, 1989). This 15-fold increase has been imputed to the practice of trading sexual favours for cocaine or crack. Since the identities of sex partners are mostly unknown, the partner notification system fails, which results in the disease spreading further.

The role of syphilis and other genital ulcer diseases in the acquisition and transmission of the human immune deficiency virus (HIV) is another distressing complication in the re-emergence of syphilis (Greenblatt *et al.*, 1988; Quinn, 1988). Furthermore, HIV-infected individuals may be serologically negative after infection with *T. pallidum* and thus remain undetected and untreated (Hicks *et al.*, 1987). One isolated case report suggests that, even when treated with penicillin, HIV-infected patients may not respond to this therapy and suffer relapses (Berry *et al.*, 1987). The shift from funding for the control of syphilis to AIDS surveillance may be yet another factor in the increase of syphilis (CDC, 1988).

The control of syphilis has until now been dependent on the serologic testing of patients, widespread screening of blood donors and contact tracing followed by penicillin treatment. The widespread screening of sera is costly, especially in a number of European countries where the relatively expensive treponemal haemagglutination assay (TPHA) is used for screening. This has led to the diminished screening in some of the European countries (e.g. Denmark). Treatment with antibiotics and improving health and hygiene nearly eradicated the venereal and non-venereal treponematoses; however, small pockets of cases have resulted repeatedly in fresh outbreaks of the disease. The fact that the great pox (syphilis), like smallpox, has no animal reservoir makes eradication of the treponematoses feasible provided an effective campaign could be mounted. The failure of previous efforts by antibiotic treatment alone to eradicate syphilis and the high costs related to serological screening, strongly suggests that a vaccine against treponematoses is necessary. How to produce such a vaccine depends largely on the knowledge of the pathogenesis of *T. pallidum* infection and host's immune response. The study of the molecular biology of *T. pallidum* will make a major contribution to understanding these aspects of the treponematoses.

4.2 THE CELL ENVELOPE OF *T. PALLIDUM*

Spirochaetes display an unusual morphology compared with other bacteria. Like all spirochaetes, treponemes have a helical form giving the bacterium its corkscrew-like morphology. The pathogenic treponemes have an average diameter of 0.15 μ and a length, as measured along the axis of the bacterium, of 5–15 μ depending on the growth conditions. Some consequences of these dimensions will be discussed below.

The cell envelope structure of bacteria determines the mechanism by which the host will destroy infecting organisms. Virtually all types of bacteria have a cytoplasmic membrane with a peptidoglycan layer. In Gram-positive micro-organisms the peptidoglycan layer is very thick and may be covered by a wax-like structure, such as seen in mycobacteria. Gram-negative bacteria possess, in addition, an outer membrane which, like the cytoplasmic membrane, consists of a lipid bilayer. Most Gram-negative bacteria possess lipopolysaccharide (LPS) associated with the outer membrane. Both the cytoplasmic membrane and the outer membrane harbour a number of proteins that can have functions such as ion transport (porins), transport of nutrients and stabilizing the cell envelope (lipoproteins). Also, numerous regulatory proteins can be found in the outer membrane. Some bacteria have a capsule consisting of polysaccharides with anti-phagocytic properties. In addition flagella and/or fimbriae may be found on the surface of bacteria. The flagella are important for motility, while the fimbriae function in the attachment of the pathogen. The spirochaete cell envelope resembles that of Gram-negative bacteria, with an inner and outer membrane. The compartment between the two membranes, the periplasmic space, harbours a number of flagella. These do not penetrate the outer membrane as do normal bacterial flagella but reside completely within the integrity of the outer membrane and are therefore called endoflagella, periplasmic flagella or axial filaments. The latter designation stems from the fact that the endoflagella start at the tips of the organism and wrap along the axis of the spirochaete. The flagella are thought to propel the cells by means of rotation, driven by a proton motive force (Macnab and Aizawa, 1984). The rotation results in a corkscrew-like motility, which, in the case of *T. pallidum*, is apparent only in a medium with a high macroscopic viscosity (N. Charon, personal communication).

Over the years the structure and composition of the outer membrane of *T. pallidum* has become one of the most controversial subjects of study. It seems obvious that components exposed on the surface of *T. pallidum* are the prime targets for antibodies. Furthermore, the structures involved in attachment of the parasite are likely to be present in the outer membrane. As the outer membrane has an important role in both the virulence and the immune response, many investigators have tried to identify these outer membrane components. Initially a number of antigens were assigned to be surface exposed or surface associated on the basis of standard separation and purification techniques. An example is the presumed surface exposure of the 47k antigen, which has been designated as the major outer membrane surface exposed antigen of *T. pallidum* (Jones *et al.*, 1984; Marchitto *et al.*, 1984; Marchitto *et al.*, 1986a). However, the validity of the experiments leading to these conclusions is now questioned.

One of the techniques used to identify surface associated antigens is the binding of monospecific or monoclonal antibody to intact *T. pallidum* under *Treponema pallidum* immobilization test (TPI) conditions (Radolf *et al.*, 1986b; Fehniger *et al.*, 1986). After incubation of these antisera with treponemes in the presence of complement for 18 hours, some antibody samples result in heavy labelling of the *T. pallidum* membrane, suggesting surface exposure of these antigens. However, several studies have shown that *T. pallidum* becomes sensitized with antibodies from the animal used for cultivation of the treponemes. This is especially noticeable if rabbits, not given exogenous glucocorticosteroids, are used for cultivation of *T. pallidum* or if treponemes are harvested after peak orchitis (Turner and Hollander, 1957). Bishop and Miller showed that preincubation of treponemes with unheated normal rabbit serum and subsequent incubation with immune rabbit serum results in killing of the treponemes (Bishop and Miller, 1976). These observations indicate that prolonged incubation in the presence of complement causes sensitized *T. pallidum* to become leaky and so creates an entrance for other antibodies that may not necessarily be directed against surface components, but can now gain access and bind to sub-surface antigens.

A second technique frequently used is the lactoperoxidase catalysed surface iodination (Morrison, 1980). Norris and Sell (Norris and Sell, 1984) studied the surface localization of T. pallidum proteins and demonstrated that slightly prolonged labelling under sub-optimal conditions can result in labelling of virtually all proteins, while[125] I-labelling under apparently optimal conditions revealed preferential labelling of a 39 kDa protein. This protein is abundant in T. pallidum, as can be concluded from the high intensity staining with Coomassie brilliant blue. Later experiments showed that this protein is the endoflagellar sheath protein, making its surface exposure questionable. Moskophidis and Müller (1984) reported the iodination of at least 13 antigens, which in light of more recent findings described below makes it very unlikely to represent surface labelling.

To identify integral membrane proteins, Penn et al. (1985) used low concentrations of Triton X-100 to solubilize the outer membrane of freshly isolated treponemes and found that under these conditions the 47k protein was the only T. pallidum antigen partially extracted. Triton X-114 extraction and phase partitioning (Bordier, 1981) is modification of the Triton X-100 treatment used to identify integral membrane proteins. Radolf and coworkers applied this elegant technique to T. pallidum, using high concentrations of Triton X-114 and showed that virtually all major immunogenic proteins could be extracted from freshly isolated T. pallidum (Radolf and Norgard, 1988; Radolf et al., 1988). The extracted antigens all partitioned in the detergent phase suggesting a hydrophobic character of these antigens. Using low concentrations of the Triton X-114 the outer membrane of T. pallidum was completely solubilized, but left the protoplasmic cylinder intact. Surprisingly, no antigens could be immunoprecipitated from the detergent phase which resulted, indicating the lack of such antigens in the outer membrane. By increasing the concentration Triton X-114 the protoplasmic cylinder became more distorted and the number of proteins solubilized increased, the 47 kDa antigen being the first antigen detected in the detergent phase. These results, similar to those obtained by Penn et al. (1985), might indicate an outer membrane location for the 47 kDa antigen, but more likely reflect the detection of minor amounts of this extremely antigenic and possibly most abundant antigen released from the cytoplasmic membrane.

Differential ultracentrifugation of disrupted ^{35}S-methionine labelled T. pallidum results in unique SDS-PAGE profiles of the soluble and insoluble fractions (Stamm and Bassford, 1985). The membrane fraction appears to contain virtually all major antigens. Similarly, Schouls and colleagues (1989a) showed that the major antigens resided in the particulate fraction of disrupted T. pallidum. Comparison of the profiles obtained by both groups, however, show a major difference in the number of protein bands detected. Schouls and coworkers analysed total protein, whereas Stamm investigated de novo synthesis of non-dividing T. pallidum cells. In vitro labelling of non-dividing cells may differ from labelling of growing cells and could result in preferential labelling of intracellular proteins. By metabolic labelling *T. pallidum* for only two hours and subsequently incubating the intact treponemes with immune rabbit serum, Stamm showed that only two antigens could be detected as 'surface exposed' (Stamm *et al.*, 1987). These proteins were shown to be endoflagellar subunits, again underlining the failure of conventional techniques to detect the exact membrane location of the *T. pallidum* proteins.

Other techniques used to evaluate surface exposure are immunofluorescence and immunoelectron microscopy of fixed treponemes. These and the above-mentioned techniques require the use of membrane perturbing agents. Considering that the outer membrane of *T. pallidum* is extremely fragile, all the previously claimed outer membrane locations are highly questionable and must be re-evaluated.

Recent freeze fracture and deep-etching electron microscopy experiments performed independently by two groups have shown that the outer membrane of *T. pallidum* has a striking paucity of intramembranous particles (IMPs) compared with other bacteria such as *E. coli* (Radolf *et al.*, 1989; Walker *et al.*, 1989). An example of a freeze-fracture showing IMPs of *T. pallidum* is depicted in Figure 4.1. IMPs are thought to represent integral membrane proteins or protein/lipid aggregates. The number of IMPs in the cultivable non-pathogenic *Treponema phagedenis* is about ten times higher than that in *T. pallidum*, but still a factor seven lower than that of *Spirochaeta aurantia* or *Escherichia coli*. *Borrelia burgdorferi* also has a limited number of IMPs, but still 20 times as many as *T. pallidum*. Because of the scarcity of particles, the outer membrane of *T. pallidum* is believed to contain only a few proteins and these are generically called treponemal rare outer membrane proteins (TROMPs). In contrast to the outer membrane, the *T. pallidum* inner membrane has numerous particles.

The paucity of surface-exposed antigens in the outer membrane of *T. pallidum* was suggested many years ago. The lack of agglutination of freshly isolated treponemes with immune serum as opposed to the agglutination observed with aged *T. pallidum* implies that the *T. pallidum* surface is inert (Turner and Hollander, 1957). As an example of more recent findings, Penn and Rhodes (1982) showed that intact, freshly harvested treponemes do not react with antibody or show radiolabelling of surface proteins, unlike surface proteins from detergent treated treponemes. The insensitivity of freshly isolated *T. pallidum* to proteases supports this finding. Treatment of whole *T. pallidum* with proteinase K or with trypsin does not result in any apparent degradation of proteins or change in morphology. Examination of Western blot profiles of protease treated treponemes shows that the major antigens remain intact except for TmpA, which appeared to be partially degraded (Cockayne, manuscript in preparation); this finding could reflect a partial exposure of this antigen and suggests a possible correlation with the TROMPs.

A reconsideration of the unusual morphology of spirochaetes makes the

Figure 4.1 Freeze-fracture electron micrograph of *T. pallidum*. The inner leaflet of the outer membrane of the fractured treponeme displays a limited number of IMPs. The freeze-fracture was performed by Dr R.Miller, Dept of Oral Biology, University of Washington, Seattle. The bar represents $1 \mu M$.

paucity of proteins in the outer membrane even more striking. The helical *T. pallidum* cell, when stretched, forms a long thin cylinder with a volume surface ratio of 1:27. Compared with the volume surface ratio of *E. coli*, which is 1:4, this suggests the presence of relatively more membrane and thus more membrane proteins in *T. pallidum* than in *E. coli*. However, freeze-fracture experiments reveal the presence of a low amount of exposed outer membrane protein suggesting that the antigens are located in the periplasmic space or in the inner membrane.

There is another remarkable feature of the *T. pallidum* outer membrane. Several lines of research have provided a body of evidence to conclude that the *T. pallidum* outer membrane has no lipopolysaccharide (LPS). Hardy and Levin (1983) were unable to detect significant *Limulus* amoebocyte lysate gelation activity supporting the concluded absence of endotoxin. Penn *et al.* (1985) showed that *T. pallidum* has very low pyrogenicity and that the outer membrane is extremely SDS susceptible, indicating the lack of smooth LPS. Furthermore, no immobilization or vesicle formation was detected when treponemes were treated with the antibiotic polymyxin B. A number of other spirochaetes such as *T. phagedenis* do have LPS (Bailey *et al.*, 1985) but others like *B. burgdorferi* also lack the LPS characteristic of Gram-negative bacteria. Leptospira possess a LPS-like component, called lipooligosaccharide (LOS), but LOS differs significantly from the normal LPS. The absence of LPS also has implications for the separation of the inner and outer membranes of *T. pallidum*. The usual procedure for separating inner and outer membranes of Gram-negative bacteria is sucrose gradient centrifugation. The outer membrane, having a higher buoyant density mainly due to the presence of LPS, forms a band in the lower part of the gradient while the cytoplasmic membrane forms the upper band. The absence of LPS does not allow this procedure to be applied to *T. pallidum*.

The data on the membrane can be used to make a model of the *T. pallidum* cell envelope as the one proposed in Figure 4.2.

4.3 THE *T. PALLIDUM* GENOME AND MOLECULAR CLONING OF GENES

Miao and Fieldsteel (1978) performed the first studies on the *T. pallidum* chromosome and found that it was 53% in GC content, contains no repeating sequences and has a genomic size of 1.4×10^7 base pairs. Miao calculated that the genomic size of *T. pallidum* was more than twice as large as that of *E. coli*. However, as a result of pulse field gel electrophoresis experiments it is now clear that the genome of *T. pallidum* is smaller than that of *E. coli* and comprises about 900kb (S. Norris, personal communication). The 900 kilobase genome potentially encodes some 500 genes from which 5% has been cloned and less than 2% sequenced. Physical or genetic maps of the genome are not available, but a project to construct such maps has been started (S. Norris, personal communication). Although Miao and Fieldsteel (1978) claimed the absence of repeating sequences, this needs to be re-evaluated using more sensitive techniques. All the cloned genes tested in Southern hybridizations are present as a single copy gene in the *T. pallidum* genome.

Norgard and Miller (1981) reported the presence of an 11 kb plasmid in *T. pallidum*. Considering the potential of such a plasmid as a cloning vehicle genetically to manipulate *T. pallidum*, it is surprising that no further

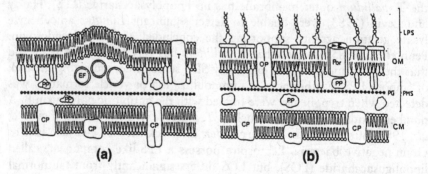

(a) **(b)**

Figure 4.2 Hypothetical structure of the cell envelope of *T. pallidum* (a) and *E. coli* (b). CM, cytoplasmic membrane; CP, cytoplasmic membrane protein; PG, peptidoglycan layer; PMS, periplasmic space; PP, periplasmic protein; EF, endoflagellum; OM, outer membrane; LPS, lipopolysaccharide; OP, outer membrane protein; Por, porin protein; T, treponemal rare outer membrane protein.

studies have been done on this extra chromosomal element. Additionally, the presence of plasmids creates the potential for plasmid mediated antibiotic resistance, a matter not to be taken lightly as can be learned from the problems with penicillin-resistant gonococci. The presence or altered composition of such a plasmid might even be responsible for the appearance of different variants of *T. pallidum*. Pulse field electrophoresis experiments have failed to show such a plasmid but the detection limit of these experiments may be too low to display a low copy number plasmid. Perhaps the circular DNA observed by Norgard and Miller was not plasmid but eukaryotic satellite DNA; the observation deserves further investigation.

A major problem in syphilis research has been the lack of purified *T. pallidum* products. The inability to cultivate *T. pallidum in vitro* limits investigators to the use of animals, mostly rabbits, to propagate the pathogenic treponemes. Cultivation in animals results in low yields of bacteria and massive contamination with host material. It was hoped that molecular cloning of *T. pallidum* genes and subsequent expression in *E. coli* would circumvent these problems. Following the first report (Stamm *et al.*, 1982), a substantial number of studies on cloned *T. pallidum* antigens have been published. At this time, 26 different genes of *T. pallidum* have been cloned and most of them expressed in *E. coli* (Table 4.1).

The selection of most recombinants has been immunological, using either syphilitic serum or monoclonal antibodies. In four cases, a DNA probe was used for detection of the gene of interest. A DNA probe based on the DNA sequence of the homologous *T. pallidum* gene was used for the selection of the *tyF1* gene *T. pallidum* subsp. *pertenue*, the causative agent of the non-venereal disease yaws. In the cloning of the endoflagellar core proteins FlaB1, FlaB2 and FlaB3, a DNA probe based on the N-terminal amino acid sequence of the flagellar subunit was used. Although a variety of cloning vehicles and different restriction enzymes for digestion of *T. pallidum* DNA were used by different investigators, the same limited set of genes was consistently selected. This arose as most investigators employed antibodies against *T. pallidum* for screening the gene libraries and so selected the major antigenic and well-expressed proteins.

Although the availability of cloned antigens has contributed in part to a better characterization of these proteins, the function of most cloned antigens is still unknown. The endoflagellar proteins clearly are involved in the motility of the treponemes. The 60 kDa, Tp4 antigen (Hindersson *et al.*, 1987) probably is the homologue of the GroEL protein, important for the correct folding of proteins. As members of the GroEL family of proteins occur unbiquitously in living organisms and share about 40% of their amino acid sequences they are called common antigens. The fibronectin binding proteins P1 and P2 might be part of the attachment mechanism of *T. pallidum*. The function of the other cloned proteins,

Table 4.1 Summary of the cloned antigens. The table indicates relative molecular weight and the antigen name used by the original authors. The cloning strategy is displayed as the cloning vehicle used and the restriction enzyme used for digestion of the *T. pallidum* DNA

Size[a]	Size[b]	Antigen	Cloning strategy	References
190	190	4D	λ charon 30 – *Sau*3A	1;2;3;4;5;6;7;8
190	190	TpF1	pHC79 cosmid – *Sau*3A	9;10;11
190	140	C1–5	λL47–1 – *Bam*HI	12
190	190	TyF1	EMBL-3 - *Sau*3A	10
89[e]	69.3	P1	EMBL-3 – *Sal*1	13
67[e]	67	67k	λ charon 30 – *Sau*3A	1;14
61	60	Tp–4	pHC79 cosmid – *Sau*3A	15;45
58	58	TpA	pHC79 cosmid – *Sau*3A	9
47	47	47k	pBR322 plasmid – *Alui/Hae*III	16;17;18;19
47	45	P6	EMBL-4 – *Eco*RI	20
47	47	TpS	pHC79 cosmid – *Sau*3A	21
44.5	44	TmpA	pHC79 cosmid – *Sau*3A	9;22;23;24;25
44.5	44.5	44.5k	pAT153 plasmid – *Sau*3A	26
41	42	42k	pAT153 plasmid – *Sau*3A	26
41[e]	41	TpF2	pHC79 cosmid – *Sau*3A	9
39	39	BMP (39k)	pBR322 plasmid – *BAm*HI	27;28;29;30
37	38	38k	λ charon 30 – *Sau*3A	1;31
37	37	37k	same clone as 38k	1;14
37	37	P2	EMBL-3 – *Sal*I	13
37[e]	37	P2e	EMBL-3 – *Bam*HI	32
37	37	FlaA	λ gtii - random DNAse	33;34
35	35	TmpC	pHC79 cosmid – *Sau*3A	9
35	29	29k	pAT153 plasmid – *Sau*3A	26
34	34	TmpB	pHC79 cosmid – *Sau*3A	9;22;23;24;25
34	37	37k	pAT153 plasmid – *Sau*3A	26
34.5	34.5	FiaB1	pUC18 plasmid – *Eco*RI/*Bam*HI	35
33	31.5	FlaB2	pUC121 plasmid – *Sau*3A	36
31	31	FiaB3	pUC18 plasmid – *Eco*RI/*Bam*HI	35
29–35	26–32	TpD	pHC79 cosmid – *Sau*3A	9;37;38
29–35	34	34k	pBR322 plasmid – *Bam*HI	39;40;41;42
29–35	29–35	29–35k	pAT153 plasmid – *Sau*3A	26
24	24	24k	pUC8 plasmid – *Bam*HI	43
24	24	TpT	pHC79 cosmid – *Sau*3A	21
24–28	19–24	TpE	pHC79 cosmid – *Sau*3A	9;38
24–28	24–28	24–28k	pAT153 plasmid – *Sau*3A	26
17	17	17k	pAT153 plasmid – *Sau*3A	26
15.5	15.5	15.5k	pAT153 plasmid – *Sau*3A	26
15.5	15	15k	pBR322 plasmid – *Aiul/Hea*III	44

[a] Consensus M_r in kDa according to Norris (Norris, 1987)
[b] M_r in kDa reported by authors
[c] Not antigenic, gene located on same plasmid as the 37k gene
[d] Cloned as truncated protein
[e] Not yet reported by Norris
[1] Walfield et al. (1982); [2] Fehniger et al. (1984); [3] Fehniger et al. (1986); [4] Coates et al. (1986); [5] Radolf et al. (1986a, b); [6] Radolf et al. (1987); [7] Borenstein et al. (1988); [8] Van Embden et al. (1983); [9] Noordhoek et al. (1989); [10] Noordhoek et al. (1990a); [11] Walfield et al. (1989); [12] Peterson et al. (1987b); [13] Rodgers et al. (1986); [14] Hinderson et al. (1987); [15] Norgard et al. (1986); [16] Chamberlain et al. (1988); [17] Hsu et al. (1989); [18] Chamberlain et al. (1989b); [19] Peterson et al. (1986); [20] Hindersson and Bangsdorf (1987); [21] Hansen et al. (1985); [22] Schouls et al. (1989a); [23] Ijsselmuiden et al. (1989); [24] Wicher et al. (1989); [25] Bailey et al. (1989); [26] Stamm et al. (1982); [27] Stamm et al. (1983); [28] Dallas et al. (1987); [29] Stamm et al. (1988); [30] Fehniger et al. (1986); [33] Isaacs and Radolf (1990); [34] Champion et al. (1990); [35] Pallesen and Hindersson (1989); [36] Hindersson et al. (1986); [37] Schouls et al. (1986b); [38] Norgard and Miller (1983); [39] Swancutt et al. (1990); [40] Swancutt et al. (1986); [41] Swancutt et al. (1990); [42] Hsu et al. (1988); [43] Purcell et al. (1989); [44] Thole et al. (1988).
[f] indicates identical antigens

however, remains obscure. This is partly due to the difficulty in creating mutated T. pallidum strains. One of the strategies to pinpoint functions is to alter or inactivate cloned genes and reintroduce the altered genes into the original organism, exchanging the altered gene for the original gene and to study the changes in the organism. The restrictions to these experiments are both technical and ethical. Technical problems are related to the necessity to use animals for the cultivation of T. pallidum, making selection of mutants difficult. Generally, the procedure for making mutants to pinpoint functions of certain genes involves the insertion of an antibiotic resistance gene. Apart from the technical difficulties, the possible development of antibiotic-resistant T. pallidum by genetic manipulation also creates an ethical problem. Recent improvements in the in vitro cultivation experiments might remedy the technical problems but not the ethical doubts. The use of non-pathogenic treponemes as acceptors of T. pallidum genes might be an alternative approach to elucidate functions of cloned proteins, although non-pathogenic treponemes are only distantly related to T. pallidum. As an example T. phagedenis has less than 5% DNA homology with T. pallidum, a difference which could interfere with this strategy.

4.4 TREPONEMA PALLIDUM LIPOPROTEINS

The finding that the outer membrane of T. pallidum contains virtually no proteins was followed by another unexpected discovery. Chamberlain et al. (1989a) found that metabolic labelling of T. pallidum with radioactive fatty acids resulted in labelling a large number of the major immunogens and they termed these antigens proteolipids. These lipid-modified T. pallidum proteins, usually called lipoproteins, have molecular masses of 47, 38, 36, 17 and 15 kDa. Independently, Schouls et al. (1989b) showed that the cloned antigens TpD, TpE, TmpA and TmpC are lipoproteins. The lipoprotein character of TpD, which is identical to the 34k protein, has recently been corroborated (Swancutt *et al.*, 1990). The four antigens described by Schouls and coworkers possess N-terminal signal sequences and the consensus lipopeptidase (signal peptidase II) cleavage site (Figure 4.3). The lipopeptidase cleavage site is a tetrapeptide and has the following preferred composition leucine-X-X-cysteine, in which X is a small relatively uncharged amino acid and the cysteine is an absolute requirement. Additionally, it has been shown that the processing of TpD, TpE and TmpA can be inhibited by the antibiotic globomycin, which specifically blocks the signal peptidase II.

Undoubtedly, the most prominent *T. pallidum* antigen in terms of humoral response is the 47k antigen. In natural and experimental syphilis, antibodies against this antigen are among the first that can be detected on Western blots and, for this reason, the 47k protein has potential for the early serodiagnosis of syphilis. The 47k antigen has been cloned and expressed in *E. coli* but expression of the cloned product is low. This probably has hampered the purification and application of this antigen to the serodiagnosis of syphilis. The DNA sequence of the 47k gene shows that the deduced amino acid sequence carries no signal sequence (Hsu *et al.*, 1989), yet evidence suggests a surface association of this antigen

```
MGRYIVPALLCVAGMGFAHAQSALQPIAEVNLFRREPVTL        BMP
MKTRNFSLVSALYVLLGVPLFVSA                        TmpB
MVGTYVVTLWGGVFAALVAG                            37k
MKKAVVLSAVALLSGVCAVADESVLIDF                    FlaA

MFDAVSRATHGHGAFRQQFQYAVEVLGEKVLSKQETEDSR        47k

MCTDGKKYHSTATSAAVGASAPGVPDARAIAAICEQLRQH        TpF1

    MNAHTLVYSGVALACAAM │LGSC│ASGAKEEAEK           TmpA
    VREKWVRAFAAVFCAML  │LIGC│SKSDRPQMGN           TmpC
       MKRVSLLGSAAIFALV│FSAC│GGGGEHQHGE           TpD
      MRKLSTGLLLWIAFISG│VTSC│KSGPPAEELV           TpE

        MKKYLLGIGLILA│LIAC│KQNVSSLDEK             OspA
         MRLLIGFALALA│LIGC│AQKGAESIGS             OspB

MIQEIMADLSGFLTYARSFRRTGSFV│LDAC│LHCRRCDGCR       LepORF1
```

Figure 4.3 N-terminal amino acid sequences of sequenced or partially sequenced cloned spirochaetal antigens. LepORF1 represents the leptospiral protein deduced from the open reading frame upstream from the *trp* genes (see text). The boxed area denotes the lipopeptidase cleavage site of the spirochaetal lipoproteins.

(Jones *et al.*, 1984; Marchitto *et al.*, 1984; Marchitto *et al.*, 1986). The lack of a N-terminal signal sequence (Figure 4.3) is very unusual for outer membrane or periplasmic proteins although most inner membrane proteins of prokaryotes do not carry a N-terminal signal sequence (Blobel, 1980; Wickner and Lodish, 1985). The region upstream from the structural 47k gene contains no characteristic ribosomal binding site or promoter sequences. The lack of consensus translational and transcriptional sequences might explain its poor expression in *E. coli*. Recently, it was shown that both the cloned and native 47k protein are lipid-modified (Chamberlain *et al.*, 1989b). The absence of an N-terminal signal sequence or potential lipopeptidase cleavage site suggests that the post-translational modification of the immunodominant 47 kDa protein might follow a pathway as yet unknown in prokaryotes.

Purcell *et al.* (1989) cloned and expressed the 15 kDa antigen of *T. pallidum* and, by showing that the 15 kDa antigen could be labelled with radioactive fatty acids, lent further support to the lipoprotein nature of this antigen. A number of the cloned antigens that are found to be lipid-modified in *E. coli* were not initially detected in the metabolic labelling of *T. pallidum* with fatty acids. This indicates that the sensitivity of the labelling technique in *T. pallidum* is not sufficiently sensitive to detect all lipoproteins, implying that an even larger number of antigenic proteins might be lipid-modified in *T. pallidum*.

The occurrence of lipoproteins is not restricted to *T. pallidum*, but can also be found in other spirochaetes. The DNA sequence of the genes encoding for the major outer membrane proteins of *B. burgdorferi* OspA and OspB indicates the presence of N-terminal signal sequences on these proteins and a typical lipopeptidase cleavage sequence (Bergström *et al.*, 1989). Preliminary labelling experiments of *B. burgdorferi* with ^{14}C-palmitic acid show that OspA and OspB are intensely labelled, indicating their lipoprotein character (Schouls, unpublished results). The deduced amino acid sequence upstream from the *trpE* and *trpG* genes of *Leptospira biflexa* (Yelton and Peng, 1989) also contains a putative cleavage site for lipopeptidase, although in this case the signal sequence is less obvious (Figure 4.3).

The finding that a large number of *T. pallidum* antigens are lipoproteins explains why the microbe is so immunogenic. Studies have shown that the attachment of a lipid moiety significantly enhances the humoral immune response to a protein (Jung *et al.*, 1985) and lipid-modification of proteins also enhances the cellular immune response (Coon and Hunter, 1975; Deres *et al.*, 1989). The mechanism for enhancement of the immune response by lipidation might be related to reduced degradation in macrophages of the lipid-modified component of the protein. The increased stability might result in a repeated exposure of the lipid-modified residue on the surface of the macrophage, which results in a reiterated presentation to lymphocytes. Also, the hydrophobic lipid anchor may be inserted better in the macrophage membrane and the antigen thus be more efficiently exposed.

The physiological implications of membrane associated lipid-modified proteins should also be considered. Lipoproteins are present among most species of bacteria and have a variety of functions. Murein-lipoprotein, proven to be present in several bacteria, performs a function in maintaining the shape and rigidity of the cell by connecting the peptidoglycan layer on the inner membrane with the outer membrane. Other lipoproteins such as β-lactamase of *Bacillus subtilis* and *Staphylococcus aureus* are enzymes necessary to survive in an environment containing penicillin derivatives. The lipid-modified secreted cloacins kill susceptible cells of closely related species by interfering with their protein synthesis. In *Klebsiella pneumoniae*, the lipoprotein pullulanase acts as a surfactant and in *Rhodopseudomonas viridis*, one of the cytochrome subunits is a lipoprotein. These few examples indicate the wide variety of functions lipoproteins fulfil in bacteria. Unfortunately, none of the identified treponemal lipoproteins has been shown to perform a specific function. The specialized motility of spirochaetes requires a highly flexible membrane structure which makes it questionable that treponemal lipoproteins have a function similar to that of murein lipoprotein.

Outer membrane proteins in most prokaryotic species have negatively charged residues in the proximal N-terminal region which are important for interaction with the membrane. Mature lipoproteins lack such a characteristic N-terminal sequence, but this lack is probably compensated by the covalent lipid-modification of the N-terminal cysteine. This N-terminus provides a hydrophobic anchor by which the otherwise hydrophilic protein becomes attached to the cell membrane and is translocated across the inner membrane. In Gram-positive bacteria which have no outer membrane, this usually results in excretion; in Gram-negative bacteria, translocated proteins mostly end up in the periplasmic space or in the outer membrane. Most of the cloned antigens of *T. pallidum* are claimed to be membrane associated. In some cases, data leading to this conclusion are derived from both the location of these antigens in *E. coli* as well as from the deduced amino acid sequences. Except for the 47k antigen, all cloned treponemal lipoproteins are produced as precursors carrying a N-terminal signal sequence which is cleaved during or after translocation (Figure 4.3). Additionally a number of the other non-lipid-modified cloned antigens (TmpB, BMP, 37k, FlaA) are also produced with a cleavable signal sequence. If, as deduced from the freeze fractures studies of *T. pallidum*, there is a paucity of outer membrane proteins then virtually all cloned antigens must be located in the periplasmic space. However, the freeze fracture technique might not detect any lipoproteins anchored with their lipid moiety in the inner layer of the outer membrane, with the hydrophilic polypeptide in the periplasmic space (Figure 4.2). Lipoproteins would therefore be outer membrane proteins, although not necessarily surface exposed.

One of the most frequently used screening tests in syphilis serology is the Venereal Disease Research Laboratory (VDRL) flocculation test. The VDRL antigen is an unstable mixture of the phospholipids cardiolipin, cholesterol and lecithin. The titre measured in the VDRL test is directly related with the activity of the syphilis infection and usually falls rapidly after successful treatment and more slowly in latent untreated infections. It has been suggested that the production of antibodies reactive with the VDRL antigen is an autoimmune reaction following tissue damage and subsequent exposure of tissue antigens to the immune system. Alternatively, the transient nature of the VDRL titre could represent a response against a true treponemal antigen cross reacting with the VDRL antigen. The existence of such a component has not been unambiguously demonstrated.

The finding that most of the major antigens in T. pallidum are lipoproteins suggests that the presence of the lipid tail on these proteins could be responsible for the positive VDRL reaction. However, VDRL antibodies do not react with any of the major T. pallidum antigens on western blots or in the fluorescent treponemal antibody – absorption test (FTA-ABS test) (Baker-Zander, personal communication). The lack of reactivity in the FTA-ABS test would argue against the presence of a cardiolipin reactive antigen on T. pallidum, although it is possible that the acetone, thereby removing the reactive component.

4.5 GLYCOSYLATION

The lipid-modification may not be the only post-translational protein modification taking place in T. pallidum. There are several indications that glycosylation of proteins also takes place. Data supporting glycosylation comes from the agglutination of T. pallidum with wheat germ agglutinin (Fitzgerald and Johnson, 1979b) and the enhanced binding to glass cover slips pretreated with concavalin A (Baseman et al., 1980). Moskophidis and Müller (1984) reported the immune precipitation of four T. pallidum antigens metabolically labelled with ^{14}C-glucosamine having molecular masses of 59, 35, 33 and 30.5 kDa respectively. Precipitation of similarly labelled T. phagedenis revealed two proteins of 34 and 33 kDa respectively. This result suggests that some of the endoflagellar core proteins are glycosylated in both T. pallidum and T. phagedenis. In the metabolic labelling experiment described by Strugnell et al. (1986), a 40 kDa ^{3}H-glucosamine labelled macromolecule, which may be similar to the 35 kDa endoflagellar protein, was conjectured to be a T. pallidum glycoprotein. Interestingly, Brahamsha and Greenberg (1988) found that the 33 kDa and the 34 kDa endoflagellar core proteins of the purified flagella of *Spirochaeta aurantia* reacted on western blots with concavalin A, a lectin that binds N-acetylglucosamine residues. By comparison, *B. burgdorferi* also seems to have a glycosylated endoflagellar subunit, amongst a number of other glycoproteins (Luft *et al.*, 1989).

Glycosylation may alter the conformation of a protein and could result in epitopes not present in non-glycosylated forms of the protein. As glycosylation does not take place in *E. coli*, antigens normally glycosylated in *T. pallidum* may not have the same antigenic characteristics as their cloned counterparts. The reduced antigenicity of the cloned endoflagellar FlaB2 subunit might be the result of the lack of glycosylation of this antigen (L. Pallesen, personal communication).

4.6 THE ENDOFLAGELLUM

Spirochaetes represent a phylogenetically distinct group of bacteria with a unique type of motility, caused by periplasmic flagella. In some spirochaetes, including *T. pallidum*, the flagella are relatively complex and consist of an inner core and an outer layer or sheath. In *T. pallidum* and *T. phagedenis* the core consists of different subunits which can be visualized on SDS-PAGE. In contrast, the endoflagellar protein of *B. burgdorferi* is detected as a single 41 kDa band on SDS-PAGE. Endoflagella of *T. phagedenis*, which are relatively easy to purify, can be used in the serology of syphilis (Strandberg Pedersen *et al.*, 1982) but some cross-reactions are observed. The non-specificity of the endoflagellum is illustrated by the cross-reactions of anti-*T. phagedenis* endoflagellum serum with a number of distantly related spirochaetes as detected by western blot (Figure 4.4). The cross-reactivity of endoflagella has been reported previously by a number of investigators (Hardy *et al.*, 1975; Nell and Hardy, 1978; Petersen *et al.*, 1981; Limberger and Charon, 1986). The extensive cross-reactivity renders the use of the endoflagellum, for purposes other than serological screening, questionable.

Norris *et al.* (1988) showed antigenic relatedness of the periplasmic flagella of *T. pallidum* and *T. phagedenis*. Purified *T. pallidum* endoflagella contain six proteins of 37, 34.5, 33, 30, 29 and 27 kDa. The flagella of *T. phagedenis* contain a major protein of 39kDa, a 37kDa protein, two major and two minor proteins in the range of 34 to 33 kDa and a 30 kDa protein. On the basis of similarities between the N-terminal sequences, Norris *et al.* (1988) classified the flagellar proteins in class A and class B proteins. The 37 kDa *T. pallidum* protein and the 39 kDa *T. phagedenis* protein are designated the class A proteins and represent the flagellar sheath. The *T. pallidum* 34.5,

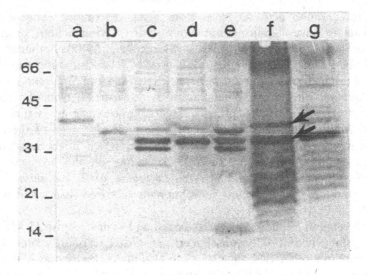

Figure 4.4 Immunoblot of distantly related spirochaetes showing the cross-reactivity of the endoflagellar proteins. The blot was incubated with anti-*T. phagedenis* endoflagellum serum. The arrows on the right-hand side indicate the positions of the 39kD sheath and of the 30–35 kD core proteins complex of *T. phagedenis*, respectively. a, *Borrelia burgdorferi*; b, *Leptospira biflexa*; c, *T. pallidum*; d, *T. denticola*; e, *T. hyodysenteriae*; f, *T. phagedenis* and g, *T. vincenti*.

33, and 30 kDa and the *T. phagedenis* complex of 34 to 33 kDa proteins form the class B proteins and represent the core of the flagellum. The 29 and 27 kDa *T. pallidum* proteins and the 37 and 30 kDa *T. phagedenis* proteins are considered unrelated to the class A or B proteins and might represent proteins from the structures involved in the attachment of the flagellum to the inner membrane. N-terminal sequencing of the class A and class B proteins of both treponemal species revealed a high degree of homology between the N-termini of flagellar proteins of *T. pallidum* and *T. phagedenis*. Furthermore, all the class B proteins showed near identity in the N-terminal sequencing. Antibodies directed against any of the class B proteins cross react with all other class B proteins irrespective of the bacterial species used, but not with class A polypeptides. Anti-class A antibodies do not react with class B but cross-reaction between *T. pallidum* and *T. phagedenis* class A polypeptides does occur. This accords with the N-terminal homology mentioned above.

Recently, Isaacs *et al.* (1989) reported the cloning and DNA sequence of the 37 kDa class A protein of *T. pallidum* and designated this protein FlaA. However, the N-terminal part of the *flaA* gene, encoding for the first 9 amino acids of the FlaA sequence as determined by Edman degradation (Norris *et al.*, 1988), was not cloned, so preventing the total amino acid sequence from being determined. The sequence appears to have significant homology with the cloned FlaA protein of *Spirochaeta aurantia* (Brahamsha and Greenberg, 1989). Brahamsha and Greenberg cloned the complete *flaA* gene of *S. aurantia* and found that the N-terminal homology between the cloned FlaA of *S. aurantia* and the *T. pallidum* FlaA starts 20 amino acids downstream from the methionine of the *S. aurantia* protein, suggesting cleavage by signal peptidase. Additionally, they observed that the molecular mass of the cloned protein was smaller then the native protein which might indicate different modification of the protein, depending on the host. Similarly, cloning the upstream region of the *flaA* gene of *T. pallidum* showed that the FlaA protein is post translationally modified (Issacs and Radolf, 1990).

Pallesen and Hindersson (1989) cloned and sequenced the 33 kDa class B flagellum subunit of *T. pallidum* and designated this core protein FlaB2. Recently, the other two flagellar class B subunits, FlaB1 and FlaB3 were cloned and expressed in *E. coli* (Champion *et al.*, 1990). The genes encoding for FlaB1 and FlaB3 are located in one operon and are coordinately transcribed from one promoter. The *flaB2* gene is probably not located in this operon. The N-termini of the cloned class B proteins and the N-terminal amino acid sequence of the 41 kDa endoflagellar protein of *B. burgdorferi* show a high degree of homology (Luft *et al.*, 1989). This homology suggests that the 41 kDa subunit represents a core protein which agrees with observations that *Borrelia* endoflagella lack sheath protein (Hovind-Hougen, 1976).

4.7 THE HEAT-MODIFIABLE OLIGOMERIC 4D ANTIGEN

A large number of reports have been generated on the immunogenic *T. pallidum* protein 4D. Three different groups have cloned this antigen (Fehniger *et al.*, 1984; Noordhoek *et al.*, 1989; Walfield *et al.*, 1989) and reported the characteristic properties of this protein which they designated 4D, TpF1 and C1–5, respectively. The 4D protein is composed of about ten non-antigenic subunits with an apparent molecular mass of 19 kDa on SDS-PAGE under normal denaturation conditions. However, if the protein preparation is not completely denatured, the 4D has an apparent molecular weight of 190 kDa and is fully reactive with serum from patients with syphilis. Treatment with proteinase K causes limited digestion of the antigen resulting in a 90 kDa degradation product which remains reactive with antibodies. The 190 kDa 4D antigen is an ordered ring structure of non-covalent polymerized 19 kDa subunits with a diameter of 10 nm, while the 90 kDa proteinase K product is a 6 nm ring structure (Fehniger *et al.*, 1986). The physical properties of this protein led to a model for the 4D molecule as depicted in Figure 4.5 (Noordhoek, personal communication). Interestingly, the properties of the 4D homologue from *T. pallidum* subsp. *pertenue*, TyF1, differs from 4D (Noordhoek *et al.*, 1989). Proteinase K treatment of latter antigen results in a digestion product with an apparent molecular mass 115 kDa rather than the 90 kDa molecule observed for 4D. Sequence analysis of the *tyF1* and *4D* genes revealed only one base pair difference between the two genes, resulting in one amino acid substitution. The large difference in apparent molecular mass on SDS-PAGE suggests that the amino acid substitution causes a conformational difference in the proteins. Efforts to exploit the minor difference between the *T. pallidum* and *T. pertenue* 4D serologically to discriminate between syphilis and yaws have until

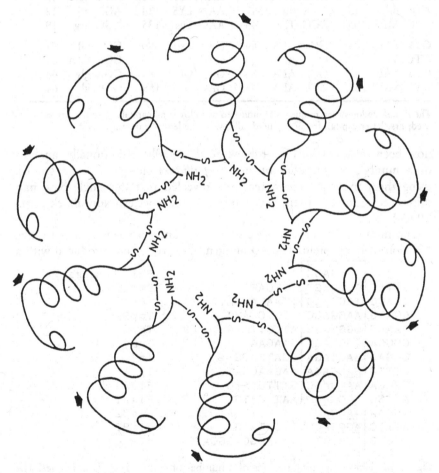

Figure 4.5 Hypothetical structure of the polymerized 4D antigen. The ordered ring structure is composed of 10 subunits of 19 kD each. The arrow indicates the site in the molecule where proteinase K presumably digests off a 5 kDa fragment.

Table 4.2 Codon usage of *T. pallidum* genes based on 11 amino acid sequences comprising 3097 amino acids in total

TTT	PHE	59	TCT	SER	66	TAT	TYR	32	TGT	CYS	6
TTC	PHE	55	TCC	SER	34	TAC	TYR	63	TCG	CYS	9
TTA	Leu	8	TCA	Ser	19	TAA	end	4	TGA	end	3
TTG	Leu	44	TCG	Ser	34	TAG	end	4	TGG	TRP	4
CTT	Leu	60	CCT	Pro	26	CAT	HIS	22	CGT	ARG	49
CTC	Leu	45	CCC	Pro	25	CAC	His	29	CGC	ARG	57
CTA	Leu	8	CCA	Pro	12	CAA	Gln	40	CGA	arg	9
CTG	LEU	47	CCG	PRO	22	CAG	GLN	129	CGG	arg	26
ATT	ILE	66	ACT	THR	37	ATT	Asn	47	AGT	Ser	29
ATC	ILE	74	ACC	THR	39	AAC	ASN	76	AGC	Ser	39
ATA	ile	15	ACA	thr	17	AAA	LYS	34	AGA	arg	12
ATG	MET	105	ACG	Thr	44	AAG	Lys	135	AGG	arg	19
GTT	VAL	75	GCT	ALA	84	GAT	Asp	80	GGT	GLY	71
GTC	Val	52	GCC	ALA	56	GAC	Asp	93	GGC	GLY	65
GTA	VAL	41	GCA	ALA	88	GAA	GLU	80	GGA	gly	45
GTG	VAL	85	GCG	ALA	135	GAG	Glu	159	GGG	gly	61

The *E. coli* codon usage frequency is indicated as follows: all capitals, very frequently used; first letter capital, frequently used; all lower-case letters, rarely used codon.

now been unsuccessful (Noordhoek *et al.*, 1990b). Additionally, analysis of a number of *T. pallidum* and *T. pallidum* subsp. *pertenue* strains, using the polymerase chain reaction, revealed that the single base pair difference is not consistent between the two subspecies (Noordhoek *et al.*, 1990a).

The multimeric ring structure of the 4D antigen suggests a porin-like function of the protein. The 4D antigen, however, is not produced with a

```
          AGGAGG                        Shine-Dalgarno
  TTTAC  GGGGAGT  TCTCATG              BMP
  TCACG  TGGGAGG  TTAACATCATG          TmpA
  GGAAG  AAGGAGC  CTCTCGATG            TmpB
  TAAAG  GGGGAGG  ATTCTTGTG            TmpC
  CCCCA  GGGGAGT  GAAGAATG             TpD
  TGGAG  GAGGAGA  CGATGATG             TpE
  GTTTG  GAGGAGA  AAGAACGATG           TpF1
  TGCGA  AAGGAGC  GTTTGAATG            FlaA
  ATTCC  ACGGAGG  AATGCATG             FlaB1
  ACCAA  AAGGAGT  AATGCTTCATG          FlaB2
  ACTTC  AGGGAGG  GTCATATG             FlaB3
  GCGCG  GGGGATC  TCGACGCAGGCATG       47k
```

Figure 4.6 Ribosomal binding sites of a number of cloned *T. pallidum* genes. The putative ribosome binding sites are located in the boxed area. The sequences indicated as the putative Shine-Dalgarno sequence in the original papers are underlined.

N-terminal signal sequence and is isolated from the soluble fraction of cell lysates.

4.8 TRANSLATIONAL AND TRANSCRIPTIONAL SIGNALS

Based on published DNA sequences of nine *T. pallidum* genes, the TpE sequence (Strugnell *et al.*, 1991) and the TmpC sequence (Schouls, 1991), a preliminary codon usage for *T. pallidum*, can be derived and is depicted in Table 4.2. The codon usage of *T. pallidum* does not grossly differ greatly from that of *E. coli* (Grosjean and Fiers, 1982), which is not completely unexpected since the GC-content of *T. pallidum* (53%) is similar to that of *E. coli*. The overall consensus is that *T. pallidum* preferentially uses a G as the wobble base, while the A is the least-used wobble base.

Most of the treponemal recombinant antigens are expressed in *E. coli* using their own transcriptional and translational signals. However, the flagellar core proteins FlaB1, FlaB2 and FlaB3 are not expressed in *E. coli*. Initially, FlaA was not expressed in *E. coli* because the region upstream from the start codon could not be cloned and the protein could only be produced as a fusion product with β-galactosidase. Recently, the region upstream of the *flaA* gene has been cloned and expressed in *E. coli* and production of the FlaA appears to be toxic for *E. coli* (Isaacs and Radolf, 1990). The class B flagellar genes were cloned along with their upstream promoter sequences but no expression in *E. coli* could be detected.

Inspection of the DNA sequence upstream from the cloned structural genes that have been published, shows that all these genes but one have a putative ribosomal binding site (Figure 4.6) (Gold *et al.*, 1981). The exception is the gene encoding for the 47k major immunogenic protein, which may have a structure with a minor similarity with the consensus ribosomal binding site (SD), although the spacing between the start codon and this putative SD sequence is large, making it doubtful that it is a good functional translational start.

The promoter sequences of the cloned *T. pallidum* genes are somewhat less evident than the SD sequences. The region upstream from the structural genes of TpD, TmpA, BMP and TpE are similar to the consensus − 10 and − 35 sequences (Rosenberg and Court, 1979). The distance between the start codon and the promoter sequences are quite similar for these four genes (Figure 4.7). At least in the case of TpD and TmpA, the lack of expression in deletion mutants has confirmed the location of the transcriptional starts (Hansen *et al.*, 1985; Schouls *et al.*, 1989b). The region upstream of the *tmpC* gene also has sequences that show similarities with the − 10 and − 35 region but the spacing between the startcodon and these regions is different from the earlier mentioned genes. Both the 4D or TpF1 and the 47k sequences display some homology with the − 10 promoter sequence. No − 35 region can be detected.

Pallesen and Hindersson (1989) speculated that transcription of the *flaB2* gene is regulated by promoters recognized by a different sigma factor. Similarly, Champion *et al.* (1990) found a putative promoter sequence for the *flaB1/flaB3* operon that resembles the sigma-28 promoter consensus sequence of *Bacillus subtilis* flagellin gene. Interestingly, the expression of the *flaA* gene appears to be controlled by an *E. coli*-like promoter, indicating the different organization of core and sheath proteins (Isaacs and Radolf, 1990).

It must be emphasized that the regions marked as promoters are speculative and are not supported by solid experimental evidence. Transcriptional terminators could not be found in any published DNA sequences but this partly arises from the lack of sufficient data on the sequences downstream from the cloned genes.

So far two polycistronic operons, the *tmpA/tmpB* and the *flaB1/flaB3* operon, have been identified. The flagellar genes are not all located in one operon. The *flaB2* gene which has very high homology (>90%) with the *flaB1* and *flaB3* genes is not part of the *flaB1/flaB3* operon. Furthermore, the *flaA* gene also is located on a different DNA segment and even has a different transcriptional start. It is unknown how these genes cooporate to assemble a complete flagellum.

4.9 RECOMBINANT TECHNOLOGY IN THE DIAGNOSIS OF SYPHILIS

The cloned antigens TmpA, TmpB, TmpC, 4D and 37k have been evaluated for use in the serodiagnosis of syphilis and the results are listed in Table 4.3. ELISAs based on both the 4D antigen and TmpA have sensitivities and specificities comparable to the classic serological tests for syphilis and so have considerable diagnostic value. Interestingly, the anti-TmpA antibody titre dropped sharply after successful treatment of patients in a fashion similar to the VDRL titre (Figure 4.8). The rapid decline in titre might be useful in monitoring the efficacy of antibiotic treatment of syphilitic patients, and contrasts with the long lasting antibody reactivity measured by tests like the TPHA and FTA-ABS. Although the sensitivity of the 37k RIA was satisfactory, the specificity was not high enough; this was not surprising considering the impurities present in the antigen preparation used for this assay. The lack of sensitivity of the TmpB and TmpC ELISA indicate that these antigens are unsuitable for serodiagnosis.

Although the use of recombinant products in the serodiagnosis may be an important step in facilitating large scale screening of sera, the detection of *T. pallidum* or its products in early lesions and in cerebrospinal fluid (CSF) might be even more important. The limit for detection of the presence of bacteria by DNA/DNA hybridization using radioactively labelled genomic DNA is about 10^4 organisms and is therefore unsuitable clinically. The recently introduced polymerase chain reaction (PCR) allows the detection

```
                        -35                    -10
Consensus              TTGACA                TATAATG

TpD   GCCGGGTGTGTGCTGCGCCGTGGAGATTTCCATTTGTTTTTCTATAATGGTGAGGA  -75
BMP   CGGCGTATTCGCGCGATTCTTGAGAAAAGCGGGGAAAAGTCGTAACTTGCCCTGCT  -69
TmpA  TCGACGCGGTGGCGAAGTCTTGAATAAGGGGGCTTTCTGGTGTACCCTCCCGGGTC  -61
TpE   CCATGAGGTGGCATGTATCTTGTCTTCATTGGGGTTCTGTGTAAAATGGGCCGCTA  -52
FlaA  ATGGGTAGATGCAGGTCAGTTGACAAAATAGGCGGTGAGCGTATACTCGAAGAGTG  -39
TmpC              CCCTTGGTTACACTGTGACCCCTTATTTTGCGCATGG  -33


TyF1  ATCGGTGAAAAAAAGCCGGGAAAAGTCCAAAAAAGACAGTGGTTATGCTCCATTTCT  -66
47k   GTGGCTCGTCTCATCATGAGACGCACTATGGCTATGCGACGCTAAGCTATGCGACT  -93
```

Figure 4.7 Putative promoter sequences of a number of cloned *T. pallidum* genes. The consensus −35 and −10 promoter regions are indicated above the sequences. The hypothesized promoter sequences in the *T. pallidum* DNA are underlined. The number on the right side of the sequence indicates the position of the −10 region relative to the startcodon of the structural gene.

of only a few treponemes in CSF or lesion exudates (Hay *et al.*, 1990; Noordhoek, personal communication). The PCR technique also allows the discrimination between the different *T. pallidum* strains based on a single base pair difference (Noordhoek *et al.*, 1990a).

4.10 PATHOGENESIS OF *T. PALLIDUM*

Bacterial pathogens are micro-organisms adapted to cause infection in a host, often by highly specialized mechanisms. Virulence factors are quite diverse and specific for different micro-organisms and hosts, yet generalizations can be made. The ability to attach and, in most cases, penetrate cells or tissue is the first prerequisite for a successful pathogen. Once in the host, the micro-organism must evade the host's immune response. Pathogenicity depends on the ability to pass these barriers and to reach the specific niche within the host. To maintain the pathogen, successful transmission to a new susceptible host is another important step in the pathogenesis of virulent micro-organisms.

T. pallidum can certainly be considered ۱s a successful pathogenic bacterium. It is therefore not surprising that this parasite displays some of these common themes in its virulence. The molecular basis for pathogenesis of some aspects of the virulence of *T. pallidum* has become clearer in the last few years.

4.10.1 The first step: adherence

T. pallidum is transmitted by direct, usually sexual, contact of mucosal tissues, thus avoiding a hazardous external environment. Therefore, the parasite has not adapted to survival outside a host and is extremely sensitive

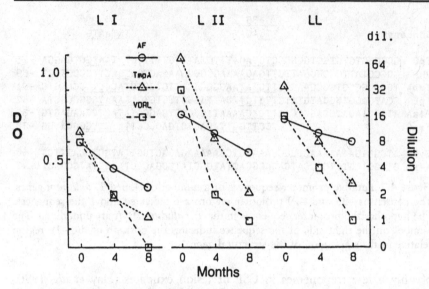

Figure 4.8 Decline of anticardiolipin (VDRL), anti-TmpA and anti-endoflagellar (AF) antibodies in syphilis patients up to eight months after penicillin treatment. The graph shows the mean values of serum samples from 27 different patients. LI, primary; LII, secondary; LL, early latent syphilis. OD, optical density of the ELISA; dil, highest dilution of serum resulting in a positive VDRL test.

to treatments that are non-physiological. The strong host dependency has frustrated investigators that have tried to cultivate *T. pallidum in vitro*. Only recently, a number of researchers have succeeded in growing *T. pallidum* on tissue culture cells (Fieldsteel *et al.*, 1981; Norris, 1982). However, the spirochaete has proven to be fastidious in its growth requirements and serial passage remains unpredictable and inconsistent (Riley and Cox, 1988).

The first major interaction between pathogen and host is attachment

Table 4.3 Sero diagnosis of syphilis with purified recombinant proteins. The table indicates the number of serum samples reactive per number of samples tested

	4D	TmpA	37k	TmpB	TmpC
Syphilis					
primary	22/27	42/55	53/78	18/24	11/24
secondary	19/19	39/39	16/16	22/25	25/25
latent	22/22	53/54	13/16	12/19	17/19
tertiary	16/28	nd	17/21	nd	nd
Yaws	11/13	14/15	nd	nd	nd
Normals	0/172	2/938	3/65	3/30	0/30

nd: not determined.

to a eukaryotic cell. Micro-organisms display multiple strategies of cell attachment. Generally only one mechanism is studied at a time, while the attachment *in vivo* could depend on several mechanisms working simultaneously. In the simplest terms, two components are involved in adherence: the adhesin and its receptor. Bacterial adhesins are typically protein structures on the surface of the microbe while receptors are usually carbohydrate structures on the surface of eukaryotic cells. In *Enterobacteriaceae* the fimbriae or pili are well known adhesins. *T. pallidum* does not possess fimbriae on its surface, although the presence of fimbriae on *Treponema pallidum* subsp. *pertenue*, the causative agent of the non-venereal treponematosis yaws, has been reported (Hovind-Hougen *et al.*, 1976).

A number of adhesins other than fimbriae have been described. Examples are the filamentous haemagglutinin from *Bordetella pertussis* and fibronectin binding proteins. Fibronectin, one of major components of the extracellular cellular matrix of eukaryotic cells, is present in large quantities on mucosal surfaces. Three treponemal proteins designated P1, P2 and P3, of which P1 and P2 have been cloned and expressed in *E. coli* (Peterson *et al.*, 1987b), are capable of binding to fibronectin (Baseman and Hayes, 1980). Fibronectin binding may be an important attachment mechanism of the pathogenic treponemes, since mucosal surfaces form the natural portal of entry for *T. pallidum*. Other pathogens, such as *Streptococcus pyogenes* and *Staphylococcus aureus*, also bind fibronectin (Beachey and Courtney, 1987; Proctor, 1987), both at the N-terminus of fibronectin although at distinct sites. In contrast, the binding site of the putative treponemal adhesins is comprised of a four amino acid sequence Arg-Gly-Asp-Ser (RGDS) in the cell binding domain of fibronectin (Thomas *et al.*, 1985a,b,c). The RGD sequence (similar to the *T. pallidum*-specific RGDS sequence) is a conserved sequence in several proteins of the extracellular matrix and blood, including fibronectin. Integrins, proteins that perform several functions such as cell attachment to extracellular matrices, bind proteins with the RGD sequence and are the receptors for invasin from *Yersinia pseudotuberculosis* (Lelong *et al.*, 1990).

The host specificity of the pathogenic treponemes is remarkable. Treponemes pathogenic for man do not cause natural infections in animals, although they can proliferate in a limited number of animal species under laboratory conditions. The host specificity may represent the necessity for, as yet unidentified, specific receptors present only on the cells of humans and those animals that can be experimentally infected with the pathogenic treponemes.

4.10.2 Dissemination and invasion

Once a pathogen has attached to the host cells, the next step is entry. There are two ways to enter the tissue: either by direct invasion of the cells, or by intercellular transfer. Many electron microscopy (EM) studies have showed the intracellular location of T. pallidum (Azar et al., 1970; Lauderdale and Goldman, 1972; Ovcinnikov and Delektorskij, 1972; Sykes et al., 1974). The above mentioned similarities with the invasin of Yersinia and the adhesin of T. pallidum, suggests that T. pallidum might invade cells. Firm evidence that intracellularity is a result of active invasion has never been provided. Instead, most recent studies confirm that T. pallidum primarily leads an extracellular life. Live T. pallidum extracted from infected rabbit testicles are able actively to penetrate a monolayer of endothelial rabbit cells in vitro through the tight junctions without altering these structures (Thomas et al., 1988). In contrast, the non-pathogenic cultivable Treponema phagedenis fails to cross the monolayer. The capacity to pass the intercellular junctions of endothelial cells might explain how treponemes enter and leave the blood stream. In a similar study, using mouse abdominal wall mounted between two culture chambers, it was also found that T. pallidum and not T. phagedenis is able to penetrate (Riviere et al., 1989). Interestingly, penetration only took place if T. pallidum was placed on the epithelial side of the abdominal wall, suggesting a cell specificity. Thomas and Comstock (1989) showed that, in the monolayer model, another pathogenic spirochaete, Borrelia burgdorferi, is also able to pass the monolayer. In contrast to T. pallidum, which passes through the junctions between the cells, B. burgdorferi is able to pass through the endothelial cell cytoplasm. A recent study with Leptospira interrogans (Thomas and Higbie, 1990) which causes a zoonotic disease, showed both intercellular passage and intracellular location of these spirochaetes in the monolayer assay.

The mechanism for tissue entry might be related to the morphology and motility of the spirochaetes. This type of motion could partially explain how T. pallidum penetrates the cells in the monolayer model and perhaps also tissue in vivo. The corkscrew-like motility, however, is not enough, since T. phagedenis, which has the same type of motility, is unable to penetrate cells. Perhaps adherence to host cells distinguishes T. pallidum from the non-pathogenic treponemes. Alternatively, T. pallidum may pass the tight junctions between cells by using secreted proteins that degrade the substance between the cells. Possibly, hyaluronidase, which degrades hyaluronic acid, the ground substance of mammalian cells (Fitzgerald and Johnson, 1979a) is such a protein. Alternatively or additionally secreted proteins might damage host cells providing free passage for the spirochaetes. A number of T. pallidum proteins, with low molecular mass, have been reported to be secreted during in vitro protein synthesis (Stamm et al., 1987). Hsu and coworkers (1988) reported the possible secretion of a 24 kDa cloned T. pallidum antigen. Disruption of tissue culture cells by T. pallidum has been observed by several researchers (Fitzgerald and Repesh, 1983; Riley and Cox, 1988). Fitzgerald showed that such cell damage did not occur if the cultured cells were incubated in high speed supernatants of viable treponemes in the presence of heat-killed treponemes, or if cells were incubated inverted in chambers with viable treponemes. These observations indicate that direct contact between T. pallidum and the host cells is required for the cytotoxicity.

4.10.3 Survival in the host

Extracellular micro-organisms generally have specialized mechanisms to acquire iron, since free iron is scarce in the body of a healthy host. Some pathogens produce siderophores for their iron uptake, others bind the mammalian iron-binding glycoproteins transferrin or lactoferrin to internalize the bound iron. T. pallidum binds lactoferrin and is able to desorb the iron from this host molecule without degrading the lactoferrin (Alderete et al., 1988). Surprisingly, T. pallidum only binds low amounts of transferrin, but acquires similar amounts of iron from transferrin as from the 100 times more efficiently bound lactoferrin.

Like a number of other spirochaetes, T. pallidum is incapable of producing long chain fatty acids required for its growth. Alderete and Baseman (1989) recently reported the capacity of T. pallidum to bind and internalize host serum lipoproteins. By analogy with iron binding, this mechanism might provide the pathogenic treponeme with its required fatty acids. A remarkable phenomenon is the enhanced binding (170 times greater) of serum lipoproteins after T. pallidum has been pretreated with sulphated proteoglycans.

When T. pallidum is treated with ruthenium red, a substance that stains polyanions, intense staining of the spirochaetes occurs (Zeigler et al., 1976). This has been confirmed by several research groups who have shown that the polyanions were mucopolysaccharides also termed proteoglycans or glycosaminoglycans (Fitzgerald et al., 1978; Fitzgerald et al., 1979; Fitzgerald and Johnson, 1979b; Fitzgerald et al., 1985). During the development of orchitis in experimental syphilis in rabbits, a mucoid material accumulates which was shown to be sulphated glycosaminoglycans and characterized as hyaluronic acid (Van der Sluis et al., 1985, 1987). Although the production of the sulphated proteoglycans initially was ascribed to T. pallidum, metabolic labelling experiments with $Na^{35}SO_4$ showed that these macromolecules are derived from the host and not the treponeme (Strugnell et al., 1984, 1986). Removal of the proteoglycan layer of T. pallidum and subsequent incubation with fresh proteoglycan results in rapid binding irrespective of the viability of the treponemes (Fitzgerald et al., 1985), further corroborating the non-treponemal origin of the proteoglycans.

The enhanced binding of host-derived lipoprotein, once T. pallidum is coated with proteoglycans, may indicate the complexity of the host–parasite interaction in T. pallidum infections; an apparently non-specifically binding host component enhances the binding of another host protein required by the parasite. Additionally, both host components might play a role in the evasion of the host immune response.

4.10.4 Evading the host defence mechanisms

Most pathogenic micro-organism's need to pass some non-specific host barriers. Since *T. pallidum* is usually sexually transmitted, it does not have to overcome most of these barriers. The vaginal pH and the presence of mucosal secretions are probably the most hazardous primary barriers for *T. pallidum*. The occurrence of penile ulcers on epithelial surfaces, however, may indicate that *T. pallidum* has the capability of passing through epithelial surfaces.

One of the first non-specific immune responses, infecting bacteria encounters, is the activation of the alternate complement pathway. Components on the surface of the micro-organism bind the complement factor C3b, which promotes opsonization by polymorphonuclear neutrophils (PMN) by C3b receptors. The polymorphs are also stimulated to produce lactoferrin to bind any free iron the pathogen might need. Other factors activated in the complement cascade induce the infiltration of more PMNs and macrophages.

The specific immune response against bacterial pathogens provides a role for both humoral and cellular immunity. Antibodies produced may block attachment of the microorganism to the cell, trigger complement mediated damage to the bacterial outer membrane, opsonize organisms, neutralize cytoxic factors or spreading factors such as hyaluronidase, block bacterial transport mechanisms and receptors and neutralize immunorepellents. Notwithstanding the important role antibody plays in the response to pathogens, most bacteria are eventually killed by macrophages or PMNs. After the initial non-specific infiltration by PMNs, T cells start triggering more phagocytes to destroy the invading organism. If the pathogen is intracellular, cytotoxic T cells may be produced and used to attack the cell harbouring the pathogen. Pathogens like *Streptococcus pyogenes*, have developed strategies to evade the phagocytosis. Others, such as *Mycobacterium tuberculosis*, are adapted to live and multiply within phagocytes.

Lukehart and coworkers described the local manifestions at the site of infection in experimental syphilis in rabbits (Lukehart *et al.*, 1980; Lukehart *et al.*, 1980). Following intratesticular infection with 10^7 *T. pallidum*, there is no noticeable change for several days, although increasing numbers of treponemes can be detected. After 3–6 days tissue damage becomes apparent and at day seven increasing infiltration with mononuclear cells, initially T cells, is observed. In contrast to infections with other pathogens such as *Neisseria gonorrhoeae*, only a limited number of polymorphs is present. Gradually the T cell population is augmented with macrophages and a minor, but increasing, number of plasma cells. The number of treponemes reaches a maximum at days 10–13, then rapidly declines and by day 17 virtually no treponemes are detectable. The decline correlates with the maximum number of infiltrating cells and with the onset of circulating antibodies, indicating that the immune response is responsible for clearing the organisms from the site of infection. The infiltrating macrophages are shown to ingest and probably degrade the treponemes. Recent experiments indicate that activated macrophages kill opsonized T. pallidum (Baker-Zander and Lukehart, personal communication). T. pallidum infection thus does not display the normal acute inflammatory process. The lack of infiltrating polymorphonuclear cells during the infection of a naive animal might be the result the lack of opsonization with complement factor C3b, simply because there is no bacterial surface component with which to react. In spite of the lack of PMNs, phagocytosis by macrophages seems to be effective and in vitro at least live and heat-killed T. pallidum are phagocytosed equally well. This implies that treponemes are unable actively to perturb ingestion by macrophages.

Although the response to T. pallidum may seem slow, it is effective enough to clear the treponemes from the infected site. However, T. pallidum has developed an ingenious strategy to prevent all of the infecting bacteria being killed. As mentioned before two events take place simultaneously: replication of bacteria at the site of infection and dissemination. Rapid dissemination makes it impossible for the host to deal with treponemes before they reach their protected niche and systemic infection is established.

Many have speculated on how T. pallidum evades the immune response. Christiansen proposed the presence of a protective layer on pathogenic treponemes which hides the antigens from the immune system (Christiansen, 1963). Indeed, many different host components are deposited on the surface of T. pallidum including serum proteins such as immunoglobulins, macroglobulin, lipoproteins, lactoferrin, proteoglycans, C3-complement factor (Alderete and Baseman, 1979) and even MHC class I molecules (Marchitto et al., 1986b). Binding of host derived fibronectin also may also result in a coating of the spirochaetes. The acquisition of this loosely associated, host-derived layer has been suggested to result in antigenic mimicry thus avoiding recognition by the host immune system. The deposition of the host material on the treponemes may represent non-specific binding but still be a virulence mechanism. Other investigators have suggested an active immunosuppression by T. pallidum (Baughn, 1983). Based on recent findings on the structure of the T. pallidum outer membrane, a number of investigators including myself currently suggest that the character of the surface of this pathogen is responsible for its exceptional pathogenicity.

4.11 THE SURFACE OF *T.PALLIDUM* AND IMMUNITY

Recently, immune rabbit serum was shown to react with the TROMPs in the outer membrane of *T. pallidum in vitro* (Blanco *et al.*, 1990). Several hours of incubation with immune serum are needed before aggregation of the rare protein particles reactive with antibodies occurs. Aggregation of the particles is required to bring the bound antibodies in close contact with each other in order to bind complement. Once complement is bound, lytic action takes place and the treponemes become leaky. The gaps created in the outer membrane then provide access for the antibodies reacting with periplasmic and cytoplasmic membrane proteins. Once antibodies directed against the endoflagellum enter the treponemes and react, immobilization of *T. pallidum* readily takes place. The immune system of the infected individual may not immediately recognize the nearly inert outer membrane of *T. pallidum*, resulting in a late and perhaps initially weak response against the outer membrane proteins. Even when a good antibody response is mounted, there may be a delay before complement mediated lysis takes place. Furthermore, opsonization with relatively few antibodies will only weakly enhance phagocytosis.

Antibody response may be needed in complement mediated lysis and opsonization but T cell infiltration may represent a more important immune response. T cells require most antigens to be processed and presented by cells such as macrophages to recognize antigen and mount a response. Therefore, treponemes must be ingested and processed by macrophages, a process that might be partially dependent on opsonization and in turn on antibody production against exposed outer membrane proteins. Once strong T cell antigens such as the flagellar proteins have been released, proliferation of *T. pallidum* specific T cells takes place, triggering and attracting more macrophages. The cascade effect results in accelerated clearing of the infecting organisms.

4.12 THE CENTRAL PARADOX

One of the most intriguing aspects of syphilis is the persistence of the infection despite high antibody levels and a cellular immune response. The antibodies produced in man or animal after infection are shown to be treponemicidal by both *in vitro* and *in vitro–in vivo* assays (Bishop and Miller, 1976). T cell responses are present and *T. pallidum* is ingested and probably killed by macrophages. The immune system seems able to eliminate the treponemes from the initial site of infection, but not all spirochaetes are cleared from the body. Apparently, the immune system also is capable of killing the massive number of treponemes during the secondary stage of syphilis yet, after healing, a number of treponemes persist for the lifetime of the host, sequestered in sites such as lymph nodes and the central nervous system. Perhaps even more surprising is the phenomenon of concomitant or infection immunity. That is, once infected, untreated individuals are resistant to re-infection with *T. pallidum*, as long as the infection persists (Magnusson *et al.*, 1956; Turner and Hollander, 1957). What could be the mechanism responsible for the escape of some treponemes? Shortly after the initial infection, dissemination takes place; the spread of spirochaetes occurs early in infection when no specific immune response has yet been mounted. After or during healing of the primary lesion, the treponemes reappear in massive numbers to cause the secondary stage. The multiplication of *T. pallidum* takes place despite the strong immune status the infected host has developed. After a few weeks, this stage of the disease also passes, again presumably as a result of the activity of the immune system. Serology and T cell proliferation assays show that patients have developed a strong humoral and cellular immune response but the treponemes persist and reappear in large numbers during secondary relapses. A possible explanation for the survival of the treponemes is that they reach immunologically privileged sites. Although attractive, this fails to explain why the treponemes are present in non-immune privileged lymph nodes. It is unlikely that the treponemes persist in the lymph nodes because of drainage of this immune-privileged site as absence of lymph drainage is one of the characteristics of such areas. Another possibility is delay in recognition due to the meagre exposure of antigens on the cell surface of *T. pallidum*. Perhaps the number of disseminated treponemes has to increase to a critical point, to provide a large enough antigenic load before the immune system can respond. We must keep in mind that infection with exogenously administered *T. pallidum* at this point, no longer results in early symptomatic infections (although gumma at the site of inoculation does occur). The phenomenon suggests that the treponemes present in the sequestered sites of an infected individual may have different properties than newly administered spirochaetes. Antigenic variation could explain this phenomenon and is discussed below. Other possible explanations include the coating of antigenic determinants with products of the immune system, such as blocking antibody or the acquisition of other host materials rendering the organism immunologically inert.

Successful pathogenic organisms must have the ability to adapt quickly to different micro-environments. Experiments with *Salmonella typhimurium* have shown that such adaptations can be extremely complex. Two dimensional gel electrophoresis protein profiles of *S. typhimurium*, metabolically labelled as free living bacteria or as intracellular organisms in macrophages, showed enhanced production of at least 34 *Salmonella* proteins and the diminished production of 136 proteins, 50 of which were not produced at all if *Salmonella* is intracellular (Buchmeier and Heffron, 1990). This complex type of regulation has some logic: why synthesize products needed for life within the cell when being extracellular or why use the type of outer membrane needed for successful infection in the skin or mucosa when already inside the host? Could it be that a repertoire of different exposed outer membrane proteins, of which the expression is regulated depending on the environment, is responsible for the escape of *T. pallidum*?

4.13 ANTIGENIC VARIATION

Antigenic variation is one of the strategies used by several parasites to evade the host immune response. In the parasite *Trypanosoma brucei*, adaptive phenotypic variation occurs as a programmed variation but is only seen under pressure of the immune response (Bloom, 1979). The rapid antigenic changes which *Borrelia hermsii*, the causative agent of relapsing fever, uses to elude the immune response indicates that at least some spirochaetes are able to adapt their antigenic make-up (Barbour, 1988). Could *T. pallidum* also be using this means of escape? Genetic rearrangements resulting in antigenic change occur in only a few organisms in the population. The altered organisms need to grow rapidly to replace the previous phenotype. Given an average generation time of 33 hours, fast growth is not characteristic of *T. pallidum* and active antigenic variation, similar to that observed in *B. hermsi* seems unlikely for *T. pallidum*. There is, however, some variation in the *T. pallidum* protein composition. Figure 4.9 shows an example. In two strains of *T. pallidum* subsp. *pertenue* the 24–28 kDa smearlike lipoprotein TpE has a reproducible lower apparent molecular mass than in some other *T. pallidum* subsp. *pertenue* or *T. pallidum* strains (Noordhoek *et al.*, 1990b). Cockayne and coworkers noted the absence of a major 34.5 kDa endoflagellar polypeptide in one of two strains of *T. pallidum* (Cockayne *et al.*, 1989). The absence of this protein had no influence on the motility of the treponemes or on the morphology of the endoflagella. A similar observation has been reported by Stamm and Bassford (1985) who found that a 35 kDa protein was absent in the *T. pallidum* street strain 14. These few examples show that there indeed is some variation in *T. pallidum*, although the modifications in protein profiles probably represents gradual changes in the antigenic make-up rather than an escape mechanism.

Observations on *T. pallidum* variants indicate that there is a risk in exclusively using *T. pallidum* strains that have been transferred from rabbit to rabbit for many years. Most treponemal research has been performed with the Nichols strain of *T. pallidum*: this strain was first isolated from the cerebrospinal fluid of a patient with recurrent secondary syphilis by Major H.J.Nichols in 1912 and has been maintained by continuous animal passage since that time. This time and the number of passages seems enough for the spirochaete to have altered its antigenic make-up, yet western blot analysis reveals no major differences compared with newly isolated street strains. Furthermore, recent laboratory accidents have shown that the Nichols strain has retained full virulence for man.

Differences in antigenic structure, such as in the endoflagellum, imply that the use of a single antigen for serodiagnosis of syphilis might result in less than optimal sensitivity. Furthermore, immunization with a single antigen which protects against a particular strain may not be protective against those which have an altered antigen or in which the antigen is lacking.

4.14 VACCINES AGAINST SYPHILIS

Ever since the discovery of *T. pallidum*, there have been numerous attempts to vaccinate rabbits with treponemal antigens and subsequently challenge the animals with live virulent *T. pallidum* (Lukehart, 1985). Except for three studies (Schobl *et al.*, 1930; Metzger *et al.*, 1969; Miller, 1973), no protection against *T. pallidum* challenge has been reported. Inspection of the literature on the vaccination trials after the successful study by Miller, reveals the lack of papers on protection studies. In the molecular biology era, only four reports on vaccination have been generated. Hindersson *et al.* (1985) showed that vaccination with purified endoflagella from *T. phagedenis* did not confer protection against challenge with *T. pallidum* in rabbits. The well-studied cloned 4D antigen was used in RIBI adjuvant to immunize rabbits, using different routes (Borenstein *et al.*, 1988). Challenge of these animals revealed a partial protection against *T. pallidum*, apparent as flat lesions with accelerated development. Wicher *et al.* (1989) showed that vaccination with the recombinant antigens TmpA, TmpB and TmpC, in spite of generating high antibody titres, did not protect guinea pigs against infection with *T. pallidum*. However, later experiments indicate that the use of a different adjuvant results in at least partial protection with TmpB (Wicher, personal communication). In an effort to stimulate the cellular response against *T. pallidum*, Strugnell and Schouls used live vaccine carriers expressing a number of recombinant antigens. Live *Salmonella typhimurium*, expressing different cloned antigens, were injected intravenously into rabbits and challenged after four weeks with virulent *T. pallidum* (Strugnell, 1989). Most antigens induced accelerated lesions similar to the result obtained with the 4D antigen, but failed to protect against infection. A recombinant vaccinia construct expressing the cloned antigen TmpA was used to infect rabbits (Schouls, unpublished results): although the rabbits developed an antibody titre against TmpA, challenge of the animals again yielded only accelerated lesions.

In conclusion, the recent knowledge on the structure of the *T. pallidum* cell envelope needs to be considered in the strategies for making an effective vaccine. TROMPs seem to be the only exposed proteins in the outer membrane of *T. pallidum* and so these rare structures should be targeted as one of the prime components of a treponemal vaccine. For maximal response, both the humoral and cellular arms of the immune system should be triggered by such a vaccine. Antibodies directed against exposed surface components could have two functions: opsonization for macrophage activity and activation of the alternative complement pathway. Once complement has perforated the outer membrane, antibodies against subsurface structures should be able to penetrate and kill the treponemes. This strategy implies that, except for the exposed outer membrane proteins, a number of other antigens such as the flagellum must be included in a vaccine.

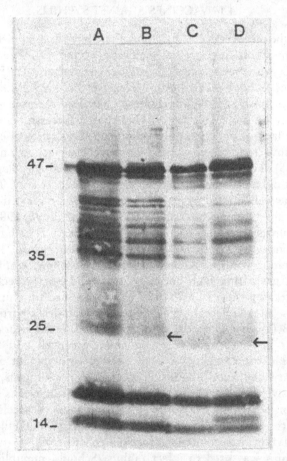

Figure 4.9 Immunoblot of *T. pallidum* Nichols strain: A, *T. pertenue* strain CDC2575; B, *T. pertenue* Gauthier strain, C and *T. pertenue* Brazaville strain, D. The western blot was incubated with serum from a yaws patient. Both the Gauthier and Brazaville strain contain a TpE antigen with a lower apparent molecular mass. The arrows indicate the position of the smear-like TpE antigens.

Unlike protective humoral immunity (opsonization, neutralization), cellular immunity is not dependent on exposed outer membrane components and the best choice of a component stimulating the cellular response should be determined in a different way. Lukehart and Baker-Zander showed that *T. pallidum* proteins purified by electro-elution from SDS-PAGE can be used to assess the T cell stimulatory capacities of individual antigens *in vitro* (Baker-Zander and Lukehart, 1988). The latter investigator also showed that using *T. pallidum* antigen preparations, production of macrophage activating factor by specific T cells can be detected (Lukehart, 1982). These techniques might be the tool to find the most potent T cell antigens of *T. pallidum* to determine which components must be included in a vaccine. Considering that the protection obtained by Metzger *et al.* (1969) and Miller (1973) was the result of long and intense vaccination schedules, it may be that the low dose of outer membrane proteins in *T. pallidum* preparations require prolonged vaccination regimens. Additionally, both Miller and Metzger used *T. pallidum* that had hardly been manipulated and therefore probably had an intact outer membrane still containing the TROMPs. More harshly treated *T. pallidum*, which certainly must have lost its outer membrane, could not confer protection. This suggested that an intact outer membrane is of major importance for achieving immunity. However, the required prolonged vaccination may also be related to the differential cellular response to individual antigens, as observed during experimental syphilis (Lukehart, personal communication). The antigens inducing protective cellular immunity may need more intense exposure to the immune system due to poor immunogenicity, inadequate processing by macrophages or, as mentioned above, paucity of the pathogenic determinant. There is another approach that might also be employed in the preparation of a vaccine: attachment is the first event in the pathogenesis of *T. pallidum* and so blocking of attachment may prevent infection. A vaccine must not only give rise to a systemic immune response but, more importantly, also must provide mucosal immunity: the evaluation of the effectiveness of this type of vaccine will be difficult as it needs a different animal model.

It should be emphasized that a safe vaccine against syphilis or against the closely related non-venereal treponematoses should not be centred on prevention of one step in the pathogenesis but should also induce immunity, which promotes elimination of treponemes, in the later stages of the infection.

ACKNOWLEDGEMENTS

The author acknowledges Dr Sheila Lukehart and Sharon Baker-Zander for their constructive criticism. I also wish to thank Cheryl Champion, Robin Isaacs, Sheila Lukehart, Sharon Baker-Zander, Alan Cockayne, Dick

Strugnell, Nyles Charon, Gerda Noordhoek, Sjoerd Rijpkema, Ruud Segers, Frans Mikx and Rob Keulers for providing unpublished data. This work was partially supported by The Netherlands Society for Scientific Research (NWO).

REFERENCES

Alderete, J.F. and Baseman, J.B. (1979) Surface associated host proteins on virulent *Treponema pallidum*. *Infect. Immun.*, **26**, 1048–56.

Alderete, J.F., Peterson, K.M. and Baseman, J.B. (1988) Affinities of *Treponema pallidum* for human lactoferrin and transferrin. *Genitourin Med.*, **64**, 359–63.

Alderete, J.F. and Baseman, J.B. (1989) Serum lipoprotein binding by *Treponema pallidum*: possible role for proteoglycans. *Genitourin Med.*, **65**, 177–82.

Azar, H.A., Pham, T.D. and Kurban, A.K. (1970) An electron microscopic study of a syphilitic chancre. *Arch. Pathol.*, **90**, 143–50.

Bailey, M.J., Penn, C.W. and Cockayne, A. (1985) Evidence for the presence of lipopolysaccharide in *Treponema phagedenis* (biotype Reiter) but not in *Treponema pallidum* (Nichols). *FEMS Microb. Lett.*, **27**, 117–21.

Bailey, M.J., Thomas, C.M., Cockayne, A. *et al.* (1989) Cloning and expression of *Treponema pallidum* antigens in *Escherichia coli*. *J. Gen. Microbiol.*, **135**, 2365–78.

Baker-Zander, S.A. and Lukehart, S.A. (1988) Development of cellular immunity to individual soluble antigens of *Treponema pallidum* during experimental syphilis. *J. Immunol.*, **141**, 4363–9.

Barbour, A.G. (1988) Antigenic variation of surface proteins of *Borrelia* species. *Rev. Infect. Dis.*, **10** (2), S399–402.

Baseman, J.B. and Hayes, E.C. (1980) Molecular characterisation of receptor binding proteins and immunogens of virulent *Treponema pallidum*. *J. Exp. Med.*, **151**, 573–86.

Baseman, J.B., Zachar, Z. and Hayes, N.S. (1980) Concavalin A-mediated affinity film for *Treponema pallidum*. *Infect. Immun.*, **27**, 260–3.

Baughn, R.E. (1983) Immunoregulatory effects in experimental syphilis, in *Pathogenesis and Immunology of Treponemal Infection* (eds R.F. Schell and D.M. Musher), Marcel Dekker, Inc. New York and Basel, 271–95.

Beachey, E.H. and Courtney, H.S. (1987) Bacterial adherence: the attachment of group A streptococci to mucosal surfaces. *Rev. Infect. Dis.*, **9** (5) S475–481.

Bergström, S., Bundoc, V.G. and Barbour, A.G. (1989) Molecular analysis of linear plasmid-encoded major surface antigens, OspA and OspB, the lyme disease spirochaete *Borrelia burgdorferi*. *Mol. Microbiol.*, **3**, 479–86.

Berry, C.D., Hooton, T.M., Collier, A.C. and Lukehart, S.A. (1987) Neurological relapse after benzathine penicillin therapy for secondary syphilis in a patient with HIV infection. *New Engl. J. Med.*, **316**, 1587.

Bishop, N.H. and Miller, J.N. (1976) Humoral immunity in experimental syphilis. II. The relationship of neutralizing factors in immune serum to acquired resistance. *J. Immunol.*, **117**, 197–207.

Blanco, D.R., Walker, E.M., Haake, D.A., Champion, C.I., Miller, J.N. and Lovett, M.A. (1990) Complement activation limits the rate of *in vitro* treponemicidal activity and correlates with antibody-mediated aggregation of *Treponema pallidum* rare outer membrane protein. *J. Immunol.*, 144, 1914–21.

Blobel, G. (1980) Intracellular protein topogenesis. *Proc. Natl Acad. Sci. USA*, 77, 1495–500.

Bloom, B.R. (1979) Games parasites play: how parasites evade immune surveillance. *Nature*, 279, 21–6.

Bordier, C. (1981) Phase separation of integral membrane proteins in Triton X-114 solution. *J. Biol. Chem.*, 256, 1604–7.

Borenstein, L.A., Radolf, J.D., Fehniger, T.E. *et al.* (1988) Immunisation of rabbits with recombinant *Treponema pallidum* surface antigen 4D alters the course of experimental syphilis. *J. Immunol.*, 140, 4023–32.

Brahamsha, B. and Greenberg, E.P. (1989) Cloning and sequence analysis of flaA, a gene encoding a *Spirochaeta aurantia* flagellar filament surface antigen. *J. Bacteriol.*, 171, 1692–7.

Buchmeier, N.A. and Heffron, F. (1990) Induction of *Salmonella* stress proteins upon infection of macrophages. *Science*, 248, 730–2.

CDC (1988) Continuous increase in infections syphillis – United States. *MMWR*, 37, 35–8.

CDC (1989) Congenital syphilis – New York City, 1986–1988. *MMWR*, 38, 825–9.

Chamberlain, N.R., Radolf, J.D., Hsu, P.L. *et al.* (1988) Genetic and physico-chemical characterisation of the recombinant DNA-derived 47-kilodalton surface immunogen of *Treponema pallidum* subsp. *pallidum*. *Infect. Immun.*, 56, 71–8.

Chamberlain, N.R., Brandt, M.E., Erwin, A.L. *et al.* (1989a) Major integral membrane protein immunogens of *Treponema pallidum* are proteolipids. *Infect. Immun.*, 57, 2872–7.

Chamberlain, N.R., DeOgny, L., Slaughter, C. *et al.* (1989b) Acylation of the 47-kilodalton major membrane immunogen of *Treponema pallidum* determines its hydrophobicity. *Infect. Immun.*, 57, 2878–85.

Champion, C.I., Miller, J.N., Lovett, M.A. and Blanco, D.R. (1990) Cloning, sequencing and expression of two Class B endoflagellar genes of *Treponema pallidum* subsp. *pallidum* encoding the 34.5 and 31.0-kilodalton proteins. *Infect. Immun.*, 58, 1697–704.

Christiansen, S. (1963) Hypothesis. Protective layer covering pathogenic trepone-mata. *Lancet*, ii: 423–4.

Coates, S.R., Sheridan, P.J., Hansen, D.S. *et al.* (1986) Serospecificity of a cloned protease-resistant *Treponema pallidum*-specific antigen expressed in *Escherichia coli*. *J. Clin. Microbiol.*, 23, 460–4.

Cockayne, A., Strugnell, R.A., Bailey, M.J. and Penn, C.W. (1989) Comparative antigenic analysis of *Treponema pallidum* laboratory and street strains. *J. Gen. Microbiol.*, 135, 2241–7.

Coon, J. and Hunter, R. (1975) Properties of conjugated protein immunogens which selectively stimulate delayed-type hypersensitivity. *J. Immunol.*, 114, 1518–22.

Dallas, W.S., Ray, P.H., Leong, J. *et al.* (1987) Identification and purification of a recombinant *Treponema pallidum* basic membrane protein antigen expressed in *Escherichia coli. Infect. Immun.*, 55, 1106–15.

Deres, K., Schild, H., WiesmGller, K.-H. *et al.* (1989) *In vivo* priming of virus-specific cytotoxic T lymphocytes with synthetic lipopeptide vaccine. *Nature*, 342, 561–4.

Fehniger, T.E., Walfield, A.M., Cunningham, T.M. *et al.* (1984) Purification and characterisation of a cloned protease-resistant *Treponema pallidum*-specific antigen. *Infect. Immun.*, 46, 598–607.

Fehniger, T.E., Radolf, J.D., Walfield, A.M. *et al.* (1986) Native surface association of a recombinant 38-kilodalton *Treponema pallidum* antigen isolated from the *Escherichia coli* outer membrane. *Infect. Immun.*, 52, 586–93.

Fehniger, T.E., Radolf, J.D. and Lovett, M.A. (1986) Properties of an ordered ring structure formed by recombinant *Treponema pallidum* surface antigen 4D. *J. Bacteriol.*, 165, 732–9.

Fieldsteel, A.H., Cox, D.L. and Moeckli, R. (1981) Cultivation of virulent *Treponema pallidum* in tissue culture. *Infect. Immun.*, 32, 908–15.

Fitzgerald, T.J., Johnson, R.C. and Wolff, E.T. (1978) Mucopolysaccharide material resulting form the interaction of *Treponema pallidum* (Nichols strain) with cultured mammalian cells. *Infect. Immun.*, 22, 575–84.

Fitzgerald, T.J. and Johnson, R.C. (1979a) Mucopolysaccharidase of *Treponema pallidum. Infect. Immun.*, 24, 261–8.

Fitzgerald, T.J., Johnson, R.C. and Ritzi, D.M. (1979) Relationship of *Treponema pallidum* to acidic mucopolysaccharides. *Infect. Immun.*, 24, 252–60.

Fitzgerald, T.J. and Johnson, R.C. (1979b) Surface mucopolysaccharides of *Treponema pallidum. Infect. Immun.*, 24, 244–51.

Fitzgerald, T.J. and Repesh, L.A. (1983) Toxic activities of *Treponema pallidum*, in *Pathogenesis and Immunology of Treponemal Infection* (eds R.F. Schell and D.M. Musher), Marcel Dekker, Inc. New York and Basel 173–93.

Fitzgerald, T.J., Miller, J.N., Repesh, L.A. *et al.* (1985) Binding of glycosamino-glycans to the surface of *Treponema pallidum* and subsequent effects on complement interactions between antigen and antibody. *Genitourin. Med.*, 61, 13–20.

Gold, L., Pribnow, D., Schneider, T. *et al.* (1981) Translation initiation in prokaryotes. *Ann. Rev. Microbiol.*, 35, 365–403.

Greenblatt, R.M., Lukehart, S.A., Plummer, F.A. *et al.* (1988) Genital ulceration as a risk factor for human immunodeficiency virus infection. *AIDS*, 2, 47.

Grosjean, H. and Fiers, W. (1982) Preferential codon usage in prokaryotic genes: the optimal codon–anticodon interaction energy and the selective codon usage in efficiently expressed genes. *Gene*, 18, 199–209.

Hansen, E.B., Pedersen, P.E., Schouls, L.M. *et al.* (1985) Genetic characterisation and partial sequence determination of a *Treponema pallidum* operon expressing two immunogenic membrane proteins in *Escherichia coli. J. Bacteriol.*, 162, 1227–37.

Hardy, P.H., Fredricks, W.R. and Nell, E.E. (1975) Isolation and antigenic characteristics of axial filaments from the Reiter treponeme. *Infect. Immun.*, 11, 380–6.

Hardy, P.H. and Levin, J.H. (1983) Lack of endotoxin in *Borrelia hispanica* and *Treponema pallidum. Proc. Soc. Exp. Biol Med.*, 174, 47–52.

Hay, P.E., Clarke, J.R., Strugnell, R.A. *et al.* (1990) Use of the polymerase chain reaction to detect DNA sequences specific to pathogenic treponemes in cerebrospinal fluid. *FEMS Microbiol. Lett.*, **68**, 233–8.

Hicks, C.B., Benson, P.M., Lupton, G.P. and Tramont E.C. (1987) Seronegative secondary syphilis in a patient infected with human immunodeficiency virus (HIV) with Kaposi sarcoma. *Annals Intern. Med.*, **107**, 492.

Hindersson, P., Petersen, C.S. and Axelsen, N.H. (1985) Purified flagella from *Treponema phagedenis* biotype Reiter does not induce protective immunity against experimental syphilis in rabbits. *Sex Transm. Dis.*, **12**, 124–7.

Hindersson, P., Cockayne, A., Schouls, L.M. and van Embden, J.D.A. (1986) Immunochemical characterisation and purification of *Treponema pallidum* antigen TpD expressed by *Escherichia coli* K12. *Sex Transm. Dis.*, **13**, 237–44.

Hindersson, P., Knudsen, J.D. and Axelsen, N.H. (1987) Cloning and expression of *Treponema pallidum* common antigen (Tp-4) in *Escherichia coli* K12. *J. Gen. Microbiol.*, **133**, 587–96.

Hindersson, P. and Bangsdorf, J. (1987) Cloning and expression of 4 *Treponema pallidum* antigens identified by monoclonal antibodies. *Abstr. Ann. Meet. Am. Soc. Microbiol.*, **D47**, 79.

Hovind-Hougen, K. (1976) Determination by means of electron microscopy of morphological criteria of value for classification of some spirochaetes, in particular treponemes. *Acta Pathol. Microbiol. Scan. Suppl. B*, **255**, 3–27.

Hovind-Hougen, K., Birch-Andersen and Skovgaard Jensen, H.J. (1976) Ultrastructure of cells of *Treponema pertenue* obtained from experimentally infected hamsters. *Acta Pathol. Microbiol. Scand.*, (B)**84**, 101–8.

Hsu, P.L., Qin, M., Norris, S.J. and Sell, S. (1988) Isolation and characterisation of recombinant *Escherichia coli* clones secreting a 24-kilodalton antigen of *Treponema pallidum*. *Infect. Immun.*, **56**, 1135–43.

Hsu, P.L., Chamberlain, N.R., Orth, K *et al.* (1989) Sequence analysis of the 47-kilodalton major integral membrane immunogen of *Treponema pallidum*. *Infect. Immun.*, **57**, 196–203.

IJsselmuiden, O.E., Schouls, L.M., Stolz, E. *et al.* (1989) Sensitivity and specificity of an enzyme linked immunosorbent assay using the recombinant DNA-derived *Treponema pallidum* protein TmpA for serodiagnosis of syphilis and the potential of TmpA to assess the effect of antibiotic therapy. *J. Clin. Microbiol.*, **27**, 152–7.

Isaacs, R.D., Hanke, J.H., Guzman-Verduzco, L. *et al.* (1989) Molecular cloning and DNA sequence analysis of the 37-Kilodalton endoflagellar sheath protein gene of *Treponema pallidum*. *Infect. Immun.*, **57**, 3403–11.

Isaacs, R.D. and Radolf, J.D. (1990) Expression in *Escherichia coli* of the 37-kilodalton endoflagellar sheath protein of *Treponema pallidum* using the polymerase chain reaction and a T7 expression system. *Infect. Immun.*, **58**, 2025–34.

Jones, S.A., Marchitto, K.S., Miller, J.N. and Norgard, M.V. (1984) Monoclonal antibody with haemagglutination, immobilisation and neutralisation activities defines an immunodominant, 47,000 mol wt., surface-exposed immunogen of *Treponema pallidum* (Nichols). *J. Exp. Med.*, **160**, 1404–20.

Jung, G., Wiesmüller, K.-H., Becker, G. *et al.* (1985) Increased production of specific antibodies by presentation of the antigen determinants with covalently coupled lipopeptide mitogens. *Angew. Chem.*, 24, 872–3.

Klein, P., Somorjai, R.L. and Lau, P.C.K. (1988) Distinctive properties of signal sequences from bacterial lipoproteins. *Prot. Eng.*, 2, 15–20.

Lauderdale, V. and Goldman, J.N. (1972) Serial ultrathin sectioning demonstrating the intracellularity of *Treponema pallidum*. *Br. J. Vener. Dis.*, 48, 87–96.

Lelong, J.M., Fournier, R.S. and Isberg, R.R. (1990) Identification of the integrin binding domain of the *Yersinia pseudotuberculosis* invasin protein. *EMBO J*, 9, 1979–1989.

Limberger, R.J. and Charon N.W. (1986) Antiserum to the 33,000-dalton periplasmic-flagellum protein of *Treponema phagedenis*: reacts with other treponemes and *Spirochaeta aurantia*. *J. Bacteriol.*, 168, 1030–2.

Luft, B.J., Jiang, W., Munoz, P. *et al.* (1989) Biochemical and immunological characterisation of the surface proteins of *Borrelia burgdorferi*. *Infect. Immun.*, 57, 3637–45.

Lukehart, S.A., Baker-Zander, S.A. and S. Sell. (1980a) Characterisation of lymphocyte responsiveness in early experimental syphilis. I. In vitro response to mitogens and *Treponema pallidum* antigens. *J. Immunol.*, 124, 454–60.

Lukehart, S.A., Baker-Zander, S.A. and Lloyd, R.M.C. (1980b) Characterisation of lymphocyte responsiveness in early experimental syphilis. II. Nature of cellular responsiveness and *Treponema pallidum* distribution in testicular lesions. *J. Immunol.*, 124, 461–7.

Lukehart, S.A. (1982) Activation of macrophages by products of lymphocytes from normal and syphilitic rabbits. *Infect. Immun.*, 37, 64–9.

Lukehart, S.A. (1985) Prospects for development of a treponemal vaccine. *Rev. Infect. Dis.*, 7, (2), S305–313.

Macnab R.M. and Aizawa, S. (1984) Bacterial motility and the bacterial flagellar motor. *Annu. Rev. Biophys. Bioeng.*, 13, 51–83.

Magnusson, H.J., Thomas E.W., Olansky, S. *et al.* (1956) Inoculation syphilis in human volunteers. *Med. (Baltimore)* 35, 33–82.

Marchitto, K.S., Jones S.A., Schell, R.F. *et al.* (1984) Monoclonal antibody analysis of specific antigenic similarities among pathogenic *Treponema pallidum* subspecies. *Infec. Immun.*, 45, 660–6.

Marchitto, K.S., Selland-Grossling, C.K. and Norgard, M.V. (1986a) Molecular specificities of monoclonal antibodies directed against virulent *Treponema pallidum*. *Infect. Immun.*, 51, 168–76.

Marchitto, K.S., Kindt, T.J. and Norgard, M.V. (1986b) Monoclonal antibodies directed against major histocompatibility complex antigens bind to the surface of *Treponema pallidum* isolated from infected rabbits or humans. *Cell Immunol.*, 101, 633–42.

Metzger, M., Michalska, E., Podwińska, J., and Smör, W. (1969) Immunogenic properties of the protein component of *Treponema pallidum*. *Br. J. Ven. Dis.*, 45, 299–303.

Miller, J.N. (1973) Immunity in experimental syphilis. VI. Successful vaccination of rabbits with *Treponema pallidum* Nichols strain, attenuated by radiation. *J. Immunol.*, 110, 1206–15.

Miao, R. and Fieldsteel, A.H. (1978) Genetics of *Treponema*: relationship between *Treponema pallidum* and five cultivable treponemes. *J. Bact.*, 133, 101–7.

Morrison, M. (1980) Lactoperoxidase-catalyzed iodination as a tool for investigation of proteins. *Meth. Enz.*, 70, 214–20.

Moskophidis, M. and Müller, F. (1984) Molecular characterisation of glycoprotein antigens on the surface of *Treponema pallidum*: comparison with nonpathogenic *Treponema phagedenis* biotype Reiter. *Infect. Immun.*, 46, 867–9.

Nell, E.E. and Hardy, P.H. (1978) Counterimmunoelectrophoresis of Reiter treponeme axial filaments as a diagnostic test for syphilis. *J. Clin. Microbiol.*, 8, 148–52.

Noordhoek, G.T., Hermans, P.W., Paul, A.N. *et al.* (1989) *Treponema pallidum* subspecies *pallidum* (Nichols) and *Treponema pallidum* subspecies pertenue (CDC 2575) differ in at least one nucleotide: comparison of two homologous antigens. *Microb. Pathog.*, 6, 29–42.

Noordhoek, G.T., Wieles, B., Van der Sluis, J.J. and Van Embden, J.D.A. (1990a) Polymerase chain reaction and synthetic DNA probes: a means of distinguishing the causative agents of syphilis and yaws? *Infect. Immun.*, 58, 2011–13.

Noordhoek G.T., Cockayne, A., Schouls, L.M. *et al.* (1990b) A new attempt to distinguish serologically the subspecies of *Treponema pallidum* causing syphilis and yaws. *J. Clin. Microbiol.*, 28, 1600–7.

Norgard, M.V. and Miller, J.N. (1981) Plasmid DNA in *Treponema pallidum* (Nichols): potential for antibiotic resistance by syphilis bacteria. *Science*, 213, 553.

Norgard, M.V. and Miller, J.N. (1983) Cloning and expression of *Treponema pallidum* (Nichols) antigen genes in *Escherichia coli*. *Infect. Immun.*, 42, 435–45.

Norgard, M.V., Chamberlain, N.R., Swancutt, M.A. and Goldberg, M.S. (1986) Cloning and expression of the major 47-kilodalton surface immunogen of *Treponema pallidum* in *Escherichia coli*. *Infect. Immun.*, 54, 500–6.

Norris, S.J. (1982) In vitro cultivation of *Treponema pallidum* in tissue culture: independent confirmation. *Infect. Immun.*, 36, 437–9.

Norris, S.J. and Sell, S. (1984) Antigenic complexity of *Treponema pallidum*: antigenicity and surface localisation of major polypeptides. *J. Immunol.*, 133, 2686–92.

Norris, S.J. *et al.* (1987) Identity of *Treponema pallidum* subsp. *pallidum* polypeptides: correlation of sodium dodecyl sulphate-polyacrylamide gel electrophoresis results from different laboratories. *Electrophoresis*, 8, 77–92.

Norris, S.J., Charon, N.W., Cook, R.G. *et al.* (1988) Antigenic relatedness and N-terminal sequence homology define two classes of periplasmic flagellar proteins of *Treponema pallidum* subsp. *pallidum* and *Treponema phagedenis*. *J. Bacteriol.*, 170, 4072–82.

Ovcinnikov, N.M. and Delektorskij (1972) Electron microscopy of phagocytosis in syphilis and yaws. *Br. J. Vener. Dis.*, 48, 224–48.

Pallesen, L. and Hindersson, P. (1989) Cloning and sequencing of a *Treponema pallidum* gene encoding a 31.3-kilodalton endoflagellar subunit (FlaB2). *Infect. Immun.*, 57, 2166–72.

Penn, C.W., Cockayne, A. and Bailey, M.J. (1985) The outer membrane of *Treponema pallidum*: Biological significance and biochemical properties. *J. Gen. Microb.*, 131, 2349–57.

Penn, C.W. and Rhodes, J.G. (1982) Surface-associated antigens of *Treponema pallidum* concealed by an inert outer layer. *Immunology*, **46**, 9–16.

Petersen, C.S., Pedersen, N.S. and Axelsen, N.H. (1981) A simple method for the isolation of flagella from *Treponema* Reiter. *Acta Pathol. Microbiol. Scand. Sect. C*, **89**, 379–85.

Peterson, K.M., Baseman, J.B. and Alderete, J.F. (1986) Isolation of a *Treponema pallidum* gene encoding immunodominant outer envelope protein P6, which reacts with sera from patients at different stages of syphilis. *J. Exp. Med.*, **164**, 1160–70.

Peterson, K.M., Baseman, J.B. and Alderete, J.F. (1987a) Cloning structural genes for *Treponema pallidum* immunogens and characterisation of recombinant treponemal surface protein, P2 (P2 star). *Genitourin. Med.*, **63**, 289–96.

Peterson, K., Baseman, J.B. and Alderete, J.F. (1987b) Molecular cloning of *Treponema pallidum* outer envelope fibronectin binding proteins, P1 and P2. *Genitourin. Med.*, **63**, 355–60.

Proctor, R.A. (1987) The staphylococcal fibronectin receptor: evidence for its importance in invasive infections. *Rev. Infect. Dis.*, **9** (4), S335–340.

Purcell, B.K., Chamberlain, N.R., Goldberg, M.S. *et al.* (1989) Molecular cloning and characterisation of the 15-Kilodalton major immunogen of *Treponema pallidum*. *Infect. Immun.*, **57**, 3708–14.

Quinn, T.C., Glasser, D., Cannon, R.O. *et al.* (1988) Human immunodeficiency virus infection among patients attending clinics for sexually transmitted diseases. *New Engl. J. Med.*, **318**, 197.

Radolf, J.D., Lernhardt, E.B., Fehniger, T.E. and Lovett, M.A. (1986a) Serodiagnosis of syphilis by enzyme-linked immunosorbent assay with purified recombinant *Treponema pallidum* antigen 4D. *J. Infect. Dis.*, **153**, 1023–7.

Radolf, J., Fehniger, T.E., Silverblatt, F.J. *et al.* (1986b). The surface of virulent *Treponema pallidum*: Resistance to antibody binding in the absence of complement and surface association of recombinant antigen 4D. *Infect. Immun.*, **52**, 579–85.

Radolf, J.D., Borenstein, L.A., Kim, J.Y. *et al.* (1987) Role of disulfide bonds in the oligomeric structure and protease resistance of recombinant and native *Treponema pallidum* surface antigen 4D. *J. Bacteriol.*, **169**, 1365–71.

Radolf, J.D. and Norgard, M.V. (1988) Pathogen specificity of *Treponema pallidum* subsp. *pallidum* integral membrane proteins identified by phase partitioning with Triton X-114. *Infect. Immun.*, **56**, 1825–8.

Radolf, J.D., Chamberlain N.R., Clausell, A. and Norgard, M.V. (1988) Identification and localisation of integral membrane proteins of virulent *Treponema pallidum* subsp. *pallidum* by phase partitioning with the nonionic detergent Triton X-114. *Infect. Immun.*, **56**, 490–8.

Radolf, J.D., Norgard, M.V. and Schulz, W.W. (1989) Outer membrane ultrastructure explains the limited antigenicity of virulent *Treponema pallidum*. *Proc. Natl Acad. Sci. USA*, **86**, 2051–5.

Riley, B.S. and Cox, D.L. (1988) Cultivation of cottontail rabbit epidermal (Sf1Ep) cells on microcarrier beads and their use for suspension cultivation of *Treponema pallidum* subsp. *pallidum*. *Appl. Environ. Microbiol.*, **54**, 2862–5.

Riviere, G.R., Thomas, D.D. and Cobb, C.M. (1989) In vitro model of *Treponema pallidum* invasiveness. *Infect. Immun.*, **57**, 2267–71.

Rodgers, G.C., Laird, W.J., Coates, S.R. *et al.* (1986) Serological characterisation and gene localisation of an *Escherichia coli*-expressed 37-kilodalton *Treponema pallidum* antigen. *Infect. Immun.*, 53, 16–25.

Rosenberg, M. and Court, D. (1979) Regulatory sequences involved in the promotion and termination of RNA transcription. *Annu. Rev. Genet.*, 13, 319–53.

Schobl, O., Tanabe, B. and Miyao, I. (1930) Preventive immunisation against treponematous infections and experiments which indicate the possibility of antitreponematous immunisation. *Phi. J. Sci.*, 35, 1–13.

Schouls, L.M., Van der Heide, J.G.J. and Van Embden, J.D.A. (1991) Characterization of the 35-kilodalton *Treponema pallidum* subsp. *pallidum* recombinant lipoprotein TmpC and antibody response to lipidated and non-lipidated *T. pallidum* antigens. *Infect. Immun.*, 59, 3536–46.

Schouls, L.M., Ijsselmuiden, O.E., Weel, J. and van Embden J.D.A. (1989a) Overproduction and purification of *Treponema pallidum* recombinant-DNA-derived proteins TmpA and TmpB and their potential use in serodiagnosis of syphilis. *Infect. Immun.*, 57, 2612–23.

Schouls, L.M., Mout, R., Dekker, J. and van Embden, J.D.A. (1989b) Characterisation of lipid-modified immunogenic proteins of *Treponema pallidum* expressed in *Escherichia coli. Microb. Pathogen.*, 7, 175–88.

Stamm, L.V., Folds, J.D. and Bassford, P.J. (1982) Expression of *Treponema pallidum* antigens in *Escherichia coli* K-12. *Infect. Immun.*, 36, 1238–41.

Stamm, L.V., Kerner, T.C. Jr, Bankaitis, V.A. and Bassford, P.J. Jr (1983) Identification and preliminary characterisation of *Treponema pallidum* protein antigens expressed in *Escherichia coli. Infect. Immun.*, 41, 709–21.

Stamm, L.V., Dallas, W.S., Ray, P.H. and Bassford, P.J. Jr (1988) Identification, cloning, and purification of protein antigens of *Treponema pallidum. Rev. Infect. Dis.*, 10 (2), S403–7.

Stamm, L.V., Hodinka, R.L., Wyrick, P.B. and Bassford, P.J. Jr (1987) Changes in the cell surface properties of *Treponema pallidum* that occur during in vitro incubation of freshly extracted organisms. *Infect. Immun.*, 55, 2255–61.

Stamm, L.V. and Bassford, P.J. (1985) Cellular and extracellular protein antigens of *Treponema pallidum* synthesized during *in vitro* incubation of freshly extracted organisms. *Infect. Immun.*, 47, 799–807.

Strandberg Pedersen, N., Sand Petersen, C., Vejtorp, M. and Axelsen, N.H. (1982) Serodiagnosis of syphilis by an enzyme-linked immunosorbent assay for IgG antibodies against the Reiter treponeme flagellum. *Scand. J. Immunol.*, 15, 341–8.

Strugnell, R.A., Handley, C.J., Lowther, D.A. *et al.* (1984) *Treponema pallidum* does not synthesise in vitro a capsule containing glycosaminoglycans or proteoglycans. *Br. J. Vener. Dis.*, 60, 8–13.

Strugnell, R.A., Handley, C.J., Drummond, L.P. and Faine, S. (1986) Characterisation of the proteoglycans synthesized by rabbit testis in response to infection by *Treponema pallidum. Austr. J. Path.*, 124, 216–25.

Strugnell, R., Schouls, L., Cockayne, A. *et al.* (1989) Experimental syphilis vaccines: Use of *aroA Salmonella typhimurium* to deliver recombinant *Treponema pallidum* antigens, in *Vaccines for Sexually Transmitted Diseases* (eds A. Meheus and RE.Spier), Butterworth & Co., London, pp. 107–13.

Swancutt, M.A., Twehous, D.A. and Norgard, M.V. (1986) Monoclonal antibody selection and analysis of a recombinant DNA-derived surface immunogen of *Treponema pallidum* expressed in *Escherichia coli*. *Infect. Immun.*, 52, 110–19.

Swancutt, M.A., Riley, B.S., Radolf, J.D. and Norgard, M.V. (1989) Molecular characterisation of the pathogen-specific, 34-kilodalton membrane immunogen of *Treponema pallidum*. *Infect. Immun.*, 57, 3314–23.

Swancutt, M.A., Radolf, J.D. and Norgard, M.V. (1990) The 34-kilodalton membrane immunogen of *Treponema pallidum* is a lipoprotein. *Infect. Immun.*, 58, 384–92.

Sykes, J.A., Miller, J.N. and Kalan, A.J. (1974) *Treponema pallidum* within cells of a primary chancre from a human female. *Brit. J. Vener. Dis.*, 50, 40–4.

Thole, J.E.R., Hindersson, P., de Bruyn, J. *et al.* (1988) Antigenic relatedness of a strongly immunogenic 65 kDA mycobacterial protein antigen with a similarly sized ubiquitous bacterial common antigen. *Microb. Pathogen*, 4, 71–83.

Thomas, D.D., Baseman, J.B. and Alderete, J.F. (1985a) Fibronectin tetrapeptide is target for syphilis spirochaete cytadherence. *J. Exp. Med.*, 162, 1715–19.

Thomas, D.D., Baseman, J.B. and Alderete, J.F. (1985b) Putative *Treponema pallidum* cytadhesins share a common functional domain. *Infect. Immun.*, 49, 833–5.

Thomas, D.D., Baseman, J.B. and Alderete, J.F. (1985c) Fibronectin mediates *Treponema pallidum* cytadherence through recognition of fibronectin cell-binding domain. *J. Exp. Med.*, 161, 514–25.

Thomas, D.D., Navab, M., Haake, D.A. *et al.* (1988) *Treponema pallidum* invades intercellular junctions of endothelial cell monolayers. *Proc. Natl Acad. Sci. USA*, 85, 3608–12.

Thomas, D.D. and Comstock, L.E. (1989) Interaction of *Borrelia burgdorferi* with host cell monolayers. *Infect. Immun.*, 57, 1324–6.

Thomas, D.D. and Higbie, L.M. (1990) *In vitro* association of *Leptosires* with host cells. *Infect. Immun.*, 58, 581–5.

Turner, T.B. and Hollander, D.H. (1957) Biology of the treponematoses. *World Health Organisation monograph series*, 35, Geneva.

Van der Sluis, J.J., Ten Kate, F.J.W., Vuzevski, V.D. and Stolz, E. (1987) Light and electron microscopy of rabbit testes infected with *Treponema pallidum* (Nichols Strain): nature of deposited mucopolysaccharides and localisation of treponemes. *Genitourin. Med.*, 63, 297–304.

Van der Sluis, J.J., Van Dijk, G., Boer, M. *et al.* (1985) Mucopolysaccharides in suspensions of *Treponema pallidum* extracted from infected rabbit testes. *Genitourin. Med.*, 61, 7–12.

Van Embden, J.D.A., Van der Donk, H.J., Van Eijk, R.V. *et al.* (1983) Molecular cloning and expression of *Treponema pallidum* DNA in *Escherichia coli* K-12. *Infect. Immun.*, 42, 187–96.

Von Heijne, G. (1989) The structure of signal peptides from bacterial lipoproteins. *Prot. Eng.*, 2, 531–4.

Walfield, A.M., Hanff, P.A. and Lovett, M.A. (1982) Expression of *Treponema pallidum* antigens in *Escherichia coli*. *Science*, 216, 522–3.

Walfield, A.M., Roche, E.S., Zounes, M.C. *et al.* (1989) Primary structure of an oligomeric antigen of *Treponema pallidum*. *Infect. Immun.*, 57, 633–5.

Walker, E.M., Zampighi, G.A., Blanco, D.R. *et al.* (1989) Demonstration of rare protein in the outer membrane of *Treponema pallidum* subsp. *pallidum* by freeze-fracture analysis. *J. Bacteriol.*, **171**, 5005–11.

Wickner, W.T. and Lodish, H. (1985) Multiple mechanisms of protein insertion into and across membranes. *Science*, **230**, 400–7.

Wicher, K., van Embden, J.D., Schouls, L.M. *et al.* (1989) Immunogenicity of three recombinant *Treponema pallidum* antigens examined in guinea pigs. *Int. Arch. Allergy. Appl. Immunol.*, **89**, 128–35.

Yelton, D.B. and Peng, S.L. (1989) Identification and nucleotide sequence of the *Leptospira biflexa* serovar *patoc trpE* and *trpG* genes. *J. Bacteriol.*, **171**, 2083–9.

Zeigler, J.A., Jones, A.M., Jones, R.H. and Kubica, K.M. (1976) Demonstration of extracellular material at the surface of pathogenic *T. pallidum* cells. *Brit. J. Ven. Dis.*, **52**, 1–8.

Molecular biology of
Candida pathogenesis

DAVID R. SOLL

5.1 INTRODUCTION

Candida albicans and related species represent an ever increasing problem not only in immunocompromised patients (Wey *et al.*, 1988) but also in patients with other predisposing conditions, as well as in apparently healthy individuals (Odds, 1988; Bodey and Feinstein, 1985). These opportunistic pathogens are capable of infecting virtually every tissue of the human body and have been responsible for an increasing number of nosocomial infections (Weber and Rutala, 1988; Isenberg *et al.*, 1989; Burnie *et al.*, 1985; Wey *et al.*, 1988). In HIV-positive individuals, oral-oesophageal candidosis represents one of the first signs of symptomatic AIDS (Greenspan, 1988; Syrjanen *et al.*, 1988), and there is preliminary evidence suggesting strain selection and evolution in these patients (Odds *et al.*, 1990). The success of *Candida* as a pathogen must stem partly from its commensal existence in the natural flora of most individuals and partly from a combination of pathogenic traits. However, for reasons not immediately obvious, but which must include the absence of a reasonable genetic system for experimental investigation (Whelan, 1987), we know little about the epidemiology, etiology and molecular pathology of *Candida*. Fortunately, in the past few years the tools of molecular genetics have been developed in *Candida* in order to attack questions of pathogenesis (Kurtz *et al.*, 1988; Kirsch *et al.*, 1990), as well as assumptions which until now have remained undocumented. This chapter attempts to define questions related to *Candida* biology, pathogenesis and epidemiology which remain unanswered, but which are now capable of being resolved.

Molecular and Cell Biology of Sexually Transmitted Diseases
Edited by D. Wright and L. Archard
Published in 1992 by Chapman and Hall, London ISBN 0 412 36510 3

5.2 GROWTH OF *CANDIDA* AS IT RELATES TO PATHOGENESIS

In the commensal condition, *Candida* is usually found in low titre and usually in the budding growth form (Odds, 1988). At sites of infection, there are usually increased cell titre and an increase in the proportion of hyphae. It is surprising that we know so little about the factors which account for increased colonization at the site of infection. *C. albicans* will grow under laboratory conditions in buffered media containing only glucose, biotin, inorganic salts and amino acids with a generation time of less than two hours (Mardon *et al.*, 1971; Land *et al.*, 1975). *C. albicans* may grow under aerobic or anaerobic conditions (Swawathowski and Hamilton-Miller, 1975), and affects its micro-environment by releasing metabolites (Saltarelli, 1973). Why is carriage more frequent in the healthy mouth than the healthy vaginal canal? Why does candida grow so densely in the vagina or during an oral cavity episode of candidosis, but remains at a very low or negligible level in the same body location in the commensal state? Are there inhibitors of growth in the vaginal fluid and saliva of healthy individuals not present in individuals suffering from candidosis, or are there less nutrients in the vaginal fluid or saliva of healthy individuals? Is there a stimulation of growth at the site of infection? Is the immune system involved in suppressing growth (Witkin *et al.*, 1986)? Does the state of the oral or vaginal mucosa dictate the level of colonization? The answer to all of these questions is unknown but experimentally approachable.

5.3 THE BIOLOGY OF DIMORPHISM AS IT RELATES TO PATHOGENESIS

Candida albicans and related species multiply in the budding form in a manner similar to, if not at times indistinguishable from, its pathogenic inocuous relative *Saccharomyces cerevisiae* (Soll, 1985a; Soll *et al.*, 1981) (Figure 5.1(a) and (b)). However, *Candida* has the added capability of growing in an elongate hyphal form (Figures 5.1(c) and (d)), and it is this capability which appears to be most fundamental to tissue penetration. When hyphae at the periphery of a colony penetrate agar, as the hyphae penetrate deeper, they release budding cells at the distal end of each hyphal compartment and these budding cells generate small secondary colonies (Figure 5.2(a) to (f)). This sequence may differ from tissue penetration by the abundant genesis of budding cells during hyphal growth, but it provides evidence of the capacity of the hyphae to move through material and to convert from one phenotype to another as it penetrates. The formation of hyphae in pathogenesis has been supported by the observation that variants defective in hyphae formation are not as virulent as their parent strains (Sobel *et al.*, 1984). However, *Candida* cells in the budding phenotype can penetrate tissue to some extent (Ray and Payne, 1988; Cole *et al.*, 1988).

Although the problem of dimorphism has been attacked over the past

30 years with vigour, we still know relatively little about the molecular mechanisms involved in the decision of an infecting cell to convert from the budding to hyphal form. Again, the reason for this is embedded in the absence of a sexual cycle in *C. albicans* or, more precisely, in the absence of a laboratory system for detailed genetics. However, some insights into the regulation of dimorphism (reviewed in section 5.4) have emerged from biological experiments (Soll, 1985a; Soll, 1986) and the standard methods of molecular biology should prove to be extremely helpful in the near future (Kurtz *et al.*, 1988; Kirsch *et al.*, 1990).

C. *albicans* grows better as a budding cell. Indeed, most mutants affecting some aspect of dimorphism usually are defective in hypha formation, not bud formation. When a normal *C. albicans* cell forms a daughter bud, the scenario is identical to bud formation in diploid *Saccharomyces cerevisiae* (Byers, 1981). F-actin granules condense at the site of bud emergence (Anderson and Soll, 1986). The cell wall and plasma membrane then bulge at this point, and condensed large actin granules or actin bars are distinguishable at the site of evagination, but still in the mother cell. At the time of evagination, a filament ring forms just under the plasma membrane

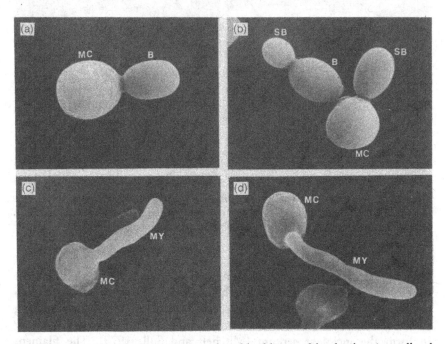

Figure 5.1 Scanning electron micrographs of budding and hypha forming cells of *Candida albicans*. (a) A mother cell (MC) with one bud (B); (B) a mother cell (MC) and mature bud (B) both with secondary buds (SB); (c) a mother cell (MC) and small hypha, or mycelium (MY); (d) a mother cell (MC) and longer hypha, or mycelium (MY).

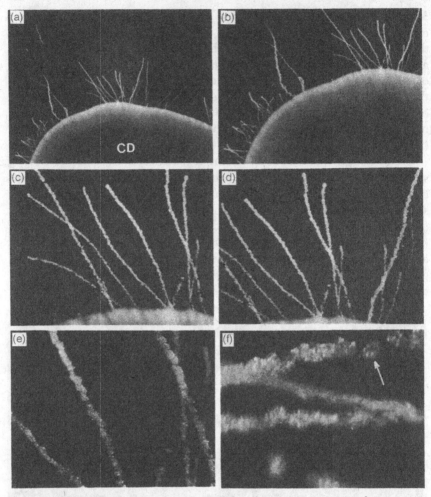

Figure 5.2 The edge of a colony of *Candida albicans* on agar. a to f, increasing magnification. Hyphae emerge from the colony into the agar, and release budding cells which form small colonies (see arrow in (f)) in the agar along the length of the hypha.

(Figures 5.3(a) and (b)) and is composed of approximately 12 filament profiles on average (Soll and Mitchell, 1983). The filaments appear to be latticed to each other and the plasma membrane. The filament ring structure is transient, and is dismantled as the chitin ring grows inwardly to form the incipient septum (Mitchell and Soll, 1979). The filament ring in budding *C. albicans* is similar to the one observed in budding *S. cerevisiae* (Byers and Goetsch, 1976). The localization and temporal dynamics are also similar. In *S. cerevisiae*, mutations in four genes result in the absence of the filament ring, CDC4, CDC10, CDC11 and CDC12

(Byers and Goetsch, 1976). These genes have been cloned from *S. cerevisiae* (Haarer and Pringle, 1987; Haarer *et al.*, 1990) and represent a gene family encoding proteins with no apparent homology to any known cytoskeletal elements. The filament ring genes are now being cloned from *C. albicans*, and probably will provide unique proteins necessary for cell division which can be targeted for new drug development. It has been suggested that the filament ring stimulates localized chitin synthesis (Soll, 1986). This localized function is reinforced by the absence of junctional localization of chitin synthesis in filament ring mutants of *S. cerevisiae* (Byers and Goetsch, 1976).

As the daughter bud grows, F-actin granules localize in the cortical cytoplasm (Anderson and Soll, 1986). F-actin filaments can sometimes be observed spanning the mother–daughter cell cytoplasm. When the daughter cell grows to roughly two-thirds of the mother cell volume, the nucleus migrates to the mother–daughter cell junction, becomes bar-belled in shape, then divides, with one of the two daughter nuclei re-entering the deep cytoplasm of the mother cell, and the other entering the daughter cell (Bedell *et al.*, 1980). Microtubules are associated with the nucleus during

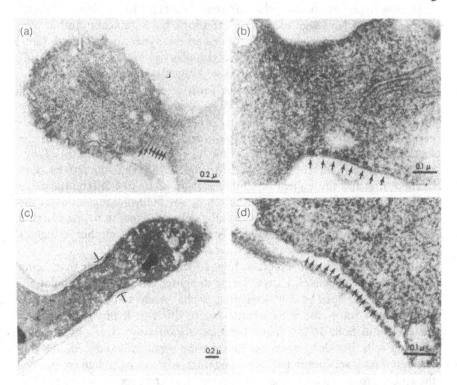

Figure 5.3 Transmission electron micrographs of budding (a)(b) and hypha forming cells (c)(d) with visible filament ring structures (pointed to with arrows).

division and no doubt play the same roles suggested in nuclear division in *S. cerevisiae* (Palmer *et al.*, 1989).

During the first two-thirds of bud growth, the wall expands due to two mechanisms, general expansion equally distributed around the wall, and apical expansion (Staebell and Soll, 1985; Soll *et al.*, 1985). General expansion accounts for roughly 30% and apical expansion 70% of wall growth. When the bud reaches roughly two-thirds of its final size, the apical zone shuts down and the wall is completed by general expansion. When the next round of budding occurs, actin granules again cluster at the point of evagination, and the cycle repeats itself.

Budding cells can be induced to form hyphae by a number of different methods (Soll, 1985a, b). Inducing media have included N-acetyl-glucosamine (Simonetti *et al.*, 1974), methionine (Mardon *et al.*, 1969), serum (Taschdjian *et al.*, 1960), carbon dioxide and sulfhydryl compounds (Nickerson and Van Rij, 1949). Simple methods have also been developed in which stationary phase cells, blocked in Gl, can be induced synchronously to form either buds or hyphae, depending upon the pH of the inducing medium (Buffo *et al.*, 1984; Mitchell and Soll, 1979).

During hypha formation, the differences in actin localization, filament ring formation, and wall expansion are subtle but significant, and in part account for the differences in cell shape and the dynamics of growth (Soll, 1986). During hypha formation, actin granules again localize at the site of evagination (Anderson and Soll, 1986). However, upon evagination, the filament ring and chitin ring do not form at the mother–daughter cell junction (Soll and Mitchell, 1983; Mitchell and Soll, 1979). Rather, they form roughly 20 minutes later, when the length of the incipient hypha, or germ tube, is on average 2 μm. They form on average 2 μm from the mother–daughter cell junction along the tube (Figure 5.3(c) and (d)). It has been suggested (Soll, 1986; Soll and Mitchell, 1983) that the difference in filament ring and subsequent septum formation is due to a differential delay in the signal for filament ring formation. If the filament ring forms at the apical growth zone of a daughter cell, the delayed signal in hypha-forming cells results in positioning along the tube rather than at the mother–daughter cell junction as in the case of bud formation.

Actin localization also differs in hypha-forming cells. Rather than being distributed in the general cortex during daughter cell growth, granules are localized in the apex of the elongating hypha (Anderson and Soll, 1986). Wall growth zones and their dynamics also differ in hypha forming cells (Staebell and Soll, 1985). The apical expansion zone is as potent as in budding cells, but the general expansion zone is radically reduced, and this difference may account in part for elongation. In contrast to bud formation, the apical expansion zone does not shut down after a specific period of time. Differences in the proportions of expansion zones may account for the difference in wall composition between bud and hypha (Chattaway *et al.*, 1968).

Nuclear migration and division during hypha formation is surprisingly similar to that during bud formation, (Soll *et al.*, 1978). The nucleus migrates to the mother-hypha junction when the hypha attains a length approximately equal to the diameter of the mother cell. The nucleus appears to divide at the junction and the two daughter nuclei then separate into the hypha and original mother cell (Soll *et al.*, 1978). There is disagreement as to whether the nucleus divides at the mother–daughter cell junction or at the site of the filament ring (Gow *et al.*, 1986).

Although the subtle temporal, spatial and quantitative differences between-bud and hypha formation may account for some aspects of the dimorphic transition in *C. albicans* and related species, we have yet to identify a gene product differentially synthesized and basic to phenotype. By 1-dimensional and 2-dimensional gel electrophoresis, it has been demonstrated that the majority of pulse-labelled polypeptides synthesized during bud and hypha formation are the same (Manning and Mitchell, 1980; Finney *et al.*, 1985; Brummel and Soll, 1982). Of the 374 polypeptides reproducibly resolved by 2D-PAGE, only one polypeptide was found to be specific for bud formation and only one for hypha formation (Finney *et al.*, 1985). Since 2D-PAGE visualizes only one-twentieth of the potential number of polypeptides which may be coded for in the *C. albicans* genome, the method by no means indicates the number of phenotype-specific polypeptides. This deficiency is accentuated by the possibility that many phenotype-specific polypeptides may be regulatory and of low abundance.

Surprisingly, although there have been several reports of antigenic differences between the developing bud and the developing hypha (e.g. Syverson *et al.*, 1975; Hopwood *et al.*, 1986; Smail and Jones, 1984; Brawner and Cutler, 1986), no genes have been cloned which are differentially expressed. The methods for cloning phenotype-specific genes indeed exist. One can synthesize cDNAs from mRNA of either bud or hypha, hybridize with saturating amounts of RNA from the alternate phenotypes, and then develop a subtraction library (Schneider *et al.*, 1988) enriched for potential phenotype-specific cDNAs. Phenotypic specificity of cloned cDNAs can then be tested by Northern blot hybridization, and the phenotype-specific cDNA clones used in turn to clone the homologous gene from a genomic library. In a second approach, one can generate a cDNA library from one phenotype, replicate plate and screen with radiolabelled cDNA from both phenotypes (Mehdy *et al.*, 1983). Clones which hybridize to phenotypic homologous cDNA and but not to phenotypic heterologous cDNA represent putative phenotype-specific mRNAs. Finally, antisera to phenotype-specific proteins can be used to screen expression libraries (Young and Davis, 1983). These approaches are being carried out in a number of *Candida* laboratories at the time this chapter is being written, and there is little doubt that if differential gene expression indeed occurs during dimorphism, phenotype-specific genes will have been cloned and reported by the time this chapter is published.

Why will phenotype-specific genes be valuable to our understanding of *Candida* pathogenesis and treatment of the disease? As argued, perhaps the most obvious virulence trait of *C. albicans* which separates it from noninfectious yeast like *S. cerevisiae* is the capacity to generate a hypha. Understanding how *C. albicans* generates a hypha and the genes involved in this process may provide us with insight into the evolution of this pathogenic process. Second, identifying gene products specific to hypha formation may provide targets for new drug development, although any essential gene product present in *Candida* but not in humans (e.g. the filament ring proteins) represents an excellent target. Finally, if we understood how a hypha forms, what signals hypha formation at the site of infection, and what molecules are involved in sensory transduction in hypha formation, prophylactic strategies for suppressing the signal-response pathways leading to tissue penetration could be developed.

5.4 HIGH FREQUENCY SWITCHING AS IT RELATES TO PATHOGENESIS

It always seemed simplistic to assume that the entire phenotypic repertoire of *C. albicans* which contributes to its success as a pathogen is the bud-hypha transition, especially when there were well over one hundred *Candida* species (Van Uden and Buckley, 1970), most of which generate hyphae, but only a few of which cause disease. Obviously, hypha formation alone is not sufficient for human pathogenesis, and other virulence traits must combine with hypha formation to generate a successful virulence profile. Unfortunately, even this more complex view of *Candida* pathogenesis turns out to be too elementary. It had been observed over 50 years ago, that *Candida* strains exhibited colony variability (e.g. Mackinnon, 1940; Ireland and Sarachek, 1968). Brown-Thomsen demonstrated, in 1968, that when *Candida* strains were maintained over long periods of time, variant colony phenotypes emerged which exhibited distinct physiologies. We now know that *C. albicans* and related species are capable of switching frequently and reversibly between a limited number of general phenotypes distinguishable by colony morphology and that these switching systems are superimposed upon the bud-hypha transition (Soll, 1990). Almost every strain of *C. albicans* and *C. tropicalis* so far tested has an induceable switching system, and no strain apparently possesses more than one switching system defined by its repertoire of switch phenotypes. The characteristics of high frequency phenotypic switching in *C. albicans* include: (1) low and high frequency modes of switching; (2) a limited number of general switch phenotypes; (3) heritability; (4) reversibility and interconvertability, and (5) induceability by ultraviolet (UV) irradiation (Soll, 1990; Slutsky *et al.*, 1985; Slutsky *et al.*, 1987; Soll *et al.*, 1987a; Soll *et al.*, 1988; Soll *et al.*, 1989; Soll and Kraft, 1988;

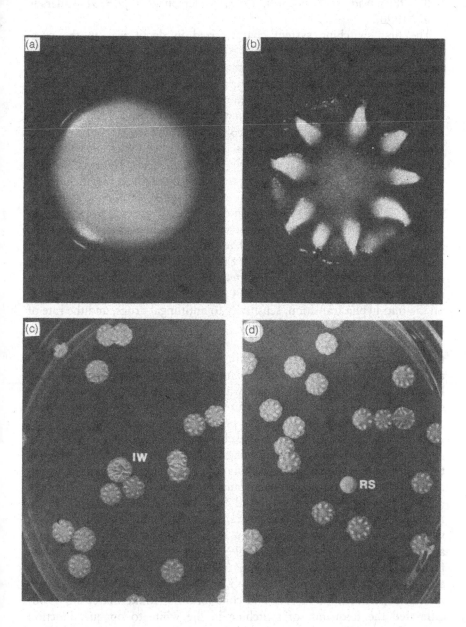

Figure 5.4 High frequency switching of colony phenotype in *Candida albicans* strain 3153A. (a) The original smooth white colony phenotype; (b) the star phenotype; (c) example of switching from star to irregular wrinkle (IW); (d) example of switching from star to revertant smooth (RS) colony phenotype.

Soll, 1989; Anderson and Soll, 1987; Anderson *et al.*, 1989; Anderson *et al.*, 1990).

The first switching system characterized in detail was in the common C. *albicans* laboratory strain 3153A (Slutsky *et al.*, 1985). Cells from this strain usually generate smooth white colonies (Figure 5.4(a)) after seven days on amino acid rich medium (Lee *et al.*, 1975) supplemented with argenine and a limiting concentration of zinc (Bergen *et al.*, 1990). However, either spontaneously or after low UV treatment, variants distinguishable by colony morphology appear (e.g. star phenotype in Figure 5.4(b)) which are heritable but which also have the feature of interconverting at relatively high frequencies (Figure 5.4(c) and (d)). The phenotypes in the switching repertoire of strain 3153A include: (1) original smooth white; (2) ring, (3) star; (4) irregular wrinkle; (5) stippled; (6) hat; (7) fuzzy, and (8) revertant smooth white (Figures 5.5). Differences in colony morphology are due to differences in the dynamics and proportions of budding cells, hyphae, pseudohyphae and hyphal branching in the domes of colonies grown on agar depleted of zinc (Slutsky *et al.*, 1985). An example of a switch from ring to star is presented in Figure 5.6. In addition to colony morphology, switching in strain 3153A affects the environmental constraints on the bud-hypha transition, sensitivity to antifungal drugs, and the rate of growth and colony size (Soll, 1989).

Perhaps the most carefully examined switching system in C. *albicans* is the white-opaque transition in strain WO-l, an isolate from the blood and lungs of a bone marrow transplant patient at the University of Iowa Hospital and clinics in Iowa City, Iowa (Slutsky *et al.*, 1987). Although the switching system observed in the white-opaque transition is found in only a minority of C. *albicans* strains, its popularity as an experimental system relies on: (1) the phase transition-like character of switching between white and opaque; (2) the ease of observing the transition using vital dyes; (3) the expression of the alternate phenotypes on virtually any agar employed, and (4) the dramatic impact this transition exerts on cellular phenotype (Slutsky *et al.*, 1987; Anderson and Soll, 1987; Anderson *et al.*, 1989; Anderson *et al.*, 1990; Rikkerink *et al.*, 1988). In the white-opaque transition, cells switch at extremely high frequencies between a standard smooth white colony-forming phenotype, a grey (or opaque) colony forming phenotype and at a lower frequency to an irregular wrinkled and two fuzzy phenotypes (Figure 5.7). Using the Luria-Debruck fluctuation experiment, Rikkerink and collaborators (Rikkerink *et al.*, 1988) estimated the frequency of switching in the white to opaque direction to be 10^{-4} to 10^{-5}, and in the opaque to white direction to be 10^{-4} at 25°C in YEPD broth. Bergen and collaborators (Bergen *et al.*, 1990) measured switching at the cellular level in microcolonies, and found the probability of a switch to be 1.7×10^{-5} in the white to opaque direction and 1.0×10^{-1} in the opaque to white direction. Since the

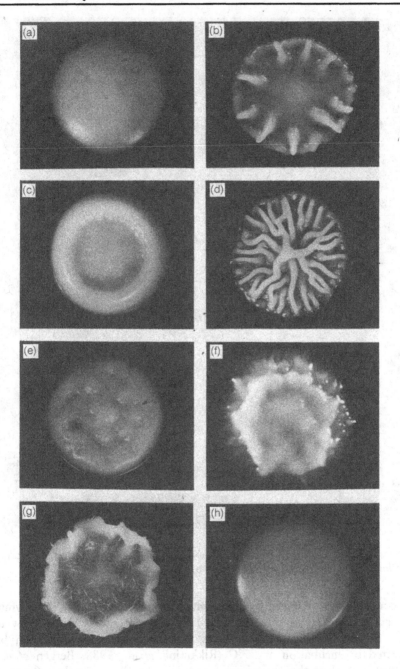

Figure 5.5 Examples of the colony morphologies in the switching repertoire of *Candida albicans* strain 3153A. (a) original smooth white; (b) star; (c) ring; (d) irregular wrinkle; (e) stippled; (f) hat; (g) fuzzy; (h) revertant smooth white.

Figure 5.6 An example of colony morphologies in switching repertoired *Candida albicans* strain 3153A. (a) original ring; (b) star.

frequency of switching in both the white to opaque and opaque to white direction can be heritably stimulated by low doses of UV (Morrow *et al.*, 1989), and since mass conversion from opaque to white can be affected by incubation at 37°C (Rikkerink *et al.*, 1988; Bergen *et al.*, 1990), it is clear that switching frequencies are dramatically influenced by environmental factors and therefore do not represent fixed rates. The unresolved paradox is the extremely high levels of spontaneous switching, induced mass conversion, with the heritability of UV-induced high frequency switching (Anderson and Soll, 1987; Rikkerink *et al.*, 1988; Bergen *et al.*, 1990)

When a cell switches from white to opaque, it undergoes a change in phenotype so extreme that even a good mycologist would probably not identify the opaque cell as in the genus *Candida*. When compared with white budding cells (Figure 5.8(a)), opaque budding cells are on average twice as large, are elongate and sometimes wider at one end than at the other and contain a prominent vacuole (Figures 5.8(b) and (c); Slutsky *et al.*, 1987). Opaque budding cells also differ from white budding cells

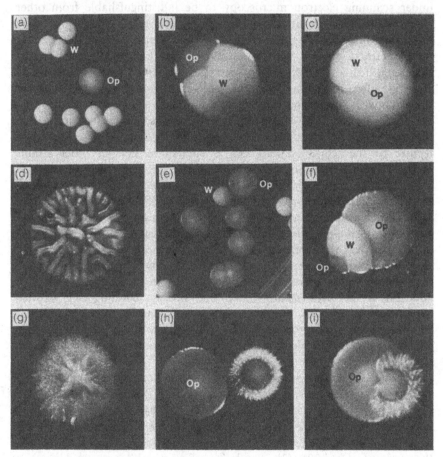

Figure 5.7 The switching system including the white-opaque transition in *Candida albicans* strain WO-1. (a) An example of a switch from white (W) to opaque (Op): (b) a white colony with an opaque sector; (c) an opaque colony with a white sector; (d) an irregular wrinkled colony; (e) an example of a switch from opaque to white; (f) an example of a switch from opaque to white to opaque in one colony; (g) the medusa colony phenotype; (h) the fried egg colony phenotype contrasted to opaque colony phenotype; (i) an opaque colony with a fried egg sector.

by: (1) the position of new buds (Slutsky *et al.*, 1987; Anderson and Soll, 1987); (2) the capacity to undergo bipolar budding (Slutsky *et al.*, 1987); (3) actin distribution during bud development (Anderson and Soll, 1987); (4) constraints on dimorphism (Anderson *et al.*, 1989); (5) lipid and sterol content (Ghannoum and Soll, 1980); (6) hydrophobicity and cohesiveness (Kennedy *et al.*, 1988); (7) the pattern of sugar assimilation (Soll, 1990) and (8) the pattern of cell wall expansion (Soll, unpublished observations). They also differ in ultrastructural architecture (Anderson and Soll, 1987; Anderson *et al.*, 1990). White cells appear under scanning electron microscopy to be indistinguishable from other *Candida* strains, exhibiting wall surfaces with no distinct topography other than bud scars (Figure 5.9(a) – (c)). In contrast, opaque cells exhibit unique surface pimples with pores from which blebs sometimes emerge (Figure 5.9(d) – (i)). When thin sections are viewed under transmission electron microscopy, the opaque pimple is obvious as a thickening of the wall with membrane bound blebs sometimes apparent at the pimple apex (Figure 5.8(b)) and on rare occasions, vesicles positioned in the cytoplasmic cortex under the wall pimple. The cytoplasmic architecture of the white budding cell is similar to that of other strains of *C. albicans*, but the architecture of the opaque cell is dominated by a large vacuole with spaghetti-like material which may very well represent collapsed vesicles (Anderson *et al.*, 1990) (Figures 5.8(b) and (c)). The role, if any, for the specialized architecture of both wall and cytoplasm in the opaque budding cell is unknown. Gene expression also differs between the white and opaque budding cell. At least one white and two opaque-specific polypeptides are readily detectable among the major polypeptides distinguishable by 2D-PAGE (Soll *et al.*, 1990). In addition, it has been demonstrated that opaque budding cells express hypha-specific antigens, and at least 3 opaque-specific antigens of 14.5, 21 and 31 kd (Anderson and Soll, 1987; Anderson *et al.*, 1990). The 14.5 kd antigen is localized in the apex of the surface pimple (Anderson *et al.*, 1990).

Although the white-opaque transition represents one of the rarer switching systems in *C. albicans*, it is the best studied (Soll *et al.*, 1990) and demonstrates how effective a switching system can be in changing phenotypes, in a reversible fashion. It has been demonstrated that in the white-opaque transition, switching affects: (1) adhesion to buccal epithelium (Kennedy *et al.*, 1988); (2) stimulation of neutrophil superoxide genesis (Kolotila and Diamond, 1990); (3) killing by intact neutrophils or cell free oxidants (Kolotila and Diamond, 1990); (4) susceptibility to major antifungal drugs (Soll *et al.*, 1990) and (5) virulence in mouse models (Ray *et al.*, in preparation). Therefore, the role of phenotype in the pathogenesis of *C. albicans* and related species like *C. tropicalis* (Soll *et al.*, 1988) may not be limited to dimorphism, but may include the capacity to switch frequently and

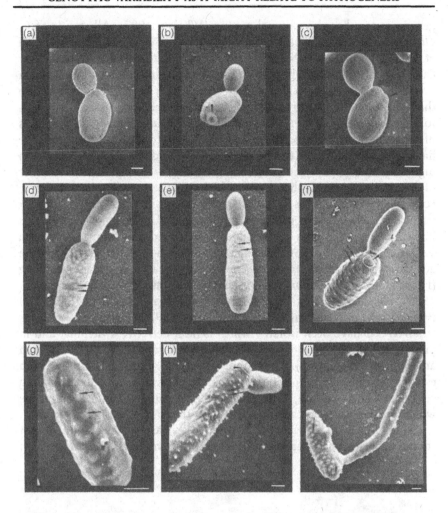

Figure 5.8 Scanning electron micrographs of white budding cells (a–c), opaque budding cells (d–h) and an opaque cell forming a hypha (i) of *Candida albicans* strain WO–1. In panels (b) and (c) bud scars are noted with arrows. In panels (D) to (H) pimples are noted with arrows. In panel (I) a pimple is noted with an arrow, and a bud scar with an arrow head.

reversibly between a number of general phenotypes specialized in different aspects of virulence.

5.5 GENOTYPIC VARIABILITY AS IT MIGHT RELATE TO PATHOGENESIS

As previously noted, the major obstacle in dissecting the cellular and molecular events involved in pathogenesis has been the absence of an effective genetic system based on sexual mating (Whelan, 1987). All strains of *C. albicans* so far examined for ploidy have been found to be diploid (Whelan *et al.*, 1980; Olaiya and Sogin, 1979; Riggsby *et al.*, 1982) and, in the few cases tested, strains contain balanced lethals (Whelan and Soll, 1982) which precludes homozygosis. The absence of a sexual cycle has apparently led to this condition. This means that there is no recombination based on meiosis followed by sexual fusion. Where, then, does *C. albicans* obtain the levels of variability one might expect of so successful and versatile a pathogen? We have observed that most strains of *C. albicans* and related species undergo high frequency phenotypic switching, providing each strain with a number of alternative general phenotypes which may serve roles in alternative ecological settings, or as a fast-change mechanism when pressured by a particular therapeutic or prophylactic treatment. Unfortunately, we do not know the molecular basis of reversible, high frequency switching. It has been suggested (Soll *et al.*, 1990; Slutsky *et al.*, 1987; Soll, 1989) that reversible genetic rearrangements are probably involved in yeast mating-type switching (Hicks *et al.*, 1977), *Salmonella* flagellin variability (Silverman and Simon, 1983), *Neisseria* antigenic variability (Swanson and Kooney, 1989), and *trypanosome* antigenic variability (Donelson, 1989). Mutants of *C. albicans* strain WO-1 have been isolated which exhibit defective switching, and in most cases, these mutants are also defective in hypha formation (Bergen and Soll, in preparation). This is not surprising since it has already been established that hypha-specific antigens are expressed in opaque budding cells (Anderson and Soll, 1987), suggesting that switching, at least in the white-opaque transition, involves the activation of hypha-specific genes.

Although we do not know the molecular basis of switching, recent studies have demonstrated that the genome of *C. albicans* can undergo high

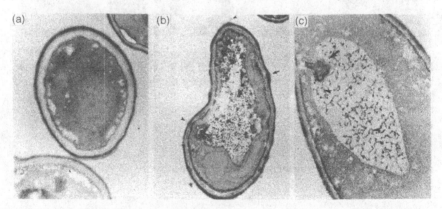

Figure 5.9 Transmission electron micrographs of white (A) and opaque (B and C) cells of *Candida albicans* strain WO-1. (A) The standard profile of a white cell; (B) A profile of an opaque cell with wall pimples and pimple blebs (noted with arrows); (c) A profile of an opaque cell's large vacuole with unusual vacuole structure.

frequency change. First, it has been demonstrated that there is extraordinary polymorphism between strains in the size of homologous chromosomes (Magee *et al.*, 1987). Second, it has been demonstrated that within a strain, there can be polymorphism for a pair of homologous chromosomes, suggesting unequal interchromosomal recombination or intrachromosomal recombination leading to increased or decreased chromosome size. Third, two moderately repetitive sequences, Ca3 and 27A, have been cloned which generate complex Northern blot patterns (Schmid *et al.*, 1990; Scherer and Stevens, 1988). Both are distributed throughout the genome and exhibit pattern polymorphisms in different strains (Soll *et al.*, 1987a; Soll *et al.*, 1988; Soll *et al.*, 1989; Schmid *et al.*, 1990; Scherer and Stevens, 1988). In the case of 27A, changes in pattern have been demonstrated over a limited number of cell divisions (Scherer and Stevens, 1988). Fourth, a telomeric sequence Ca7 has been cloned which is present on most, if not all, chromosomes (Soll *et al.*, 1987a; Hicks *et al.*, in preparation). This sequence exhibits rapid recombination in a single strain over a limited number of divisions (Soll *et al.*, 1987a; Soll, unpublished observations). Fifth, a sequence has recently been cloned from *C. albicans*, which may have the capacity to recombine reversibly and at high frequency with tandem sequences, is located on a single chromosome, and may contain a sequence for its own recombinase gene (Srikantha *et al.*, in preparation). This is the first time such a gene has been cloned from a eukaryote. To have uncovered so many genomic variability systems without the aid of an effective genetic system indicates that we have only scratched the surface of high frequency genetic variability in *C. albicans*. This is indeed an ominous suggestion since it predicts that *C. albicans* has the capacity rapidly to evolve more virulent phenotypes with the correct selective pressures. Indeed, these pressures may have found a perfect evolutionary target in AIDS strains. AIDS patients are frequently treated prophylactialy with antifungal drugs because of high carriage and frequent episodes of oral-oesophageal candidosis (Epstein, 1989; Schidt and Pindborg, 1987). There are already preliminary indications that AIDS strains may represent a select group (Odds *et al.*, 1990), and that recurrence of oral-oesophageal candidosis in an AIDS patient is usually due to the same strain (J. Greenspan, personal communication; Odds *et al.*, in preparation). These findings possibly imply that *C. albicans* is a dynamic pathogen with the capacity for rapid genetic and phenotypic change.

5.6 OTHER PHENOTYPIC TRAITS AS THEY RELATE TO PATHOGENESIS

In addition to hypha formation, phenotypic switching and the capacity to live as a commensal in the healthy individual, a number of other potential virulence factors have been identified and analysed.

5.6.1 Adhesion

Adhesion to mucosa must represent an important factor in tissue penetration (e.g. Calderone *et al.*, 1985; Persi *et al.*, 1985). In many instances, the site of infection is coated with yeast (Kennedy and Volz, 1985) and numerous pictures have been published of cells in the budding form adhering to surface epithelium with germ tubes penetrating the tissue (e.g. Howlett and Squire, 1980). Differences in the level of adherence have been demonstrated between species (Ray *et al.*, 1984), between strains within a species (McCourtie and Douglas, 1984), in mutants (Calderone *et al.*, 1985), between the bud and hyphal forms (Anderson *et al.*, 1989; Kimball and Pearsall, 1980), and as a result of high frequency switching (Kimball and Pearsall, 1980). Adhesiveness may also vary as a result of changes in the host mucosa. Putative adhesions on the *Candida* surface have been characterized (Douglas, 1985; Lee and King, 1983) and specific sugars have been demonstrated to compete with *Candida* cells for binding to epithelium (Collins-Lech *et al.*, 1984; Centeno *et al.*, 1983), although in a very careful review, Odds points to conflicting results between these studies (Odds, 1988). It has been assumed that mannoproteins function as *Candida* adhesions (Douglas, 1985; Lee and King, 1983). Unfortunately, *in vitro* assays of adhesion do not usually take into account molecules in saliva, the digestive tract, or vaginal fluid which may first coat the cell prior to adhesion. It has been demonstrated that fibronectin binds to C. *albicans* (Skerl *et al.*, 1984) and may act as one of many adherence interphases.

The problem of adhesion has been reviewed in more detail (Odds, 1988; Kennedy, 1987; Rotrosen *et al.*, 1986), and deserves more concentrated research. Adhesion to materials other than mucosal cell surfaces (Klotz *et al.*, 1985; Critchley and Douglas, 1985) has become extremely important in hospital practice, where the risk of systemic infections in immunosuppressed patients resulting from catheterization remains high. The problem of adhesion must be dealt with in models which most faithfully mimic the *in vivo* situation. Only in such models can molecules which interfere with adhesion be identified and then tested in prophylactic therapies.

5.6.2 Excreted enzymes

We do not know how hyphae penetrate tissue, but some reasonable assumptions lead to models with predictive value. Since hyphae adhere quite strongly to mucosal cells (Anderson *et al.*, 1989; Kimball and Pearsall, 1980) and growth is primarily at the apex (Staebell and Soll, 1985), the long surface of an expanding hypha most likely remains fixed to the supporting tissue as the tip of the hypha expands into the tissue. Adhesion between hyphal surface and tissue probably acts to anchor the penetrating hypha. It has been proposed that two excreted enzymes, an acid protease (Staib, 1965) and a phospholipase (Pugh and Cawson, 1977), facilitate tissue penetration, but proof is awaited. Mutants which are defective in acid protease secretion are less virulent (MacDonald and Odds, 1983; Kwon-Chung et al., 1985), acid protease can be detected in infected tissue by immunofluorescent labelling (MacDonald and Odds, 1980), and the level of secreted acid protease follows roughly the rank order of species virulence. However, several problems exist in a final interpretation of the role of secreted acid protease. First, the pH of human saliva is too high for effective enzyme activity (Germaine and Tellefson, 1981), and even the pH of vaginal fluid is higher than the measured optimal range of the enzyme (Ruchel, 1981). Second, variants have been found which are not high-level secretors but are, none the less, virulent when injected into mice through the tail vein (Ray and Payne, unpublished observations). The acid protease secreted in the medium may participate in cell wall maturation of the yeast, or another less conspicuous process. It is likely that, as in the case of all the proposed virulence factors of C. albicans, no single phenotypic characteristic makes or breaks the pathogenic capacity of the organism. More likely, combinations of factors, which may be controlled by high frequency switching, together dictate the virulent nature of the organism in a particular environmental niche. It has recently been demonstrated that switching can affect the level of acid protease secretion more than 20 fold in the white-opaque transition of C. albicans strain WO-1 (Ray et al., in preparation).

Recently, the acid protease gene was cloned and partially sequenced (Lott et al., 1989). This will allow detailed analysis of the regulation of this gene under inducing conditions, as a function of switching, as a function of dimorphism and as functions of strain and species. Results of these studies may provide insights into methods for inhibiting secretion or synthesis at sites of infection. Site-specific mutagenesis (Kelley et al., 1987; Kurtz et al., 1986) or overproduction will provide us with more definitive information on the role of this enzyme, both in the cell biology of C. albicans and in pathogenesis.

Less is known about phospholipase. It is secreted into the medium (Banno et al., 1985) and at least in one study appears to correlate with virulence (Barrett-Bee et al., 1985). Possible coregulation with acid protease has not been demonstrated, and there has been no report that the gene for excreted phospholipase has been cloned.

5.7 THE APPLICATION OF MOLECULAR GENETIC TOOLS TO QUESTIONS OF CANDIDA ETIOLOGY AND EPIDEMIOLOGY

The tools for investigating virulence traits at the molecular level are rapidly being developed (Soll, 1990; Kurtz et al., 1986), but they have not yet been successfully applied to expand our understanding of Candida pathogenesis.

In contrast, the application of molecular genetic methods to questions of etiology and epidemiology has been effective and immediately informative and in the remaining portion of this chapter unanswered questions in this area and the development of new methods for obtaining answers will be considered.

5.8 THE BASIC QUESTIONS WHICH REMAIN UNANSWERED AND WHY

Because *C. albicans* and related species live as commensals in a number of body sites, it has long been assumed that when *Candida* causes disease, it is not the result of an infection transmitted from one individual to another, but rather the conversion of the carried organism from a commensal to a pathogenic state. Underlying this assumption is the more basic assumption that *Candida* infections are primarily due to changes in the host physiology. There is no argument that in cases of immunodeficiency, patients are far more prone to *Candida* infection (Odds, 1988). For example, AIDS patients exhibit increased frequencies of oral-oesophageal candidosis (Greenspan, 1988; Syrjanen *et al.*, 1988), and female AIDS patients in addition exhibit increased frequencies of vaginal candidosis (Rhoads *et al.*, 1987). Leukaemics and immunosuppressed patients also exhibit dramatically increased incidences of candidosis (Bodey, 1984). These latter patients exhibit increased susceptibility to systemic candidosis (Reingold *et al.*, 1986), although this does not seem to be the case in AIDS patients (Lane and Fauci, 1987). Other physiological and physical conditions which may predispose individuals to candidosis or increased colonization include: (1) diabetes mellitus (Odds, 1988; Sonck and Somersalo, 1963); (2) pregnancy (Frerich and Gad, 1977); (3) infancy (Winner and Hurley, 1964); (4) burns (Law *et al.*, 1972); (5) oral contraception (Jackson and Spain, 1968); (6) antibiotic treatment (Silverman and Simon, 1983); (7) denture-ware (Budtz-Jorgensen, 1978); (8) corticosteroids (Johnston *et al.*, 1967); (9) leukoplakias of the oral cavity (Krogh *et al.*, 1987), and (10) catheterization (Williams *et al.*, 1971). In all of these situations, it is a condition of the host which predisposes the patient to candidosis, and the etiological agent is considered a uniform pathogen, opportunistic and pervasive, waiting to take advantage of the situation. However, this interpretation is too simple. As already described, *C. albicans* and related species exhibit strain variation in every virulence trait so far tested (Odds, 1988), leading to the conclusion that some strains will be more virulent than others, depending on host conditions. In addition, *Candida* is capable of genomic evolution by a host of mechanisms in addition to point mutation, suggesting that virulence traits may also be able to evolve at a rapid rate. Questions still to be answered in relation to strain origin in this disease are important. (1)

Is the source of an infecting strain the commensal strain of the individual prior to infection? Alternatively, is the genotypic range of commensals in a healthy population similar to the genotypic range of infecting strain in a population suffering from candidosis? (2) Is there only one commensal strain at different body locations of a healthy individual, or are there multiple strains? (3) Is there selection or evolution of particular strains with particular phenotypes during the course of a prolonged infection? (4) Are recurrent infections due to single or multiple strains? (5) Are there genetic and phenotypic differences between strains from different geographical locations? (6) Are there particular strains, based on genotypic relatedness, associated with particular kinds of candidosis such as oral candidosis, disseminated candidosis, vaginal candidosis? The reason many questions remain partially or completely unanswered has been the absence, until recently, of methods for assessing the genetic relatedness of strains. In this case, it has been the tools of molecular genetics which have provided us with the methods for assessing strain relatedness, and in combination with computer systems, with methods for performing large epidemiological and evolutionary studies of *Candida*.

5.9 THE RANGE OF METHODS USED TO ASSESS GENETIC RELATEDNESS BETWEEN *CANDIDA* STRAINS

In the past, strains of C. *albicans* and, to a lesser extent related species, have been biotyped by phenotype. Biotyping methods have included: (1) serotyping (Hasenclever and Mitchell, 1961; Hasenclever et al., 1961); (2) sensitivity to yeast killer factor (Polonelli et al., 1983); (3) assimilation patterns for sugars and nitrogen sources (Odds and Abbott, 1980); (4) streak morphology (Hunter et al., 1988); (5) resistance to chemicals (McCrieght and Warnock, 1982; Odds and Abbott, 1980), and (6) isoenzyme polymorphism (Berchev and Izmirov, 1967). More complex biotyping methods, pioneered by Odds and associates, employed multiple phenotypic characteristics, thus generating a phenotypic profile of each strain (Odds and Abbott, 1980; 1983). Unfortunately, these methods all run the risk of grouping genetically unrelated strains with the same phenotype and separating genetically related strains with different phenotypes, a real possibility when one considers the dramatic effects of high frequency reversible switching on cell physiology and phenotype (Soll, 1989; 1990).

For the genetic assessment of relatedness between *Candida* strains, a number of methods have been used in recent years, including: (1) polymorphisms in mitochondrial DNA (Olivo et al., 1987; Wills et al., 1984); (2) ethidium bromide-stained restriction fragment polymorphisms (Scherer and Stevens, 1988; Soll et al., 1987b; Cutler et al., 1988); (3)

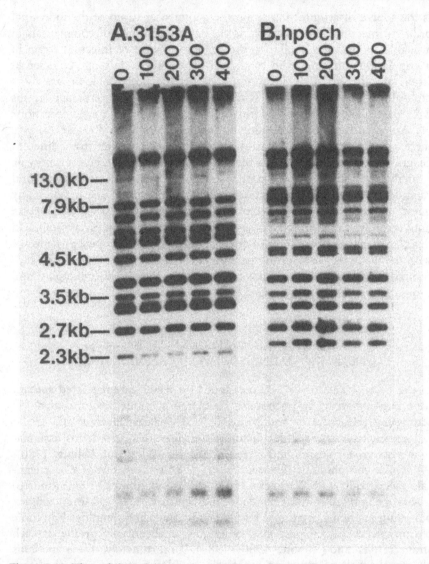

Figure 5.10 The stability of the Southern blot hybridization pattern with probe Ca3 in *Candida albicans* strains 3153A and hp6ch over 400 generations. Molecular sizes are noted to the left of the blots.

karyotypes using OFAGE, TAFE or FIGE (Magee and Magee, 1987; Snell *et al.*, 1987; Merz *et al.*, 1988); (4) polymorphisms in Southern blot hybridization patterns probed with *S. cerevisiae* ribosomal DNA (Magee *et al.*, 1987), and (5) polymorphisms in Southern blot hybridization patterns probed with moderately repetitive genomic sequences (Soll *et al.*, 1987b; 1988; Schmid *et al.*, 1990; Scherer and Stevens, 1988). Although all of these systems can be employed for epidemiological and etiological studies of limited scope, it appears that the last method is the most sensitive for discriminating between strains (Schmid *et al.*, 1990; Soll, 1990), and when combined with new computer methods for dendrography (Schmid *et al.*, 1990) may prove to be a paradigm for evolutionary, etiological and epidemiological studies of infectious fungi.

5.9.1 Fingerprinting with moderately repetitive sequences

The effectiveness of a fingerprinting probe depends on four characteristics: (1) the probe should generate a pattern which is complex; (2) the pattern should be stable within a strain over extended periods of growth; (3) the pattern should vary between separate strains, and 4) the difference in pattern should reflect the degree of genetic relatedness. So far, two moderately repetitive sequences, 27A (Scherer and Stevens, 1988) and Ca3 (Schmid *et al.*, 1990) appear to fulfil these requirements. Both are dispersed throughout the genome and both are species specific (Scherer and Stevens, 1988; Soll *et al.*, 1987a; 1988). Both produce Southern blot hybridization patterns which are relatively stable over a number of generations in a single strain (Figure 5.10) but vary between strains (Figure 5.11) (Schmid *et al.*, 1990; Scherer and Stevens, 1988).

The general method for Southern blot hybridization, the basis of fingerprinting with probes, is presented in Figure 5.12. In brief, DNAs of individual strains are purified (Scherer and Stevens, 1988) and digested with a particular endonuclease (e.g. EcoRI). The resultant fragments are separated electrophoretically in agarose gels, blotted into nitrocellulose membranes and hybridized with the radioactive probe (the endonuclease digestion map of Ca3 is presented in Figure 5.13). After hybridization, the blots are autoradiographed and the hybridization patterns analysed.

5.9.2 Analysing fingerprints: the automated dendron system

For small epidemiological or etiological studies, 'eye-balling' the similarity of patterns is usually sufficient (e.g. Soll *et al.*, 1987a; 1988; 1989; Scherer and Stevens, 1988; Fox *et al.*, 1989). However, for broader epidemiological studies involving large numbers of strains, computer-assisted techniques are necessary. Recently, an automated fingerprinting system, 'Dendron', was developed at the University of Iowa. This system provides not only auto-mated comparisons of strains, but also automatically generates dendrograms based on similarity values, and permits large retrospective analyses of strain relatedness based on the ever-growing database of strain fingerprints (Schmid *et al.*, 1990). This system will be described in some detail since the technology can be applied to most organisms in the same manner.

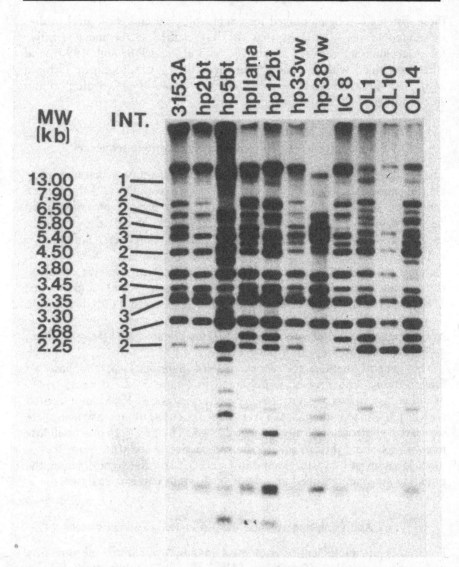

Figure 5.11 The differences in Southern blot hybridization patterns with probe Ca3 between 11 strains. Molecular sizes and relative band intensities are noted to the left of blot.

A general scheme of the Dendron system is presented in Figure 5.14. In the original Dendron study (Schmid *et al.*, 1990), band densities were categorized into four categories: 0, no band; 1, light band; 2, medium band; and 3, heavy band. The similarity between two strains A and B; the S_{AB} value, was then calculated by the following formula:

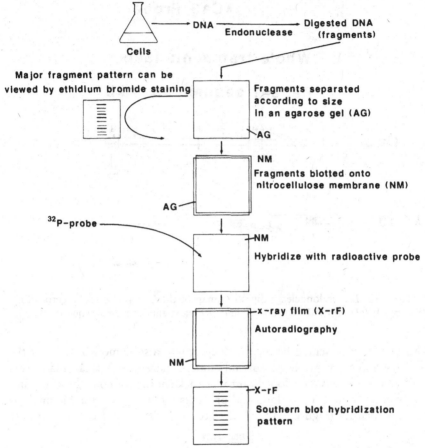

Figure 5.12 General scheme for fingerprinting with the Southern blot hybridization method.

$$S_{AB} = \frac{\displaystyle\sum_{i=1}^{k} (a_i + b_i - |a_i - b_i|)}{\displaystyle\sum_{i=1}^{k} (a_i + b_i)}$$

where a_i and b_i are the intensities of band i in patterns a and b, respectively, and k equals the total number of bands. If the positions and intensities of the bands in pattern A and B are identical, the S_{AB} value will equal 1. If the positions are nonidentical for all bands, the S_{AB} will equal 0. In this original study (Schmid *et al.*, 1990), it was demonstrated that S_{AB} values

λCa3 Probe

Whole fragment: 15kb

Repeat sequences: 11kb

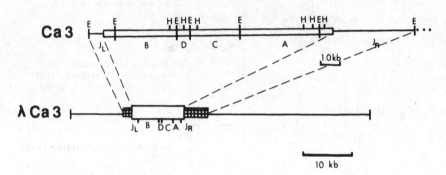

Figure 5.13 The endonuclease digestion map of the Ca3 probe for fingerprinting. (E) EcoRI; (H) HinDIII. (B) (D) (C) and (A) fragments of repeat sequence.

above 0.96 represented perfect identity. It was also demonstrated that the average S_{AB} value for 46 presumably independent *C. albicans* isolates was 0.69. In the newly developed, automated Dendron system, the computer scans each Southern blot lane and provides a continuous graphic analysis of pixel density (Figure 5.15). Peaks are integrated and the same formula

Dendron

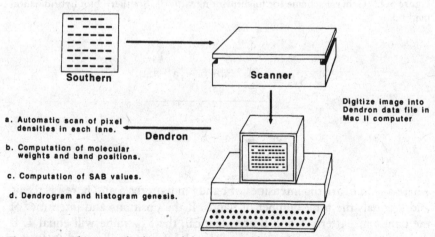

Figure 5.14 The general scheme of the automated Dendron system.

employed to assess S_{AB} values. Again, perfect identity is obtained at values of 0.96 or greater (Vawter *et al.*, in preparation).

The power of Dendron is in the capacity to analyse the relatedness of large numbers of strains. As each strain is analysed, the digitized Ca3 hybridization pattern is stored on an optical disc, and the pattern analysis stored in the data file. As each new pattern is analysed, it is compared not only with the other patterns on the 14 lane Southern blot, but also with all patterns previously analysed. Therefore, retrospective analysis of strains grouped for whatever reason can be performed. Dendron automatically generates dendrograms for all or any subset of strains analysed and stored in its file. In Figure 5.16, a dendrogram is presented for 59 strains obtained from healthy individuals (HMH and hp strains), patients with oral candidosis (OL strains) and immunocompromised patients (IC and WO strains).

Although the Dendron system represents a powerful tool for large epidemiological studies, some precautions in interpreting results are warranted. First, because evolution of the Ca3 pattern is not understood (Schmid *et al.*, 1990), one cannot be confident in interpreting S_{AB} values in terms of genetic

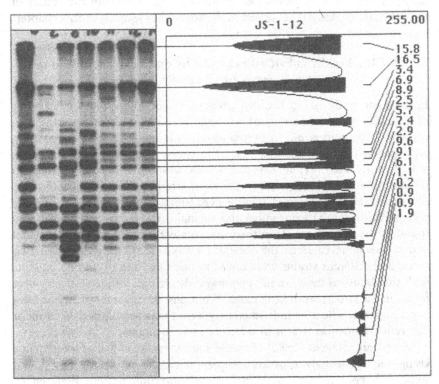

Figure 5.15 Dendron printout of the digitized Southern blot on left, and automated analysis of pixel density and molecular size on right.

relatedness. Indeed, it has already been demonstrated that there is an upper limit of 37 Ca3-containing EcoRI fragment sizes, and some fragments exhibit higher hybridization intensity than others (Schmid *et al.*, 1990). If changes in pattern are due to rearrangements at preferred sites, one might expect divergence in related strains with time, and a continuous decrease in S_{AB} value. However, rearrangements at a limited number of sites can also lead to pattern convergence. It is therefore clear that for a Dendron-type system to be effective in determining genetic relatedness in large epidemiological studies, the sequence of Ca3 and the rules of rearrangement in the Ca3 gene family must be elucidated.

5.10 PRELIMINARY ANSWERS TO EPIDEMIOLOGICAL AND ETIOLOGICAL QUESTIONS USING DNA FINGERPRINTING

Molecular techniques for fingerprinting strains have been applied in a number of studies on the epidemiology and etiology of *Candidosis*. In the majority of cases, the Dendron system could not be applied since it has only been recently developed (Schmid *et al.*, 1990), but the results of the previous studies are reviewed in the remaining sections of this chapter.

5.10.1 Is a healthy individual colonized by one, or by more than one, strain of *Candida*?

In a recent study of 50 healthy, predominantly young women, carriage of *Candida* at one or more of 17 body locations was demonstrated in roughly 73% (Soll *et al.*, in preparation). Employing DNA fingerprinting with probe Ca3 and cloning from the tested body sites, the following was demonstrated: (1) in five individuals with simultaneous carriage in the mouth and anal-rectal areas, the same strain was found in both areas in three cases, and different strains were found in two cases; (2) in three patients with simultaneous vulva and vaginal carriage, the same strain was found in the two areas; (3) in three patients with simultaneous anal-rectal and vulva-vaginal carriage, the same strain was found in the two areas in two cases, and different strains were found in one case; and (4) in two patients with simultaneous mouth and vulva-vaginal carriage, different strains were found in the two areas in both cases. From this limited analysis, it has been concluded that a healthy individual may have the same commensal strain or different commensal strains in different body locations.

Scherer and Stevens (1988) obtained similar results in five of six patients. Using the moderately repetitive sequence 27A as a probe, they found some difference in the Southern blot hybridization pattern between *C. albicans* strains isolated from different body locations. Many of the pattern differences they found were minor, suggesting that the changes in 27A

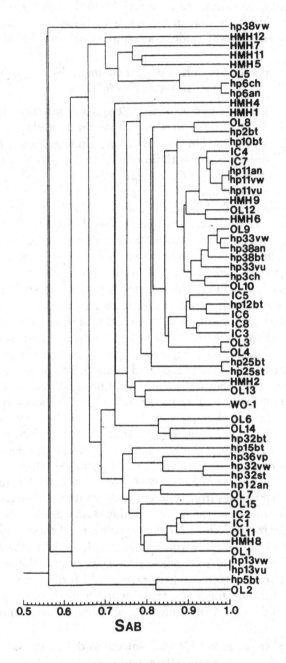

Figure 5.16 A dendrogram for 59 strains generated by Dendron. HMH, hp; strains from healthy individuals; OL; strains from patients with oral candidosis; IC, WO; strains from immunocompromised patients.

159

organization may have occurred in one strain in the colonized individual. They noted in this study that in three sets of isolates with 27A pattern differences, the ethidium bromide-strained patterns were similar.

5.10.2 Is an immunocompromised patient colonized by one, or more than one, strain of *Candida*?

In the most detailed study to date of a single systemic infection, species and strain variability was assessed in a bone marrow transplant over a two month period, using a combination of colony cloning, colony phenotype assessment, and fingerprinting with probe Ca3 and Ct13−8, a moderately repetitive sequence specific for *C. tropicalis* (Soll *et al.*, 1988). The results of this study were indeed unexpected (Figure 5.17 for temporal and spatial dynamics of colonization). During the course of the disease, two strains of *C. albicans* and three strains of *C. tropicalis* were distinguished by switching repertoire, fingerprints and sugar assimilation patterns. When first surveyed, a strain of *C. albicans* with a smooth colony phenotype was cloned from urine. Two days later, a strain of *C. tropicalis* with an irregular edge colony phenotype was cloned from the blood. One day later, a second *C. tropicalis* strain with a fuzzy phenotype was cloned from skin blisters. Two days later, both the irregular edge and fuzzy *C. tropicalis* strains were cloned from the blood. Amphotericin B cleared the blood of the fuzzy *C. tropicalis* strain, but not the irregular edge *C. tropicalis* strain. Subsequent treatment with 5-fluorocytosine removed the irregular edge strain. The patient then ceased oral clotrimazole treatment on her own volition; resulting immediately in colonization of the throat by a second smooth white *C. albicans* strain, a new *C. tropicalis* strain with a bumpy halo phenotype, and the irregular edge and fuzzy *C. tropicalis* strain originally colonizing the blood (Figure 5.17).

The extraordinary complexity of colonization observed in the preceeding study was not observed for 20 immunocompromised patients tested for relatedness with the probe 27A (Fox *et al.*, 1989). In this latter study, it was demonstrated that isolates from different body locations exhibited the same fingerprint. Unfortunately, the isolates in this study were not cloned from the sites of colonization. Therefore, initial culturing in these studies would result in enrichment of strains which grew fastest on the media employed, and would therefore not identify mixed colonization. Since the isolates were stocked over a seven year period, storage would further select for single strains. It therefore remains to be tested whether multiple strains are frequently responsible for systemic infections in immunocompromised hosts.

5.10.3 Are recurrent *Candida* infections due to the same or to different strains?

In the most detailed study to date, *Candida* strains, colony phenotype, and the levels of colonization were monitored through three successive episodes of vaginal candidosis in one patient using a combination of cloning from 14 body locations, fingerprinting with probe Ca3, and an analysis of switching (Soll *et al.*, 1989). At the time of the initial vaginal infections, three different strains of C. *albicans* were isolated, one from the oral cavity, one from under the breast and one from the vulva-vaginal, anal-rectal areas. Twenty-two days after completion of clotrimazole treatment, colonization was low to negligible in all body locations, but 87 days after the first infection, a second infection appeared. The same two strains were evident in the mouth and vagina, respectively. However, the colony phenotype of the vaginal strain had switched to small sized, irregular edge. Seven days after completion of butoconazole treatment, colonization in all body locations was negligible, but 32 days after the second infection, a third infection appeared. Again, the same two strains colonized the mouth and vagina, respectively. However, in the third infection, the colony phenotype had switched back to the medium sized, smooth white phenotype of the first infection. Therefore, the same strain caused three recurrent episodes of vaginal *candidosis* and underwent switching between episodes and drug therapies. Additional studies have resulted in the same fundamental observation that the same strain is usually responsible for recurrent infections.

In two recent studies of AIDS patients with oral-oesophageal candidosis using Ca3 fingerprinting, it has also been demonstrated that the same strains are responsible for recurrent infections after antifungal drug therapy (Schmid *et al.*, in preparation; Greenspan, personal communication). In one of these two studies involving nine AIDS patients in a self-help group in Leicester, there was an extraordinary level of genetic related-ness between strains demonstrated by similar S_{AB} values (Figure 5.18), suggesting that either these AIDS patients were infected by a common

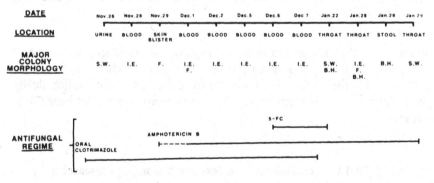

Figure 5.17 A diagram of the dates of isolation, major colony phenotypes on *Candida* agar, and antifungal drug regimens during the course of a systemic infection in a bone marrow transplant patient. The dates are presented only when *Candida* was detected. S.W., smooth white; I.E., irregular wrinkle; F., fuzzy; bumpy halo. The dashed line represents time between a low test dose and a high therapeutic dose of amphotericin B.

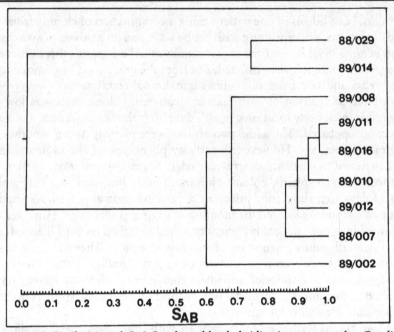

Figure 5.18 Similarities of Ca3 Southern blot hybridization patterns for *Candida* isolates from nine AIDS patients. The dendrogram was generated from the similarity values between the strains through the Dendron software package. DNA from the nine isolates were digested with the endonuclease EcoRI, fragments were separated by electrophoresis, and the blots were probed with the moderately repetitive sequence Ca3. Band intensities and positions were scored and S_{AB} values determined. An S_{AB} value of 0.96 or greater represents absolute identity, and a value of 0.0 absolute nonidentity. The average similarity value of *C. albicans* strains is 0.69. Strains 88/29 and 89/14 did not type as the species *C. albicans* according to Ca3 specificity.

strain which then proceeded to evolve rapidly, as was suggested in Scherer and Stevens' analysis of strain variability in patients (Scherer and Stevens, 1988), or that these individuals selected from their environment genetically related strains with presumably a common propensity for colonizing AIDS patients.

5.10.4 Can candidosis be a sexually transmitted disease?

In two studies using fingerprinting methods, isolates from the genitalia of sexual partners exhibited identical fingerprints (Stevens *et al.*, 1990; Soll *et al.*, unpublished observations). Therefore, there is little doubt that the organism can be transferred between individuals, but there is less proof that one partner can cause an infection in the other partner. In several studies, treating of the male sexual partner with an antifungal drug had no significant effect on the recurrence of vaginal candidosis (Buch and Christensen, 1982; Bisschop *et al.*, 1986; Calderon-Marquez, 1987), suggesting that although partners may share a common genital strain, the disease is due to the endogenous strain of the infected female. However, there is evidence suggesting that penile candidosis can indeed result from sexual intercourse with a female with vaginal candidosis (Thin *et al.*, 1977; Catterall *et al.*, 1966; Chakraborty and Datta, 1979).

5.11 DISCUSSION

Because *Candida* is a common commensal in the natural flora of the oral-oesophageal, gastrointestinal, anal-rectal, and genital areas of healthy humans, its status as an infectious disease has never received much attention. It has been assumed that the endogenous commensal is the source of the infecting strain. However, *Candida* can be transferred between individuals and can cause disease after transference, and is responsible for nosocomial infections (Weber and Rutala, 1988; Wey *et al.*, 1988).

In this review, we have considered those phenotypic characteristics of C. *albicans* which promote virulence, and the molecular tools now being used to analyse these traits. It has been noted that although molecular genetics has only begun to pay off in the cloning of genes involved in pathogenesis, molecular genetic techniques have recently been applied with success to etiological and epidemiological questions. With the assistance of computer programs to analyse DNA fingerprints and generate dendograms based on genetic relatedness, many of the most basic questions related to the origin of the infecting agent and strain specificity are now in the process of being answered.

In view of the capacity for rapid phenotypic and genotypic change in *Candida* and related species, rapid evolution of the organism may possibly lead to increased states of virulence. This may be especially true when continuous prophylactic treatment is applied to patients with suppressed immune systems, as in AIDS patients. Such strains may prove to be more infectious among subpopulations with shared physiological or immunological disorders. It is therefore inperative that we understand how virulence traits evolve in *Candida*, which genes are involved in virulence, and how high frequency genetic and phenotypic switching occur. It should also be evident from the preceding discussion that few answers will be obtained without a careful combination of biological and molecular genetic techniques.

ACKNOWLEDGEMENT

The research performed in the author's laboratory and reported in this review was funded by Public Health Service grant AI23922 from the National Institute of Health.

REFERENCES

Anderson, J.M. and Soll, D.R. (1986) Differences in actin localization during bud and hypha formation in the yeast *Candida albicans*. *J. Gen. Microbiol.*, **132**, 2035–47.

Anderson, J.M. and Soll, D.R. (1987) The unique phenotype of opaque cells in the 'white-opaque transition' in *Candida albicans*. *J. Bacteriol.*, **169**, 5579–88.

Anderson, J.M., Mihalik, R. and Soll, D.R. (1990) Ultrastructure and antigenicity of the unique cell wall 'pimple' of the *Candida* opaque phenotype. *J. Bacteriol.*, **172**, 224–35.

Anderson, J.M., Cundiff, L., Schnars, B. *et al.*, (1989) Hypha formation in the white-opaque transition of *Candida albicans*. *Infect. Immun.*, **57**, 458–67.

Anderson, M.L. and Odds, F.C. (1985) Adherence of *Candida albicans* to vaginal epithelia: significance of morphological form and effect of ketoconazole. *Mykosen.*, **28**, 531–40.

Banno, Y., Yamada, T. and Nozawa, Y. (1985) Secreted phospholipases of the dimorphic fungus, *Candida albicans*; separation of three enzymes and some biological properties. *Sabouraudia*, **23**, 47–54.

Barrett-Bee, K., Hayes, Y., Wilson, R.G. and Ryley, J.F. (1985) A comparison of phospholipase activity, cellular adherence and pathogenicity of yeasts. *J. Gen. Microbiol.*, **131**, 1217–21.

Bedell, G.W., Werth, A. and Soll, D.R. (1980) The regulation of nuclear migration and division during synchronous bud formation in released stationary phase cultures of the yeast *Candida albicans*. *Exptl. Cell Res.*, **127**, 103–13.

Berchev, K. and Izmirov, I. (1967) Isoenzymes of some oxido-reductases in the *Candida* genus as a basis of species identification after electrophoresis. *Experent.*, **23**, 961–2.

Bergen, M., Voss, E. and Soll, D.R. (1990) Switching at the cellular level in the white-opaque transition of *Candida albicans*. *J. Gen. Microbiol*, in press.

Bisschop, M., Merkus, J., Scheygrond, H. and Van Cutsem, J. (1986) Co-treatment of the male partner in vaginal candidosis: a double-blind randomized control study. *Br. J. Obstet. Gynecol.*, **93**, 79–81.

Bodey, G.P. (1984) Candidiasis in cancer patients. *Am. J. Med.*, **77(4D)**, 13–19.

Bodey, G.P. and Feinstein, V. (1985) *Candidiasis*. Raven Press, New York.

Brawner, D.L. and Cutler, J.E. (1986) Ultrastructural and biochemical studies of two dynamically expressed cell surface determinants on *Candida albicans*. *Infect. Immun.*, **51**, 337–60.

Brown-Thomsen, J. (1968) Variability in *Candida albicans* (Robin) Berkhout I. Studies on morphology and biochemical activity. *Hereditas*, **60**, 355–98.

Brummel, M. and Soll, D.R. (1982) The temporal regulation of protein synthesis during synchronous bud or mycelium formation in the dimorphic yeast *Candida albicans*. *Develop. Biol.*, **89**, 211–24.

Buch, A. and Christensen, E.S. (1982) Treatment of vaginal candidosis with natamycin and effect of treating the partner at the same time. *Acta Obstet. Gynecol. Scand.*, **61**, 393–6.

Budtz-Jorgensen, E. (1978) Clinical aspects of *Candida* infection in denture wearers. *J. Am. Dent. Assoc.*, **96**, 474–9.

Buffo, J., Herman, M. and Soll, D.R. (1984) A characterization of pH-regulated dimorphism in *Candida albicans*. *Mycopath.*, **85**, 21–30.

Burnie, J.P., Odds, F.C., Lee, W. *et al.* (1985) Outbreak of systemic *Candida albicans* in intensive care unit caused by cross infection. *Brit. Med. J.*, **290**, 746–8.

Byers, B. (1981) Cytology of the Yeast Life Cycle, in *The Molecular Biology of the Yeast Saccharomyces: Life Cycle and Inheritance* (eds J.N. Strathern, E.W. Jones and J.R. Broach). Cold Spring Harbor Laboratory, New York. pp. 59–96.

Byers, B. and Goetsch, L. (1976) A highly ordered ring of membrane-associated filaments in budding yeast. *J. Cell Biol.*, **69**, 717.

Byers, B. and Goetsch, L. (1976) Loss of the filamentous ring in cytokinesis-defective mutants of budding yeast. *J. Cell Biol.*, **70**, 35a.

Calderon-Marquez, J.J. (1987) Intraconazole in the treatment of vaginal candidosis and the effect of treatment of the sexual partner. *Rev. Infect. Dis.*, **9**, 143–5.

Calderone, R.A., Cihlar, R.L., Lee D.D. *et al.* (1985) Yeast adhesion in the pathogenesis of endocarditis due to *Candida albicans*: studies with adherence-negative mutants. *J. Infect. Dis.*, **152**, 710–15.

Cattarall, R.D. (1966) Urethritis and balanitis due to *Candida*, in *Symposium on Candida Infections* (eds H.I. Winner, and R. Hurley), pp. 113–18. Livingstone, Edinburgh.

Centeno, A., Davis, C.P., Cohen, M.S. and Warren, M.M. (1983) Modulation of *Candida albicans* attachment to human epithelial cells by bacteria and carbohydrates. *Infect. Immun.*, **39**, 1354–60.

Chakraborty, A.K. and Datta, A.K. (1979) Candidal balanoposthitis in male. *Indian J. Dermatol.*, **24**, 23–30.

Chattaway, F.W., Holmes, M.R. and Barlow, A.J.E. (1968) Cell wall composition of the mycelial and blastospore forms of *Candida albicans*. *J. Gen. Microbiol.*, **51**, 367–76.

Cole, G.T., Sesham, K.R., Pope, L.M. and Yancey, R.J. (1988) Morphological aspects of gastrointestinal tract invasion by *Candida albicans* in the infant mouse. *J. Med. Vet. Mycol.*, **26**, 173–86.

Collins-Lech, C., Kalbfleisch, J.H., Franson, T.R. and Sohnle, P.G. (1984) Inhibition by sugars of *Candida albicans* adherence to human buccal mucosal cells and corneocytes *in vitro*. *Infect. Immun.*, **46**, 831–4.

Critchley, I.A. and Douglas, L.J. (1985) Differential adhesion of pathogenic *Candida* species to epithelial and inert surfaces. *FEMS Microbiol. Letts.*, **28**, 199–203.

Cutler, J., Glee, P.M. and Horn, H. (1988) *Candida albicans* and *Candida stellatoidea*-specific DNA fragments. *J. Clin. Microbiol.*, **26**, 1720–4.

Donelson, J.E. (1989) *DNA Rearrangements and Antigenic Variation in African Trypanosomes*, in *Mobile DNA* (eds D.E. Berg and M.M. Howe). ASM Press, Washington, D.C., pp. 763–81.

Douglas, L.J. (1985) Adhesion of pathogenic *Candida* species to host surfaces. *Microbiol. Sci.*, **2**, 243–7.

Epstein, J.B. (1989) Oral and pharyngeal candidiasis: topical agents for management and prevention. *Postgrad. Med.*, 85, 257–69.

Finney, R., Langtimm, C.J. and Soll, D.R. (1985) The programs of protein synthesis accompanying the establishment of alternative phenotypes in *Candida albicans*. *Mycopath.*, 91, 3–15.

Fox, B.C., Mobley, H.L.T. and Wade, J.C. (1989) The use of a DNA probe for epidemiological studies of candidiasis in immunocompromised hosts. *J. Infect. Dis.*, 159, 488–93.

Frerich, W. and Gad, A. (1977) The frequency of *Candida* infections in pregnancy and their treatment with clotrimazole. *Curr. Med. Res. Opin.*, 4, 640–4.

Germaine, G.R. and Tellefson, L.M. (1981) Effect of pH and human saliva on protease production by *Candida albicans*. *Infect. Immun.*, 31, 323–6.

Ghannoum, M.A. and Soll, D.R. (1980) Variation of lipid and sterol contents in white and opaque phenotypes of *Candida albicans*. *J. Med. Vet. Mycol.*, in press.

Gow, N.A.R., Henderson, G. and Gooday, G.V. (1986) Cytological interrelationships between the cell cycle and duplication cycle of *Candida albicans*. *Microbios.*, 47, 97–105.

Greenspan, D. (1988) Oral manifestations of HIV infection, in *Perspectives on Oral Manifestations of AIDS*. PSG Publishing Company, Inc. Littleton, pp. 38–48.

Haarer, B.K. and Pringle, J.R. (1987) Immunofluorescence localization of the *Saccharomyces cerevisiae* CDC12 gene product to the vicinity of the 10-nm filaments in the mother-bud neck. *Mol. Cell Biol.*, 7, 3678–87.

Haarer, B.K., Ketcham, S.R., Ford, S.K. *et al.* (1990) The *Saccharomyces cerevisiae* CDC3, CDC10, CDC11 and CDC12 genes encode a family of related proteins. Submitted for publication.

Hasenclever, H.F. and Mitchell, W.O. (1961) Antigenic studies of *Candida* I. Observation of two antigenic groups in *Candida albicans*. *J. Bacteriol.*, 82, 570–3.

Hasenclever, H.F., Mitchell, W.O. and Loewe, J. (1961) Antigenic studies of *Candida* II. Antigenic relation of *Candida albicans* group A and group B to *Candida stellatoidea* and *Candida tropicalis*. *J. Bacteriol.*, 82, 574–7.

Hicks, J.B., Strathern, J.N. and Herskowitz, I. (1977) The cassette model of mating type interconversion, in *DNA Insertion Elements, Plasmids and Episomes* (eds A.J. Bukkaria, J.A. Shapiro and S.L. Althya). Cold Spring Harbor Lab, Cold Spring Harbor, New York, pp. 457–62.

Hopwood, V., Poulain, D., Fortier, B. *et al.* (1986) A monoclonal antibody to a cell wall component of *Candida albicans*. *Infect. Immun.*, 54, 222–7.

Howlett, J.A. and Squire, C.A. (1980) *Candida albicans* ultrastructure: colonization and invasion of oral epithelium. *Infect. Immun.*, 29, 252–60.

Hunter, P.R., Fraser, C.A.M. and MacKenzie, D.W.R. (1988) Morphotype markers of virulence in human candidal infections *J. Med. Microbiol.*, 28, 85–91.

Ireland, R. and Sarachek, A. (1968) A unique minute-rough colonial variant of *Candida albicans*. *Mycopathol. Mycol. Appl.*, 35, 346–60.

Isenberg, H.D., Tucci, V., Cintron, F. *et al.* (1989) Single-source outbreak of *Candida tropicalis* complicating coronary bypass surgery. *J. Clin. Microbiol.*, 27, 2426–8.

Jackson, J.L. and Spain, W.T. (1968) Comparative study of combined and sequential antiovulatory therapy on vaginal moniliasis. *Am. J. Obstet. Gynecol.*, **101**, 1134–5.

Johnston, R.D., Chick, E.W., Johnston, N.S. and Jarvis, M.A. (1967) Asymptomatic quantitative increase of *Candida albicans* in the oral cavity: predisposing conditions. *South. Med. J.*, **60**, 1244–7.

Kelley, R., Miller, S.M., Kurtz, M.B. and Kirsch, D.R. (1987) Directed mutagenesis in *Candida albicans*: one-step gene disruption to isolate ura3 mutants. *Molec. Cell Biol.*, **7**, 199–207.

Kennedy, M.J. (1987) Adhesion and association mechanisms of *Candida albicans*, in *Current Topics in Medical Mycology*, vol. 2 (Ed. M.R. McGinnis). Springer-Verlag, New York, pp. 73–169.

Kennedy, M.J., Rogers, A.L., Hasselman, L.R. *et al.* (1988) Variation in adhesion and cell surface hydrophobicity in *Candida albicans* white and opaque phenotypes. *Mycopath.*, **102**, 149–56.

Kennedy, M.J. and Volz, P.A. (1985) Ecology of *Candida albicans* gut colonization: inhibition of *Candida* adhesion, colonization, and dissemination from the gastrointestinal tract by bacterial antagonism. *Infect. Immun.*, **49**, 654–68.

Kimball, L.H. and Pearsall, N.N. (1980) Relationship between germination of *Candida albicans* and increased adherence to human buccal epithelial cells. *Infect. Immun.*, **28**, 464–8.

Kirsch, D.R., Kelly, R. and Kurtz, M.B. (1990) The Genetics of *Candida*. CRC Press, Boca Raton, Florida.

Kolotila, M.P. and Diamond, R.D. (1990) Effects of neutrophils and in vitro oxidants on survival and phenotypic switching of *Candida albicans* WO-1. *Infect. Immun.*, **58**, 1174–9.

Klotz, S.A., Drutz, D.J. and Zajic, F.E. (1985) Factors governing adherence of *Candida* species to plastic surfaces. *Infect. Immun.*, **50**, 97–101.

Krogh, P., Holmstrup, P., Thorn, J.J. *et al.* (1987) Yeast species and biotypes associated with oral leukoplakia. *Oral. Surg.*, **63**, 48–54.

Kurtz, M.B., Cortelyon, M.W. and Kirsch, D.R. (1986) Integrative transformation of *Candida albicans*, using a cloned *Candida* ADE2. *Molec. Cell Biol.*, **6**, 142–9.

Kurtz, M.B., Kirsch, D.R. and Kelly, R. (1988) The molecular genetics of *Candida albicans*. *Microbiol. Sci.*, **5**, 58–63.

Kwon-Chung, K.J., Lehman, D., Good, C. and Magee, P.T. (1985) Genetic evidence for role of extracellular proteinase in virulence of *Candida albicans*. *Infect. Immun.*, **49**, 571–5.

Land, G.A., McDonald, W.C., Stjernholm, R.L. and Friedman, L. (1975) Factors affecting filamentation in *Candida albicans*: Changes in respiratory activity of *Candida albicans* during filamentation. *Infect. Immun.*, **12**, 119–27.

Lane, H.C. and Fauci, A.S. (1987) Infection complications of AIDS, in *AIDS: Modern Concepts and Therapeutic Challenges* (Ed. S. Brodes), Marcel Dekker, Inc., New York, pp. 185–203.

Law, E.J., Kim, O.J., Stieritz, D.P. and MacMillan, B.G. (1972) Experience with systemic candidiasis in the burned patient. *J. Trauma.*, **12**, 543–52.

Lee, K.L., Buckley, H.R. and Campbell, C.C. (1975) An amino acid rich liquid synthetic medium for the development of mycelial and yeast forms of *Candida albicans*. *Sabouraud.*, **13**, 148–53.

Lee, J.C. and King, R.D. (1983) Adherence mechanisms of *Candida albicans*, in *Microbiology 83* (Ed. D. Schlessenger). ASM, Washington DC 269–72.

Lott, T.J., Page, L.S., Boiron, P. *et al.* (1989) Nucleotide sequence of the *Candida albicans* aspartyl proteinase gene. *Nuc. Acid. Res.*, 17, 1779–80.

MacDonald, F. and Odds, F.C. (1980) Inducible proteinase of *Candida albicans* in diagnostic serology and in the pathogenesis of systemic candidiosis. *J. Med. Microbiol.*, 13, 423–35.

MacDonald, F. and Odds, F.C. (1983) Virulence for mice of a proteinase-secreting strain of *Candida albicans* and a proteinase-deficient mutant. *J. Gen. Microbiol.*, 129, 431–8.

Mackinnon, J.E. (1940) Dissociation in *Candida albicans*. *J. Infect. Dis.*, 66, 59–77.

Magee, B.B. and Magee, P.T. (1987) Electrophoretic karyotypes and chromosome numbers in *Candida* species. *J. Gen. Microbiol.*, 133, 425–30.

Magee, B.B., D'Souza, T.M. and Magee, P.T. (1987) Strain and species identification by restriction fragment length polymorphisms in the ribosomal DNA repeat of *Candida* species. *J. Bacteriol.*, 169, 1639–43.

Manning, M. and Mitchell, T.G. (1980) Analysis of cytoplasmic antigens of the yeast and mycelial phases of *Candida albicans* by two-dimensional electrophoresis. *Infect. Immun.*, 30, 484–95.

Mardon, D.W., Balish, E. and Phillips, A.W. (1969) Control of dimorphism in a biochemical variant of *Candida albicans*. *J. Bacteriol.*, 100, 701–7.

Mardon, D.W., Hurst, S.K. and Balish, E. (1971) Germ tube production by *Candida albicans* in minimum liquid culture medium. *Can. J. Microbiol.*, 17, 851–6.

McCourtie, J. and Douglas, L.J. (1984) Relationship between cell surface composition, adherence and virulence of *Candida albicans*. *Infect. Immun.*, 45, 6–12.

McCreight, M.C. and Warnock, D.W. (1982) Enhanced differentiation of isolates of *Candida albicans* using a modified resistogram method. *Mykosem.*, 25, 589–98.

Mehdy, M.C., Ratner, D. and Firtel, R. (1983) Induction and modulation of cell-type-specific gene expression in *Dictyostelium*. *Cell*, 32, 763–71.

Merz, W.G., Connelly, C. and Hieter, P. (1988) Variation of electrophoretic karyotypes among clinical isolates of *Candida albicans*. *J. Clin. Microbiol.*, 26, 842–85.

Mitchell, L. and Soll, D.R. (1979) Commitment to germ tube or bud formation during release from stationary phase in *Candida albicans*. *Exptl. Cell Res.*, 120, 167–79.

Mitchell, L.H. and Soll, D.R. (1979) Temporal and spatial differences in septation during synchronous mycelium and bud formation by *Candida albicans*. *Exptl. Mycol.*, 3, 298–309.

Morrow, B., Anderson, J., Wilson, E. and Soll, D.R. (1989) Bidirectional stimulation of the white-opaque transition of *Candida albicans* by ultraviolet irradiation. *J. Gen. Microbiol.*, 135, 1201–8.

Nickerson, W.J. and Van Rij, N.J. (1949) The effect of sulphydryl compounds, penicillin, and cobalt on the cell division mechanism of yeast. *Biochem. Biophys. Acta*, 3, 461–75.

Odds, F.C. and Abbott, A.B. (1980) A simple system for the presumptive identification of *Candida albicans* and differentiation of strains within the species. *Sabouraud.*, **18**, 301–18.

Odds, F.C. and Abbott, A.B. (1983) Modification and extension of tests for differentiation of *Candida* species and strains. *Sabouraud.*, **21**, 79–81.

Odds, F.C. (1988) *Candida and Candidosis: a Review and Bibliography*, Baillière Tindale, London.

Odds, F., Schmid, J. and Soll, D.R. (1990) Epidemiology of *Candida* infections in AIDS, in *Mycoses in AIDS Patients* (Ed. H.V. Bossche). Plenum Press, New York. In press.

Olaiya, A.F. and Sogin, S.J. (1979) Ploidy determination of *Candida albicans*. *J. Bacteriol.*, **140**, 1043–9.

Olivo, P.D., McManus, E.J., Riggsby, W.S. and Jones, J.M. (1987) Mitochondrial DNA polymorphism in *Candida albicans*. *J. Infect. Dis.*, **156**, 214–15.

Palmer, R.E., Koval, M. and Koshland, D. (1989) The dynamics of chromosome movement in the budding yeast *Saccharomyces cerevisiae*. *J. Cell Biol.*, **109**, 3355–66.

Persi, M.A., Burnham, J.C. and Duhring, J.L. (1985) Effects of carbon dioxide and pH on adhesion of *Candida albicans* to vaginal epithelial cells. *Infect. Immun.*, **50**, 82–90.

Polonelli, L., Archibusacci, C., Sestito, M. and Morace, G. (1983) Killer system: a simple method for differentiating *Candida albicans* strains. *J. Clin. Microbiol.*, **17**, 774–80.

Pugh, D. and Cawson, R.A. (1977) The cytochemical localization of phospholipase in *Candida albicans* infecting the chick chorio-allantoic membrane. *Cell Mol. Biol.*, **22**, 125–32.

Ray, T.L., Digre, K.B. and Payne, C.D. (1984) Adherence of *Candida* species to human epidermal corneocytes and buccal mucosal cells: correlation with cutaneous pathogenicity. *J. Invest. Dermatol.*, **83**, 37–41.

Ray, T.L. and Payne, C.D. (1988) Scanning electron microscopy of epidermal adherence and cavitation in murine candidiasis: a role for *Candida* acid proteinase. *Infect. Immun.*, **56**, 1942–9.

Ray, T.L., Payne, C.D. and Soll, D.R. (1989) Linkage of acid proteinase expression to phenotype switching in *Candida albicans* strain-WO-1. *Clinical Res.*, **37**, 895.

Reingold, A.L., Lu X.D., Plikaytis, B.D. and Ajello, L. (1986) Systemic mycoses in the United States, 1980–1982. *J. Med. Vet. Mycol.*, **24**, 433–6.

Rhoads, J.G., Wright, C., Redfield, R.R. and Burke, D.S. (1987) Chronic vaginal candidiasis in women with human immunodeficiency virus infection. *JAMA*, **257**, 3105–6.

Riggsby, W.S., Torres-Bauza, L.J., Wills, J.W. and Townes, T.M. (1982) DNA content, kinetic complexity, and the ploidy question in *Candida albicans*. *Mol. Cell. Biol.*, **2**, 853–62.

Rikkerink, E.H.A., Magee, B.B. and Magee, P.T. (1988) Opaque-white phenotype transition: a programmed morphological transition in *Candida albicans*. *J. Bacteriol.*, **170**, 895–9.

Rotrosen, D., Calderone, R. and Edwards, J.E. (1986) Adherence of *Candida* species to host tissues and plastic surfaces. *Rev. Infect. Dis.*, **8**, 73–85.

Ruchel, R. (1981) On the renin-like activity of *Candida* proteinases and activation of blood coagulation *in vitro*. *Biochem. Biophys. Acta*, **659**, 99–113.

Saltarelli, C.G. (1973) Growth stimulation and inhibition of *Candida albicans* by metabolic by-products. *Mycopathol. Mycol. Appl.*, 51, 53–63.

Scherer, S. and Stevens, D.A. (1987) Application of DNA typing methods to the epidemiology and taxonomy of *Candida* species. *J. Clin. Microbiol.*, 25, 675–9.

Scherer, S. and Stevens, D.A. (1988) A *Candida albicans* dispersed, repeated gene family and its epidemiological applications. *Proc. Natl Acad. Sci. (USA)*, 85, 1452–6.

Schidt, M. and Pindborg, J.J. (1987), AIDS and the oral cavity. *Int. J. Oral Maxill. Surg.*, 16, 1–14.

Schmid, J., Voss, E. and Soll, D.R. (1990) Computer-assisted methods for assessing strain relatedness in *C. albicans* by fingerprinting with the moderately repetitive sequence Ca3. *J. Clin. Microbiol.*, in press.

Schneider, C., King, R.M. and Phillipson, L. (1988), Genes specifically expressed at growth arrest of mammalian cells. *Cell*, 54, 787–93.

Silverman, M. and Simon, M. (1983) Phase variation and related systems, in *Mobile Genetic Elements* (Ed. J. Shapiro). Academic Press, New York, pp. 537–57.

Simonetti, N., Stripolli, V. and Cassone, A. (1974), Yeast-mycelial conversion induced by N-acetyl-D-glucosamine in *Candida albicans*. *Nature (London)*, 250, 344–6.

Skerl, K.G., Calderone, R.A., Segal, E. *et al.* (1984) *In vitro* binding of *Candida albicans* yeast cells to human fibronectin. *Can. J. Microbiol.*, 30, 221–7.

Slutsky, B., Buffo, J. and Soll, D.R. (1985) High frequency switching of colony morphology in *Candida albicans*. *Science*, 230, 666–9.

Slutsky, B., Staebell, M., Anderson, J. *et al.* (1987) White-opaque transition: a second high-frequency switching system in *Candida albicans*. *J. Bacteriol.*, 169, 189–97.

Smail, E.H. and Jones, J.M. (1984) Demonstration and solubilization of antigens expressed primarily on the surfaces of *Candida albicans* germ tubes. *Infect. Immun.*, 45, 74–81.

Snell, R.G., Herman, I.F., Wilkins, R.J. and Conner, B.E. (1987) Chromosomal variations in *Candida albicans*. *Nuc. Acid Res.*, 15, 3625.

Sobel, J.D., Muller, G. and Buckley, H.R. (1984) Critical role of germ tube formation in the pathogenesis of candidal vaginitis. *Infect. Immun.*, 44, 576–80.

Soll, D.R. (1985a) *Candida albicans*, in *Fungal Dimorphism: With Emphasis on Fungi Pathogenic to Humans* (Ed. P. Staniszlo). Plenum Press, New York, pp. 167–95.

Soll, D.R. (1985b) The role of zinc in *Candida* dimorphism. *Current Top. Med. Mycol.*, 1, 258–85.

Soll, D.R. (1986) The regulation of cellular differentiation in the dimorphic yeast *Candida albicans*. *Bioessays*, 5, 5–11.

Soll, D.R. (1989) High Frequency Switching in Candida, in *Mobile DNA* (eds M.M. Howe and D.E. Berg). ASM Press, Washington, D.C., pp. 791–7.

Soll, D.R. (1990) Dimorphism and high frequency switching in *Candida albicans*, in *The Genetics of Candida* (eds D.R. Kirsch, R. Kelley and M.B. Kurtz). CRC Press Inc., Boca Raton, Florida, pp. 147–76.

Soll, D.R. (1990) In *Fungal Spores and Disease Initiation in Plants and Animals* (ed. G.T. Cole). Plenum Publishing Corp., New York, in press.

Soll, D.R., Anderson, J.M. and Bergen, M. (1990) The developmental biology of the white-opaque transition in *Candida albicans*, in *Candida albicans: Cellular and Molecular Biology* (Ed. R. Prasad). Springer-Verlag, in press.

Soll, D.R. and Kraft, B. (1988) A comparison of high frequency switching in *Candida albicans* and *Dictyostelium discoideum. Develop. Genet.*, 9, 615–28.

Soll, D.R. and Mitchell, L.H. (1983) Filament ring formation in the dimorphic yeast *Candida albicans. J. Cell Biol.*, 96, 486–93.

Soll, D.R., Bedell, G., Thiel, J. and Brummel, M. (1981), The dependency of nuclear division on volume in the dimorphic yeast *Candida albicans. Exptl. Cell Res.*, 133, 55–62.

Soll, D.R., Galansk, R., Isley, S. *et al.* (1989) Switching of *Candida albicans* during successive episodes of recurrent vaginitis. *J. Clin. Microbiol.*, 27, 681–90.

Soll, D.R., Herman, M.A. and Staebell, M.A. (1985) The involvement of cell wall expansion in the two modes of mycelium formation of *Candida albicans. J. Gen. Microbiol.*, 131, 2367–75.

Soll, D.R., Langtimm, C.J., McDowell, J. *et al.* (1987a) High frequency switching in *Candida* strains isolated from vaginitis patients. *J. Clin. Microbiol.*, 25, 1611–22.

Soll, D.R., Slutsky, B., MacKenzie, S. *et al.* (1987b) In *Oral Mucosa Diseases: Biology, Etiology and Therapy* (eds. I. MacKenzie, C. Squier and E. Dabelsteen). Laegerforeningens Folarg, Copenhagen, pp. 52–9.

Soll, D.R., Staebell, M., Langtimm, C. *et al.* (1988) Multiple *Candida* strains in the course of a single systemic infection. *J. Clin. Microbiol.*, 26, 1448–59.

Soll, D.R., Stasi, M. and Bedell, G. (1978) The regulation of nuclear migration and division during pseudomycelium outgrowth in the dimorphic yeast *Candida albicans. Exptl. Cell Res.*, 116, 207–15.

Sonck, C.E. and Somersalo, O. (1963) The yeast flora of the anogenital region in diabetic girls. *Arch. Derm.*, 88, 846–52.

Staebell, M. and Soll, D.R. (1985) Temporal and spatial differences in cell wall expansion during bud and mycelium formation in *Candida albicans. J. Gen. Microbiol.*, 131, 1467–80.

Staib, F. (1965) Serum-proteins as nitrogen soruce for yeast-like fungi. *Sabouraud*, 4, 187–93.

Stevens, D.A., Odds, F.C. and Scherer, S. (1990) Application of DNA typing methods to *Candida albicans* epidemiology and correlations with phenotype. *Rev. Infect. Dis.*, 12, 258–66.

Swanson, J. and Kooney, J.M. (1989) Mechanisms for variation of pili and outer membrane protein II in *Neisseria gonorrhea*, in *Mobile DNA* (eds D.E. Berg and M.M. Howe), ASM Press, Washington, D.C., pp. 743–61.

Swawathowski, M. and Hamilton-Miller, J.M. (1975) Anaerobic growth and sensitivity of *Candida albicans. Microbios. Lett.*, 5, 61–6.

Syrjanen, S., Valle, S.-L., Antonen, J. *et al.* (1988) Oral candidal infection as a sign of HIV infection in homosexual men. *Oral Surg.*, 65, 36–40.

Syverson, R.E., Buckley, H.R. and Campbell, C.C. (1975) Cytoplasmic antigens unique to the mycelial or yeast phase of *Candida albicans. Infect. Immun.*, 12, 1184–8.

Taschdjian, L.L., Burchall, J.J. and Kozian, P.J. (1960) Rapid identification of

Candida albicans by filamentation on screen and screen substitutes. *Am. J. Dis. Child.*, **99**, 212–15.

Thin, R.N., Leighton, M. and Dixon, M.J. (1977) How often is genital yeast infection sexually transmitted? *Brit. Med. J.*, **2**, 93–4.

Van Uden, N. and Buckley, H. (1970) *Candida Berkhout*, in *The Yeasts* (Ed. J. Lodder). North Holland, Amsterdam, pp. 893–1087.

Weber, D.J. and Rutala, W.A. (1988) Epidemiology of nosocomial fungal infections. *Curr. Top. Med. Mycol.*, **2**, 305–37.

Wey, F.B., Mori, M., Pfaller, M.A. *et al.* (1988) Hospital acquired candidemia: attributable mortality and excess length of stay. *Arch. Intern. Med.*, **148**, 2642–5.

Whelan, W.L. (1987) The qualities of medically important fungi. *CRC Crit. Rev. Microbiol.*, **21**, 99–170.

Whelan, W.L. and Soll, D.R. (1982) Mitotic recombination in *Candida albicans*: recessive lethal alleles linked to a gene required for methionine biosynthesis. *Mol. Gen. Genet.*, **187**, 477–82.

Whelan, W.L., Partridge, R.M. and Magee, P.T. (1980) Heterozygosity and segregation in *Candida albicans. Mol. Gen. Genet.*, **180**, 108–13.

Williams, R.J., Chandler, J.G. and Orloff, M.L. (1971) *Candida* septicaemia. *Arch. Surg.*, **103**, 8–11.

Wills, J.W., Lasker, B.A., Sirotkin, K. and Riggsby, W.S. (1984) Repetitive DNA of *Candida albicans*: nuclear and mitochondrial components. *J. Bacteriol.*, **157**, 918–24.

Winner, H.I. and Hurley, R. (1964) *Candida albicans*. Churchill, London.

Witkin, S.S., Hirsch, J. and Ledger, W.J. (1986) A macrophage defect in women with recurrent *Candida* vaginitis and its reversal by prostaglandin inhibitors. *Am. J. Obstet. Gynecol.*, **155**, 790–5.

Young, R.A. and Davis, R.W. (1983) Efficient isolation of genes by using antibody probes. *Proc. Natl Acad. Sci. (USA)*, **80**, 1194–8.

Molecular analysis of *Trichomonas vaginalis* surface protein repertoires

JOHN F. ALDERETE, ROSSANA ARROYO, DON C. DAILEY,
JEAN ENGBRING, MOHAMMAD A. KHOSHNAN, MICHAEL W.
LEHKER and JIM MCKAY

I don't mind a parasite, but
I do object to a cut-rate one.
Humphrey Bogart,
Casablanca

6.1 INTRODUCTION

Important reasons necessitate interest and support for basic research on the sexually transmitted parasite, *Trichomonas vaginalis* (Table 6.1). Surprisingly, this protozoan remains one of the most poorly investigated infectious agents yet it is the most prominent parasite causing an illness in developed countries. In all world societies trichomoniasis causes an economic and emotional burden equal to other devastating pathogens of microbial etiology.

The parasite is a flagellated protozoan responsible for a world-wide sexually transmitted disease (Ackers, 1982; Honigberg, 1978; Krieger, 1981; Müller, 1983; Rein and Chapel, 1975). Yearly estimates of numbers of patients with trichomoniasis in the United States alone range from four million to as high as ten million. Fifty percent of all patients will go

Molecular and Cell Biology of Sexually Transmitted Diseases
Edited by D. Wright and L. Archard
Published in 1992 by Chapman and Hall, London ISBN 0 412 36510 3

undiagnosed using the standard procedure of visualization of wet-mount preparations (Spence *et al.*, 1980). Alternative diagnostic methods, such as culturing of the parasite from vaginal swabs, are expensive, time consuming and not readily available.

Trichomoniasis is a non-self-limiting, gender-discriminating disease of women. Infection of men with *T. vaginalis* will mostly be asymptomatic and self-limiting. A major problem with this disease is the broad spectrum of symptomatology, ranging from relatively asymptomatic carriers who are reservoirs for the parasite to patients with a foul-smelling bloody discharge, severe inflammation and discomfort. For reasons unknown relatively asymptomatic women can become highly symptomatic and host or parasite factors which help explain the susceptibility or resistance to *T. vaginalis* infection and/or trichomoniasis remain undefined. The differentiation among trichomonal isolates, based on pathogenicity levels inherent to individual isolates, has not been satisfactorily established (Krieger *et al.*, 1990), even using rodent animal models (Honigberg *et al.*, 1966) which do not mirror the human infection. This absence of an animal model remains a major impediment toward studying the extent and nature of sequelae of *T. vaginalis* infection. Recent epidemiologic evidence points toward a possible predisposition of certain women with trichomoniasis for HIV infection (Laga *et al.*, 1989; 1990). Finally, although nitroimidazoles used to treat trichomoniasis in humans are extremely efficacious, the drugs have toxic side-effects and drug resistance by trichomonal isolates has been documented (Müller *et al.*, 1980).

Trichomonas vaginalis is a highly evolved parasite capable of surviving in the adverse host environment. To colonize and parasitize the vaginal epithelial surface of susceptible hosts, this pathogen must overcome the barriers of fluid flow and the mucus bathing the target cells. Furthermore, the organism must evade the immune and nonimmune microbicidal molecules in vaginal secretions, which otherwise may eliminate the organism. Finally,

Table 6.1 Summary of important reasons for interest in and support of basic research on *T. vaginalis*

One of the most common, clinically recognized sexually transmitted disease agents
Tremendously varied symptomatology and diagnosis is suboptimal
Adverse side-effects of drug and emergence of drug refractory isolates
Absence of animal model and extent of sequelae unknown
Mental health of parasitized women
Economic burden imposed on all societies through drug costs and medical expenses
Predisposition to HIV infection
Adverse pregnancy outcome

---

the parasites must be capable of growth and multiplication within the host environment, which is nutrient-limiting, prohibiting optimal metabolic activity of the organism.

This chapter will summarize research which elucidates some of the special features of the cell biology of the *T. vaginalis*-host interrelationship that allows for parasitism and survival in the human host. The molecular analysis of the parasite surface and the delineation of the roles of surface protein repertoires of *T. vaginalis* in the overall biology and virulence of this sexually transmitted parasite will be emphasized, since the parasite surface molecules immediately and directly interact with the host environment. This includes our work on the phenotypic variation systems of four trichomonad surface protein repertoires. Data are presented to begin to illustrate how the environment regulates biological properties of this parasite, such as protein-antigen composition and regulation of cytadherence. The role of the vaginal micro-environment on virulence of *T. vaginalis* and the mechanism(s) by which this organism is able to survive and multiply in the nutrient-limiting mucosal surface of the vagina should receive enhanced attention.

6.2 ANTIGENIC DIVERSITY AND THE IMMUNOGEN REPERTOIRES

6.2.1 Protein/epitope phenotypic variation

Initially, to understand the extensively reported antigenic heterogeneity among *T. vaginalis* isolates (Kott and Adler, 1961; Lanceley, 1958; Krieger *et al.*, 1985b; Su-Lin and Honigberg, 1983; Teras, 1966) we performed one- and two-dimensional gel electrophoresis for the analysis of total proteins of intrinsically labelled trichomonads (Alderete, 1983a; Alderete *et al.*, 1986). Surprisingly, data revealed a conservation rather than diversity in total protein patterns among the isolates.

Immunochemical surface characterization of a common laboratory isolate of *T. vaginalis* was performed using a combination of assays with pooled sera from female trichomoniasis patients and experimental animals (Alderete *et al.*, 1985). Data indicated an antibody response during natural and experimental infection that was directed toward a group of high molecular weight surface proteins. With the pooled patient sera, which contained antibody to numerous high molecular weight surface proteins (Alderete *et al.*, 1985), it was possible to evaluate several other isolates using a whole cell-radioimmunoprecipitation assay (Alderete, 1983b; 1987; Alderete *et al.*, 1985). In this assay, antibody from patients' sera was allowed first to bind to live, iodinated organisms followed by solubilization and precipitation of immune complexes. We found that isolates could be divided into two categories. In one case some isolates possessed a repertoire of antibody-binding surface proteins readily detected on autoradiograms. Other isolates lacked the antibody-binding immunogens, appearing as if they had naked surfaces.

A subsequent and intriguing finding was that the repertoire of immunogens was synthesized by all isolates (Alderete *et al.*, 1985); however, only trichomonads of some isolates were able to express the immunogens on the parasite surface (Alderete, 1983b; Alderete *et al.*, 1985). This clarified the paradoxical observations concerning the antigenic similarities (Alderete, 1983a) with the highly divergent surface immunogen profiles (Alderete *et al.*, 1985). Recently, cDNAs encoding the immunogens described below were shown to hybridize with genomic DNA of all isolates tested, strongly suggesting that all isolates have the potential to express the same protein antigens.

Equally surprising, indirect immunofluorescence using the pooled patient sera demonstrated the heterogeneous nature of parental trichomonal populations. Those isolates, which gave strong reactions by the whole cell-radioimmunoprecipitation assay mentioned above, comprised subpopulations of parasites in which some were fluorescent and some were non-fluorescent with the patient sera (Alderete *et al.*, 1985; Alderete, 1987a). This earlier work also revealed that changing proportions of fluorescent and non-fluorescent parasites occurred during *in vitro* cultivation and this now required clarification.

The generation of monoclonal antibodies (MAbs) reactive to *T. vaginalis* surfaces was fortuitous, since one MAb (C20A3) gave cytofluorometric patterns among the various trichomonal isolates identical to those detected with the pooled sera from patients (Alderete *et al.*, 1985; Alderete *et al.*, 1986). This MAb was a key reagent for confirming the property of protein phenotypic variation based on the surface expression of immunogens (Alderete *et al.*, 1986b), as well as the existence of two types of parental populations of *T. vaginalis* parasitizing humans (Table 6.2) (Alderete *et al.*, 1986b; 1986; 1987). Type I isolates consist of trichomonads that synthesize but do not undergo phenotypic variation for the surface expression of the immunogens. At the present time, approximately 40% of women are infected with Type I isolates. Type II isolates comprise trichomonads capable of phenotypic variation. Only one-third of women with Type II isolates harbour both fluorescent and non-fluorescent trichomonads in the vagina but the numbers of fluorescent trichomonads are small in comparison with MAb non-fluorescent parasites (Alderete *et al.*, 1987). Thus the *in vivo* environment favours or selects for *T. vaginalis* organisms lacking the expression of the surface immunogen repertoire although on *in vitro* cultivation the parasites readily revert to the opposite phenotype. This property was also in part experimentally demonstrated by showing the isolation of only non-fluorescent trichomonads after subcutaneous inoculation into mice hindquarters with only MAb C20A3 fluorescent parasites

Table 6.2 Differentiation of T. vaginalis isolates using monoclonal antibody (from Alderete et al., 1987b)

Reaction designation[a]	Fluorescence phenotype with MAb[b]	
	C20A3[c]	DM126[c]
Type I	−[d]	−
Type I	−	−
Type II	+/−	−
Type II	+/−	+/−

[a] Reaction designations were arbitrary and based on the fluorescence reactions with only the monoclonal antibody C20A3. Type I and Type II populations of parasites are as described in the text, and the MAbs have been previously reported (Alderete et al., 1987)
[b] Indirect immunofluorescence was performed on live organisms as described before (Alderete et al., 1986b)
[c] C20A3 and DM126 are monoclonal antibodies that recognize trichomonad proteins with relative molecular weights of 270 000 and 230 000, respectively
[d] The plus symbol refers to surface expression and antibody binding to live organisms. The minus sign refers to the absence of surface expression of the immunogen recognized by the C20A3 MAb and inaccessibility of epitope binding by the DM126 MAb. Only the C20A3 MAb negative phenotype parasites are capable of host cell parasitism and cytotoxicity, as reported (Alderete et al., 1986b)

(Alderete, 1987). Nonetheless, that 90% of women make antibody to the C20A3-reactive immunogen (Alderete et al., 1987) showed the molecule was synthesized regardless of the type or subpopulation designations of the infecting T. vaginalis isolate.

Finally, there was the delineation of three subpopulations of parasites among the Type II isolates (Alderete et al., 1986b). In addition to the non-fluorescent organisms some, but not all, fluorescent trichomonads were killed in a complement-independent fashion after exposure to the C20A3 MAb (Alderete, 1986). The lysis of these parasites expressing the C20A3 immunogen might indicate an important biofunctionality for the molecule; however, T. vaginalis without surface immunogen are equally capable of in vitro and in vivo growth and multiplication.

The complexity of the surface immunogens with regard to antibody accessibility was further reinforced in these earlier studies by the identification of another high molecular weight protein by a MAb called DM126 (Table 6.2). The protein was found on the surface of parasites of all isolates (Alderete et al., 1987), but, flow cytofluorometry with DM126 again indicated the existence of homogenous, non-fluorescent and heterogeneous populations, as seen for the MAb C20A3. The accessibility of the immunogen to antibody binding was highly variable despite its presence on the trichomonal surface. This suggests that this protein was undergoing conformational changes, possibly relating to the immediate micro-environment of the parasite. This property was termed epitope phenotypic variation.

More recently, the availability of monospecific antiserum to the purified immunogen has revealed the predominance of this molecule throughout trichomonal surfaces. Of particular interest, however, was the refractoriness of the parasites to complement-mediated killing in the presence of this monospecific antiserum. Although there is no precise explanation for this property, it illustrates the overall resistance of the protozoan to killing by antibody against common, stable immunogens which might otherwise be targets for vaccine development strategies. These immune evasion capabilities certainly will allow the parasite to survive the host antibody responses.

6.2.2 Phenotype specific virulence relationships

The differentiation of individual trichomonads on the basis of C20A3 reactivity (Table 6.2) and the ability to perform fluorescence-activated cell sorting of Type II isolates allowed for the first time the evaluation of a precise phenotype and a virulence property. Under controlled experimental conditions only non-fluorescent parasites produced cytotoxicity of HeLa cells (Alderete *et al.*, 1986b; Alderete, 1987) when compared with the positive phenotype counterparts. Because it was shown previously that host cell cytopathogenicity was contact-dependent under similar experimental conditions (Alderete and Pearlman, 1984; Alderete and Garza, 1985), this observation prompted investigation into the identity of possible adhesin proteins on the parasites mediating cytoadherence (Alderete and Garza, 1988), and as summarized in section 6.5. The data indicate an inverse relationship between surface expression of immunogens and trichomonad adhesins required for host cell killing. This led to the concept of alternating expression of the immunogens and adhesin proteins on the *T. vaginalis* surface (Alderete, 1988). The identification of the adhesins by using HeLa and vaginal epithelial cells as experimental host targets confirmed the concept (Alderete and Garza, 1988b; Alderete *et al.*, 1988).

6.2.3 The proteinase immunogens are also differentially expressed

The involvement of cysteine proteinases of *T. vaginalis* during initial host recognition and binding events (Arroyo and Alderete, 1989, as summarized in section 6.5.2 below) led to our analysis of the parasite proteinases by using two-dimensional-substrate SDS-PAGE in which gelatin was copolymerized with acrylamide for the second dimension (Neale and Alderete, 1990). Our results confirmed and extended those of others (Coombs and North, 1983; Lockwood *et al.*, 1987) on the complexity of the proteinase composition of pathogenic human trichomonads. Interestingly, a subset of the trichomonad proteinases are released into culture supernatants during *in vitro* growth

(Lockwood *et al.*, 1988) and, more importantly, these and other cysteine proteinases were observed in vaginal wash material from women with trichomoniasis (Alderete *et al.*, 1991b). Data from immunoprecipitation experiments using antisera from immunized mice and rabbits with trichomonal extracts clearly demonstrated the highly immunogenic nature of the trichomonal proteinases (Neale and Alderete, 1990). The results are being extended to the use of sera from patients with trichomoniasis (Alderete *et al.*, 1991a).

There are also quantitative and qualitative differences between proteinase patterns among isolates. This is further complicated by the ability of the trichomonads differentially to express the proteinases *in vivo* and during *in vitro* cultivation (Figure 6.1). Changes in the expression of proteinase activities were not coordinated with the immunogens described above (Table 6.2). These data, with the ability of the parasite to respond to environmental signals both *in vitro* and *in vivo*, underscore the complexity of the host–parasite interrelationship. These proteinases have not been previously characterized as major immunogens or virulence determinants and represent an important consideration for future research with this sexually transmitted parasite.

6.2.4 Molecular biology of a phenotypic variation immunogen

Recombinant DNA techniques were used by us for additional analysis of the immunogens that undergo phenotypic variation. We hope that this will tell us about the processes that govern this important biological trait. A cDNA library of *T. vaginalis* constructed in bacteriophage λgt11 was screened with MAb C20A3, and a 396 bp cDNA clone encoding the epitope of P270 was isolated (Dailey and Alderete, 1991). The cDNA encodes for an approximately 15kDa protein which has a half-life of only 30 minutes in *E. coli*. Sequence analysis of the cloned cDNA fragment revealed the following features for epitope and native high molecular weight protein. The insert has a 49 bp direct repeat at its ends (Figure 6.2), and contains two additional internal direct repeats of 10 bp and 7 bp. Both copies of the 10 bp repeat are in the same translational reading frame as that encoding the epitope, resulting in the reiteration of the tripeptide, TEL, in the protein. Furthermore, the 49 bp direct repeat results in the reiteration of 17 amino acids in the recombinant protein.

Southern blot analysis using this cDNA insert as a probe determined that the clone represented a juncture of two copies of a gene segment tandemly repeated up to 12 times within the gene encoding for the immunogen. This is consistent with what is known about the protein biochemistry, showing the immunogen contains multiple copies of the epitope recognized by the MAb C20A3 (Alderete and Neale, 1989). Furthermore, this cDNA probe did not identify any restriction fragment length polymorphisms (RFLPs) among trichomonal isolates, demonstrating the stability of this gene and possibly the surrounding DNA sequence. The hydrophilicity plot of the deduced amino acid sequence from the open reading frame encoding the C20A3 reactive epitope predicted the amino acid sequence DREGRD as the antigenic determinant (Hopp and Woods, 1981).

It has been shown that surface protein containing repetitive epitopes (Chulay *et al.*, 1987; Kemp *et al.*, 1990) may regulate or influence the host immune response (Dintzis *et al.*, 1976; Dintzis *et al.*, 1983). Competition immunoblot experiments between patient sera and MAb C20A3 demonstrated that all the patients' antibody response is directed against the tandemly-repeated epitope on this phenotypically varying immunogen (Dailey and Alderete, 1991). This immunogen may divert the immune response away from other parasitic targets, essential to the parasite's survival. Interestingly, unlike the repetitive epitopes in other model systems which undergo dramatic alterations (Chulay *et al.*, 1987; Kemp *et al.*, 1990), the C20A3-repetitive epitope has remained unchanged over the past 10 years of serial *in vitro* culture of numerous *T. vaginalis* isolates (Alderete *et al.*, 1986; 1987; Alderete and Neale, 1989). It is likely that this prolonged

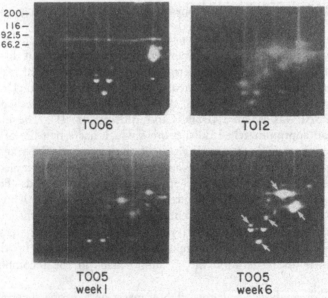

Figure 6.1 Two-dimensional substrate-SDS-PAGE of fresh clinical isolates of *T. vaginalis*. Differentially expressed proteinase activities of fresh isolate T005 after six weeks of *in vitro* passage are shown (arrows). The pattern of isolate T012 was identical to that of the common laboratory isolate NYH 286 (from Neale and Alderete, 1990).

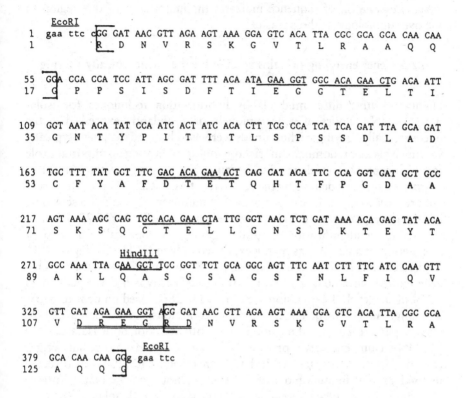

Figure 6.2 Sequence analysis of the cDNA clone encoding the immunodominant epitope of the phenotypic variation immunogen (Alderete *et al.*, 1986b). (a) The DNA sequence of the cDNA was obtained following subcloning of the insert into pUC19, and DNA sequence analysis was performed with the PC/Gene DNA software (IntelliGenetics). The cDNA sequence contained directly repeated termini of 49 nucleotides (bracketed region). Additional internal repeats of 7 bp and 10 bp are denoted as underlined and small text. An additional, imperfect third copy of the 10 bp repeat containing two adenosine insertions is located at nucleotide 178. The open reading frame encoding the protein was confirmed by directly sequencing the ends of the cDNA in the recombinant phage clone. The single letter code denoting the amino acids are listed below each codon. Hydrophilicity analysis (Hopp and Wood, 1981) was used to identify the amino acid sequence of the best fit antigenic determinant (DREGRD); (b) Diagramatic representation of the structure of the phenotypically varying immunogen. Previous data demonstrating autodigestion of the immunogen suggested the epitope recognized by monoclonal antibody C20A3 (*) had a central location (Alderete and Neale, 1989). Moreover, Southern blot analysis using the cDNA as a probe estimated that there were approximately 12 copies of the repeated epitope sequence. Data indicate that the cDNA represents a junction of two of the repeated DNA segments and contains a single C20A3-reactive epitope (Dailey and Alderete, 1991).

181

exact preservation of sequence indicates the importance of this region to the overall biology of the parasite.

6.2.5 Genes encoding proteins specific for the phenotypically varying isolates

Highly sensitive differential cDNA hybridization techniques for isolation of cDNA clones of genes uniquely or abundantly expressed in the phenotypically varying trichomonads were used because the phenotypically varying isolates predominate in trichomoniasis, indicating an important role for this property in virulence, and because the phenotype distinctions of *T. vaginalis* have been defined by the reactivity of MAb C20A3 and patient sera with respect to the surface expression of immunogens. Differential screening between a representative Type I (IR 78) and Type II (NYH 286) isolate resulted in the identification of a full length cDNA clone, called p1, that by Northern blot analysis was highly expressed only in NYH 286 (Figure 6.3). (Alderete *et al.*, 1990). Longer exposure of X-ray films revealed the possible gene expression among the Type I isolate IR 78, which corresponded to <1% of the level of expression seen for NYH 286, based on densitometric scanning. As a control under the same conditions, an unrelated cDNA, termed p24, reacted to the same extent with RNA of both isolates. The p1 cDNA clone encodes a protein (P55) with a relative molecular weight (M_r) of 55 000. Consistent with the Northern blot data, P55 was detected in total protein immunoblots of Type II but not Type I isolates, further confirming the specific expression of the gene in Type II isolates. The data shows how this sophisticated technology may isolate specifically expressed genes, depending upon the present and future phenotype distinctions of the pathogenic trichomonads.

Moreover, data indicates that this is a multicopy gene present in all isolates (Figure 6.4), including Type Is (Table 6.2). RFLPs are also apparent for this gene among the different isolates but an individual RFLP does not correlate with the phenotypic variation of immunogens among trichomonads. Therefore, transcription from this gene is controlled by a specific regulatory process, and the absence of transcriptional product in Type I isolates is not simply a result of spontaneous changes in the DNA, observed through either mutations or recombinations. What is responsible for regulating the transcription of one or more of these gene copies is not known but, clearly, the multicopy nature of the gene is significant. It may be that differential expression of the multicopy genes in some isolates of *T. vaginalis* is responsible for a new aspect of phenotypic variation that has yet to be reported for this or other pathogenic protozoa.

Although the *in vivo* significance of P55 remains to be determined, 55% of sera from patients detected this recombinant protein (Alderete *et al.*, 1990), consistent with the numbers of patients infected with Type II

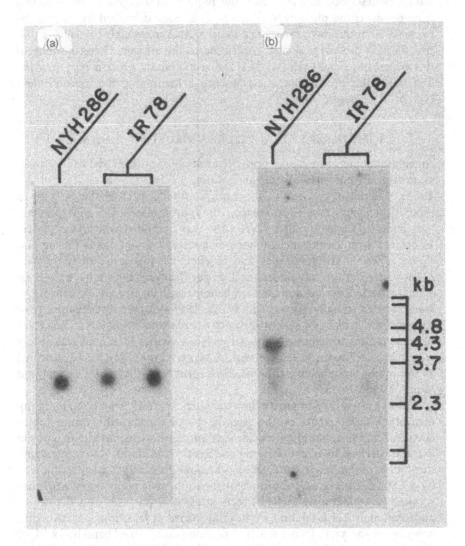

Figure 6.3 Northern blot analysis of total RNA from representative Type I (IR 78) and Type II (NYH 286) isolates of *T. vaginalis* using a cDNA found to be specific for the phenotypically varying isolates (Table 6.2). The hybridization of 15 μg RNA from both isolates (lanes 1 and 2) and 75 μg RNA of isolate IR 78 (lane 3) was performed using a ^{32}P-labelled cDNA, called p24, which reacted with both isolates (part A) and with a ^{32}P-labelled cDNA, called p1, uniquely reacting with only isolate NYH 286 (part B). The preparation of RNA was as previously described (Chirgwin *et al.*, 1979). RNA was electrophoresed through a 1% agarose gel containing formaldehyde and blotted onto Nytran. Hybridization was for 18 hr at 42°C. Blots were washed in 0.1X SSC for 1 hr at 65°C (Maniatis *et al.*, 1982). Autoradiography was done for 24 hr at 70°C with intensifying screens.

parasites described earlier. Given the highly immunogenic nature of this gene product; the phenotype distinctions already described by us and the presence of antibody in the sera of patients, it is reasonable to propose that this 55 kDa protein may also contribute to the antigenic heterogeneity of this organism. Finally, this cDNA and recombinant protein represent the first molecular probes potentially useful in identifying the more virulent type of *T. vaginalis* isolates.

6.3 THE DOUBLE-STRANDED RNA VIRUS OF *T. VAGINALIS*

An emerging literature on the presence of double-stranded (ds) RNA viruses in protozoan parasites (Wang and Wang, 1985a, 1986a, b; Tarr *et al.*, 1988; Widmer *et al.*, 1989) illustrates another aspect of the biology of these organisms. For *T. vaginalis*, an approximately 35 nm diameter icosahedral double-stranded RNA virus was identified which was closely associated with membrane fractions of infected trichomonads (Wang and Wang, 1986b). The molecular and biochemical properties of these viral particles are not yet fully understood, as purification methods have not been standardized. Although a single copy genome with an estimated size of 5.5 kb was reported initially (Wang and Wang, 1985a), other investigators have shown the existence of several distinct populations of ds RNA molecules of various sizes, resembling those of ds RNA viruses in animal cells (Flegr *et al.*, 1988). Finally, a protein with a M_r of 85 000 (85K) was identified as the major component of the ds RNA virus for *T. vaginalis* (Wang and Wang, 1986b).

We have attempted to purify the viral particles away from contaminating membrane components of the parasite to perform further immunobiochemical characterization of specific viral proteins. Viral particles in parasite extracts, derived by freeze-thawing and gentle sonication, were centrifuged through a combined sucrose cushion to remove gross cellular debris followed by a CsCl density gradient ultracentrifugation. Our preliminary work has resulted in isolation of ds RNA-containing viral particles at a density of approximately 1.4 g/ml, and we have identified at least four putative viral proteins, as shown in Figure 6.5. The predominant stained band had a M_r of 80K, consistent with a previously reported value (Wang and Wang, 1986b) other minor protein bands were reproducibly identified with M_rs of 135K, 18K and 16K. Furthermore, sera from animals experimentally infected with *T.vaginalis* Type II organisms possessed antibody to these putative viral proteins.

It will be important to establish the composition of the viral particles by a variety of ways for several reasons. An earlier report described the presence of virus only in *T. vaginalis* isolates which undergo phenotypic variation for immunogens and adhesins (Wang *et al.*, 1987). More recent investigations with the same monoclonal antibody used earlier (Alderete

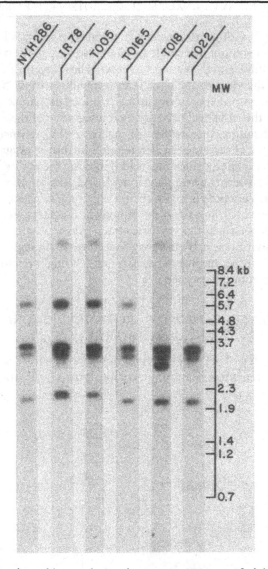

Figure 6.4 Southern blot analysis of genomic DNA purified by an established procedure (Wang and Wang, 1985b) from several isolates of *T. vaginalis*. Representative isolates IR 78, T005 and T01605 are Type I, while NYH 286, T018 and T022 are Type II (Table 6.2). In this experiment 10 μg of DNA was first digested to completion with Eco RI and electrophoresed through 0.8% agarose before blotting onto Nytran. The blot was hybridized with ^{32}P-labelled p1 cDNA shown in Figure 6.3. After 18 hr at 42°C, the blot was washed with 0.1% SSC for 1 hr at 42°C (Maniatis *et al.*, 1982). Autoradiography was performed as in Figure 6.3.

et al., 1986b) but with a more extensive number of isolates has revealed that some isolates without detectable ds RNA also vary phenotypically. The availability of antibody to the viral proteins will allow investigations into the contribution of the virus to the overall antigenic diversity of this parasite. Expression of these proteins in organisms without detectable ds RNA or viruses might further support the notion that integration of the viral genome has occurred. Therefore, the presence of nuclear DNA sequence homology with the viral RNA, as suggested earlier, requires further investigation (Wang and Wang, 1985a; 1986b), since this would indicate the ability of the virus genome to integrate into the host nucleic acids and possibly affect important regulatory genes. Very little is really known about the viruses of pathogenic protozoa and their contribution to the biology of the parasites. The finding of retrotransposon elements in trypanosomes (Aksoy et al., 1990) shows that reverse transcriptase activities may indeed exist among protozoa and the examination for this enzyme in parasite extracts might reinforce the idea of viral integration into host genomes.

6.4 BATCH VERSUS CHEMOSTAT GROWN TRICHOMONADS AND ANTIGENIC DIVERSITY

Nutrient limitation and pH may represent important environmental signals, which bear on the protein-antigenic-composition of the trichomonads.

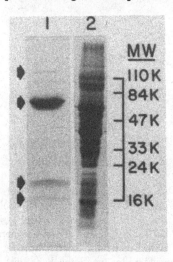

Figure 6.5 Coomassie-brilliant blue stained proteins of purified double-stranded RNA virus of *T. vaginalis* NYH 286 (lane 1) after SDS-PAGE on 10% acrylamide. Arrows denote proteins reproducibly detected under similar experimental conditions, showing the possibly complex protein composition of the virus particles. Lane 2 presents a typical stained pattern of trichloroacetic acid-precipitated proteins of *T. vaginalis* as described before (Alderete, 1983a).

Therefore, an approach to resolving some of the important issues regarding antigenic heterogeneity is to grow *T. vaginalis* under conditions similar to those *in vivo*. We began cultivation of these parasites using continuous flow culture conditions (Tempest, 1970) to study how changes in physiological conditions affect the expression of factors involved in establishment and maintenance of *T. vaginalis* infections (Lehker and Alderete, 1990b).

We found that parasites could be maintained stably and nutrient limitation led to the finding that trichomonads could achieve a generation time approaching 200 hours. This contrasted with the four to six hour generation time for batch-grown organisms. The pH was found to affect cell density and it was possible to differentiate trichomonal isolates on the basis of survival at pH <5.0. Indeed, although most patients have a vaginal pH value of approximately 5.7, some patients maintain a more normal acidic pH of approximately 4.5. Thus, observations that some isolates grow and multiply at pH of 4.5 may be relevant to the pathogenesis.

Finally, test tube grown trichomonads showed major differences in total protein profiles, when compared with those parasites cultivated at pH 4.5. Undoubtedly, these differences in immunochemical characterization reflect *in vivo* environmental signals that the pathogens encounter, and it will lead to evaluation these altered chemostat-grown parasites for their surface immunogen profiles. Axenic cultures of fresh isolates are achieved by investigators only after the parasite has undergone numerous divisions. The antigenic profile of these organisms may resemble trichomonads grown in culture for long periods of time, which differs from parasites grown under *in vivo*-like, continuous-flow conditions (Lehker and Alderete, 1990b).

6.5 CYTOADHERENCE AND THE TRICHOMONAD ADHESIN REPERTOIRE

Trichomonas vaginalis is a mucosal parasite which must traverse the mucous layer to parasitize the vaginal epithelium. Little is known regarding chemotactic signals, given by host fluids or the vaginal epithelium, which guide trichomonads toward the squamous vaginal epithelial cells (VECs). Host cell anchoring, however, certainly protects the parasite from being cleared by the constant fluid secretions of the vagina.

Experiments, summarized in Table 6.3, demonstrate the highly specific nature of *T. vaginalis* attachment to host epithelial cells in monolayer cultures (Alderete and Pearlman, 1984; Alderete and Garza, 1985) and VECs (Alderete *et al.*, 1988; Arroyo *et al.*, 1992). Although it has not yet been possible to obtain genetic mutants to certain determinants or traits, it was noteworthy in these earlier experiments that the nonpathogenic human trichomonad of the oral cavity, *T. tenax*, was unable to parasitize cell monolayers under the same conditions.

Table 6.3 Evidence for the highly specific nature of *T. vaginalis* cytadherence

---

Time, temperature, and pH dependence of trichomonal parasitism of epithelial cells

Saturation binding kinetics implicating a limited number of host cell receptor sites for parasite recognition

Competition of cytadherence does not occur with non-pathogenic *T. tenax* trichomonads

Trichomonad surface proteins involved in cytadherence are removed by trypsinization and regeneration of host cell attachment occurs after incubation of live, trypsinized parasites in growth medium

Protein synthesis inhibition by cycloheximide treatment of trypsinized parasites prevents regeneration of cytadherence

---

Trypsinization of live organisms was highly effective at abolishing cytadherence, showing the existence of trichomonad surface proteins as mediators of host parasitism (Alderete and Garza, 1985a). The ability to prevent the regeneration of expression of putative adhesins by inhibiting protein synthesis of the trypsinized organisms further reinforced the idea that parasite-derived molecules were involved in the host cell recognition and binding events. It became necessary, therefore, to try and identify the specific *T. vaginalis* surface adhesin proteins.

6.5.1 The adhesin protein repertoire

A variety of data suggests that four trichomonad proteins as shown in Figure 6.6 are adhesins mediating *T. vaginalis* cytadherence to epithelial cells (Alderete and Garza, 1988; Arroyo and Alderete, 1990). The evidence which has been accumulated is summarized in Table 6.4. For example, Figure 6.6 shows results of a representative experiment on the inhibition of trichomonal cytadherence by antibody directed to each of the adhesins. A mixture of adhesin proteins derived from a ligand assay described below inhibited parasite attachment to host cells in a competition experiment (Figure 6.6).

A ligand assay developed by others to identify microbial ligands involved in host cytadherence (Baseman and Hayes, 1980; Baseman *et al.*, 1982) was modified to demonstrate the specific binding of trichomonad proteins to fixed epithelial cells (Alderete and Garza, 1988). Data indicated that the four proteins were of parasite origin (Figure 6.6) and were distinct gene products. For example, disulfide bonding among the candidate adhesins was not evident and pulse-chase experiments did not reveal the existence of a higher molecular weight precursor adhesin protein. Inactivation of proteinases before or during detergent extraction followed by the ligand assay identified the same four proteins, again showing no evidence for proteolytic cleavage of a higher sized, parental adhesin protein. It is germane that incubation of the detergent extract at 37°C for ten minutes without proteinase inhibitors resulted in greatly diminished amounts of bound adhesins. These data were in concordance with the lability of adhesins to proteolysis and their removal from the trichomonal surface following trypsinization (Alderete and Garza, 1985; 1988).

6.5.2 Role of proteinases in cytoadherence

A novel finding was the requirement of cysteine proteinase activity for *T. vaginalis* cytadherence (Arroyo and Alderete, 1989). A reduction in the level of host cell attachment by as much as 80% was achieved by cysteine proteinase inhibitors either with pretreatment of parasites or addition to the host cell-trichomonad mixture. The absence of adverse effects by inhibitors on trichomonal metabolism, mobility and viability during these experiments provide strong evidence for the external surface action of the inhibitors on trichomonad proteinases (Arroyo and Alderete, 1989). A key feature, important to possible future pharmacological targeting and interference with proteinase action, was that adherent trichomonads readily detached following addition of proteinase inhibitor, indicating the requirement for continuous proteinase activity for tissue parasitism. The decreased cytoadherence corresponded with a diminished cytotoxic effect, underlining the contact-dependent nature for host cell killing, which has been previously reported (Alderete and Pearlman, 1984; Alderete and Garza, 1985; Krieger *et al.*, 1985a).

A challenge regarding future dissection of the molecular events of *T. vaginalis*: epithelial cell recognition and binding is two-fold. First, to identify the important proteinase or proteinases involved in cytoadherence, from the highly complex proteinase composition of these organisms (Figure

Table 6.4 Evidence for four *T. vaginalis* surface proteins as adhesins mediating cytoadherence

---

Enrichment of four trichomonad proteins bound to glutaraldehyde-fixed epithelial cells from a complex, total protein detergent extract

Binding to fixed host cells fulfils biochemical criteria of receptor-ligand interactions, e.g. saturation, time, temperature and competition binding kinetics

Surface location of adhesins demonstrated by removal by trypsinization of live organisms and extrinsic labelling

Purified adhesins inhibit trichomonal attachment to host cells

Concentration-dependent inhibition of trichomonal cytodherence by antibody to adhesins

Parasite origin of adhesins shown by intrinsic (^{35}S) labelling

Higher amounts of adhesin proteins corresponds to greater levels of cytodherence

---

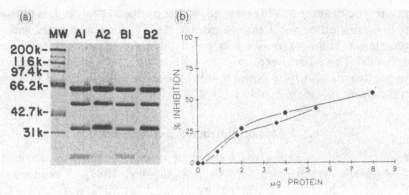

Figure 6.6 Demonstration of the synthesis of the adhesins by *T. vaginalis* (a) and inhibition of cytoadherence with antibody to adhesins (B); (a) Coomassie brilliant blue-stained protein patterns (lanes A1 and B1) and fluorograms (lanes A2 and B2) of [^{35}S]methionine-labelled trichomonads after SDS-PAGE of adhesins of fresh isolates (a) and long-term grown isolates (b). Electrophoretic analysis of proteins was performed as described before (Alderete, 1983a). MW refers to molecular weight standards (k = 1000); (b) the extent of inhibition was measured by the cytoadherence assay performed as described previously (Arroyo and Alderete, 1989). Circles show the representative inhibition kinetics of [^{3}H]thymidine labelled trichomonads, which were first treated with increasing amounts of Protein A-Sepharose purified IgG obtained from antiserum to the 65 kDa adhesin protein seen in part A. Antibody was added to parasites for 15 minutes at 37°C followed by addition of the antibody-parasite mixture to the HeLa cells in monolayer cultures. Diamonds show the inhibition obtained by blocking of the host cell receptors when increasing amounts of the four adhesins as seen in part A were first added to fixed HeLa cell monolayers for 18h at 4°C before measuring the ^{3}H-trichomonad-HeLa cell interaction. For this experiment, adhesins were obtained from the fixed HeLa cells using a ligand assay (Alderete and Garza, 1988). Adhesins were first renatured by extensive dialysis before addition to the fixed HeLa cells. Unrelated proteins, such as bovine serum albumin, used in blocking experiments as controls did not inhibit parasite recognition and binding to epithelial cells.

6.1). Secondly, the mechanism by which the proteinase activity promotes cytodherence must be delineated, although data indicate that the proteinase activity is directed toward the parasite and not to the host cell surface (Arroyo and Alderete, 1989).

6.5.3 Iron modulation of *T. vaginalis* cytoadherence

Iron, an essential element for growth of many pathogens (Griffith, 1985; Weinberg, 1978), is not readily available *in vivo*. For example, the level of free iron in serum is kept low by transferrin while on mucosal surfaces most iron is sequestered by lactoferrin (Masson *et al.*, 1966). Thus, the availability of iron for microorganisms is restricted in both host environments, and it appears that pathogens capable of competing for iron with host iron-binding proteins have enhanced virulence properties (Griffith, 1985).

Iron was found to play a role in regulating the level of trichomonal cytoadherence, as seen in Figure 6.7 (Lehker and Alderete, 1990a; 1991). In the presence of excess iron in the growth medium, parasite attachment to epithelial cells was much greater than when grown in complex medium, previously found to be iron-limiting (Gorrell, 1985), or when trichomonads were incubated overnight in the presence of 2,2-dipyridal, an iron chelator. Section 6.6 discusses the importance of iron to various biological properties of *T. vaginalis* in addition to the regulation of cytodherence. Iron, as a regulatory element is becoming apparent by the numerous sources of host molecules from which trichomonads can acquire iron (section 6.6.3 and Figure 6.9).

The elevation of cytoadherence levels of long-term grown isolates, such as isolate NYH 286 shown in Figure 6.7, to attachment values detected for fresh isolates is especially noteworthy. These data illustrate that environmental signals or conditions, discussed earlier in section 6.4, are important to virulence expression. The differences between fresh and long-term grown *T. vaginalis* organisms, as summarized in Table 6.5, contribute to our understanding of trichomonal virulence and pathogenesis. Clearly, placing trichomonads from the human host in an artificial, complex medium results in an alteration of a variety of properties and has

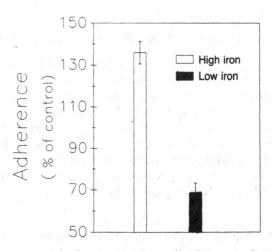

Figure 6.7 Comparison of adherence to HeLa cells of *T. vaginalis* NYH 286 grown in high or low iron medium. Parasites were radiolabelled with [³H] thymidine in TYM-serum medium (Diamond, 1957; Alderete and Garza, 1985) replete with iron or iron-deficient. After 24 hours of growth, washed parasites were added to HeLa cell monolayers in individual 96-well tissue culture plates and incubated for 15 min at 37°C as described elsewhere (Arroyo and Alderete, 1989). Associated radioactivity was measured by scintillation spectroscopy.

Table 6.5 Properties of trichomonads from fresh isolates when compared with long-term grown isolates[a]

Morphology More heterogeneity in size and shape of parasites. More often they are small and highly motile with occasional large round, multiflagellated forms. Many are ameboid

Generation time Typically a 70 hr generation time after isolation, which stabilizes to that seen for long-term grown cultures having 4–6 hr generation times after 1 week of isolation

Cytoadherence and Adhesins Levels of cytodherence are up to ten times greater than that of long-term grown isolates. Corresponding higher amounts of adhesin proteins are present

Cytotoxicity Using a standardized assay of a monolayer culture of host cells and a 1:5 ratio of parasites to host cells, fresh isolates kill cells within a 15 to 60 min time period compared with several hours for long-term grown cultures

Immunogen phenotype Most fresh isolates do not surface express immunogens. However, in vitro growth results in phenotypic variation and surface expression of immunogens in about one-half to two-thirds of the fresh isolates

[a]Fresh isolates are defined as parasites examined immediately after isolation or grown in complex medium for ≤3 days prior to measurement of cytoadherence, cytotoxicity and phenotypic variation.

contributed to discrepancies in the literature among investigators studying this parasite.

6.6 THE NUTRIENT ACQUISITION PROTEIN REPERTOIRES

Successful host parasitism by *T. vaginalis* is related to the uptake of nutrients like lipids and iron. Earlier reports by us showed highly specific binding of host proteins by trichomonad receptors (Peterson and Alderete, 1982, 1983, 1984a, b, c) and that biological properties resulting from host protein acquisition appeared to contribute to the well-being of *T. vaginalis* organisms. This initial work was carried out in a step-wise fashion to delineate the true significance of the trichomonal-host protein interactions. Evidence was obtained on the existence of a repertoire of *T. vaginalis* surface proteins involved in nutrient acquisition by this parasite.

6.6.1 Apoprotein CIII receptors for lipoprotein uptake

Lipids represent a class of host molecules essential for trichomonal growth and multiplication (Holz *et al.*, 1987; Lund and Shorb, 1962). The pathogenic human trichomonads appear incapable of *de novo* lipid synthesis and may not convert or retroconvert certain long chain fatty acids or cholesterol (Holz *et al.*, 1987; Roitman *et al.*, 1978), components essential for trichomonad membranes and therefore survival.

The *T. vaginalis* parasites have specific mechanisms to acquire lipids from mucous secretions (Peterson and Alderete, 1984a,c). Acquisition of lipoproteins was found to occur by a trichomonad surface protein of M_r >250 000. The receptor bound apoprotein CIII, a protein found in the various lipoprotein subfractions. Detailed investigations using each of the subfractions revealed an ability of the parasites to internalize the lipoproteins and accumulate the lipids into trichomonad membranes while discarding the apoproteins. In fact, the trypticase-yeast extract complex medium (Diamond, 1957) supplemented with purified lipoprotein subfractions gave growth and multiplication at rates and levels equal to medium supplemented with serum, a source of lipids in the trichomonal growth medium (Peterson and Alderete, 1984a,c).

This earlier work showed an important *in vivo* source of lipids for *T. vaginalis*. It is easy to envisage how quantitative and qualitative differences in the composition of lipoproteins in the mucous secretions among humans would be important to virulence and pathogenesis. For example, differences in lipoproteins among patients might affect the antigenic profiles of the infecting parasite or alter host susceptibility or resistance to infection. Finally, levels of lipoproteins in vaginal mucous secretions may not be sufficient to reproduce the optimal growth and multiplication of the infecting trichomonads, seen in test tube cultures. In consequence, *T. vaginalis* have evolved an ability to utilize alternative lipid sources, as described below.

6.6.2 The erythrocyte-binding receptor proteins for lipids

Observations regarding the interactions between erythrocytes and *T. vaginalis* have not received a great deal of attention. Vaginal wet mounts from patients, for example, rarely show trichomonads with bound erythrocytes. Nonetheless, haemagglutination and haemolysis by trichomonads is an important alternative nutrient acquisition system for this parasite (Lehker *et al.*, 1990).

Two immunologically cross-reactive but distinct trichomonad surface proteins mediate the specific binding of erythrocytes by *T. vaginalis* (Figure 6.8IIB). These erythrocyte adhesins were identified using a modified ligand assay (Table 6.4) using glutaraldehyde-fixed erythrocytes incubated with a combined SDS-Triton X-100 detergent extract (Lehker *et al.*, 1990). Mouse IgG to the RBC adhesins inhibited erythrocyte binding and the presence of antibody in sera from patients was demonstrated, indicative of an *in vivo* relevance of these molecules. The level of erythrocyte binding among individual isolates, although reproducible over several generations in test tube cultures, was also found to vary widely over a period of weeks, if not days, exhibiting a phenotypic variation in the expression of the erythrocyte adhesins similar to that described earlier for immunogens. In addition, there was no relationship between variations in this property and those already described for immunogens, adhesins and proteinases. This showed the emergence, in *T. vaginalis*, of another class of independently regulated, biofunctional molecules.

The contact-dependent nature of haemolysis was recently demonstrated by us (Lehker *et al.*, 1990; Dailey *et al.*, 1990) and the previously reported rapid haemolysis (Krieger *et al.*, 1983), which occurs soon after erythrocyte attachment by parasites, probably explains the few reports describing trichomonal-erythrocyte associations. Nonetheless, the efficiency by which *T. vaginalis* organisms can internalize erythrocytes was apparent by showing the rapid engorgement by trichomonads of fixed erythrocytes (Figure 6.8IA). This erythrocyte uptake continues unabated until lysis of the parasite occurs under these *in vitro* conditions.

One significant aspect of the erythrocyte binding was demonstrated by the growth and multiplication of *T. vaginalis* in medium supplemented with purified lipids or unfixed erythrocytes. Levels equal to those seen for lipoprotein- or serum-supplemented medium were readily achieved (Lehker *et al.*, 1990). Epithelial cells were previously found to be unable to provide lipids for these parasites (Peterson and Alderete, 1984c), showing the highly specific sources of cholesterol and fatty acids (Nelson, 1967) required by this parasite.

6.6.3 Host iron-binding protein receptors for iron uptake

Trichomonas vaginalis requires iron levels of >200 μM for maximal metabolic (hydrogenosomal) activity (Gorrell, 1985). The high iron requirement by trichomonads contrasts greatly with most bacterial pathogens, which need from 0.4 μM to 4 μM iron (Griffith, 1987). The importance of iron for survival of the parasites *in vitro* and possibly *in vivo* was supported by showing that an overall reduction in cell density occurred during test tube culture when 2,2-dipyridal was added to the growth medium (Figure 6.9). This growth inhibitory effect was easily reversed by the addition of exogenous ferrous ammonium sulphate but not by other trace metals.

How this parasite acquires its iron from the host, despite the absence of free iron in normal body fluids as mentioned earlier, prompted us to look for the presence of lactoferrin receptors on the parasite surface (Peterson and Alderete, 1984b). Indeed, trichomonads were found to possess specific lactoferrin receptors with a K_d of 1 μM (Peterson and Alderete, 1984b). A heterodimer receptor protein complex was identified by lactoferrin-affinity chromatography and other assays. Binding of lactoferrin but not transferrin, the iron-binding protein of plasma, resulted in iron accumulation as measured by ^{59}Fe uptake when compared with the molar binding of ^{125}I-lactoferrin.

The significance of specific lactoferrin binding by *T. vaginalis* was reinforced by stimulation of parasite growth and multiplication by lactoferrin-iron (Figure 6.9), as well as by stimulation of a key metabolic enzyme (Peterson and Alderete, 1984b), under iron-limiting conditions. Apolactoferrin and transferrin were unable to allow for similar trichomonal growth. Nonetheless, these exciting observations were tempered by the known variations in lactoferrin iron concentrations which occur *in vivo* (Cohen *et al.*, 1987). Thus, the high iron requirement for trichomonal metabolism (Gorrell, 1985) and cytodherence as shown earlier may be adversely affected by the limiting amounts of lactoferrin in vaginal secretions.

Haemolysis of erythrocytes would possibly make available yet another source of iron for *T. vaginalis* and receptor proteins unique from those which bind lactoferrin were indeed implicated in haemoglobin associations

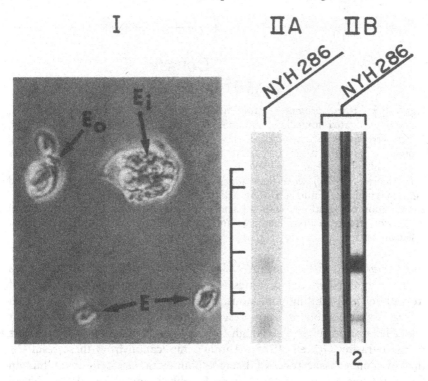

Figure 6.8 Internalization of glutaraldehyde-fixed erythrocytes by live *T. vaginalis* isolate NYH 286 (I) and identification of the erythrocyte binding protein adhesins (II) E_o refers to adherent erythrocytes outside the parasite compared with internalized erythrocytes (E_i) and unbound erythrocytes (E). IIA shows autoradiograms of iodinated trichomonad proteins from a detergent extract which bound to fixed erythrocytes. IIB shows immunoblot data using control, prebled mouse serum (lane 1) compared with mouse antiserum generated to the 27.5-kDa erythrocyte-binding protein. Antiserum to the 12.5 kDa protein gave identical results as that for the 27.5 kDa protein (from Lehker *et al.*, 1990).

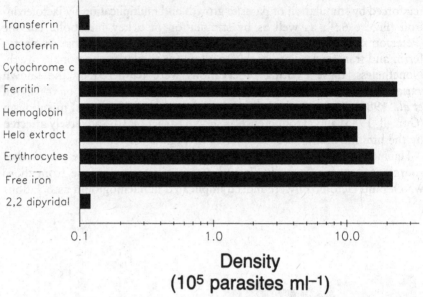

Density
(10^5 parasites ml^{-1})

Figure 6.9 Growth promoting properties of various iron-binding proteins for *T. vaginalis* incubated in medium containing 0.3 mM 2,2-dipyridal, an iron chelator. Tubes were seeded at an initial density of 1×10^5 organisms ml^{-1}. Iron-binding proteins (equivalent to 25 μM iron), free iron (250 μM), human erythrocytes (10%; vol:vol) or HeLa cell extract (2 $\times$ 10^6 cell equivalents/ml) (Peterson and Alderete, 1984c) were then added to the culture medium and incubated until maximum cell numbers were achieved. Cell densities were determined by enumeration of organisms with a haemacytometer. Note the absence of growth in the iron-chelated culture medium control and the presence of transferrin as the sole iron source.

(Lehker *et al.*, 1990). Figure 6.9 also shows the ability of trichomonads to achieve excellent cell densities with haemoglobin as a sole source of iron under iron-limiting conditions. The exact form of haemoglobin iron utilized by trichomonads is not presently known, although it is possible that intact haem iron might be enough for the various metabolic requirements of the parasite (Gorrell, 1985). Another implication from these results is the immediate association of haemoglobin with haptoglobin in human fluids (Helms *et al.*, 1984; Muller–Eberhard and Liem, 1974; Nagel and Gibson, 1967). Thus, haemoglobin may be immediately sequestered following erythrocyte lysis by *T. vaginalis*, consistent with the presence of haemoglobin receptors (Lehker *et al.*, 1990), or alternatively, organisms possess additional receptors for haptoglobin. The latter could further contribute to the complexity of nutrient iron acquisition, as the haptoglobin varies among humans (Muller–Eberhard and Liem, 1974). In this case it is intriguing to speculate that the susceptibility and/or resistance to infection, or the level of symptomatology due to growth and multiplication of the parasite, might be related to the type of haptoglobin present in patients and recognized by *T. vaginalis*. A role of haem compounds and haptoglobin in microbial virulence has been documented (Helms *et al.*, 1984).

These data show the high metabolic demand for host iron by *T. vaginalis* (Gorrell, 1985) and for optimal cytoadherence as discussed earlier (section 6.5.3 and Figure 6.7) (Lehker and Alderete, 1990b; Lekher *et al.*, 1991). This indicates that iron sources other than lactoferrin and haemoglobin might be advantageous to the parasite. Figure 6.9 also shows that cytochrome c, ferritin, and even HeLa cell extracts represent alternative iron sources. There is no evidence for cytochrome and ferritin binding by the lactoferrin or haemoglobin receptors, indicating that *T. vaginalis* possesses a repertoire of surface receptors for host iron proteins. During our initial studies on the lactoferrin receptors, an idea for control of trichomoniasis through pharmacological or vaccine interference strategies was the starvation of parasites for this important nutrient by interference with lactoferrin binding. These recent observations (Figure 6.9), however, illustrate the difficulties in achieving this goal, at least for the iron-acquisition system of *T. vaginalis*.

6.7 CONCLUSIONS

This brief overview of ongoing research activities has summarized three aspects of the biology of the parasite and the host–parasite interaction, namely the antigenic diversity, the cytadherence property and nutrient acquisition. The highly evolved and complex nature of the interrelationship between host and parasite is evident by dynamics in protein-antigen phenotypic variations, and the various protein repertoires are summarized in Table 6.6. The phenotypic variation of trichomonad protein repertoires can either be coordinated, such as for the immunogens and epithelial cell adhesins, or be independent of one another, as seen for the proteinases and erythrocyte-binding proteins.

The ability of trichomonads to survive without expression of major immunogens, as presented in Table 6.2, does not indicate an absence of an important biofunctionality for these molecules. Other biofunctional proteins such as the epithelial cell- and erythrocyte-binding adhesins also undergo phenotypic variation under certain circumstances.

Table 6.6 The *T. vaginalis* surface protein phenotypic variation systems

Immunogens
Epithelial cell-binding adhesins
Proteinases
Erythrocyte-binding adhesins
Iron-acquisition receptors

The studies on the nutrient acquisition systems of *T. vaginalis* are important not just for our understanding of the mechanisms of parasite survival in the nutrient limiting environment of its host, but may also lead to interference strategies, apart from the development of vaccines to trichomonad proteins, like the adhesins. The future for pharmacological targeting also appears possible with the observation that continuous proteinase activity is necessary for initial and sustained host parasitism.

REFERENCES

Ackers, J.P. (1982) Immunology of amebas, giardia, and trichomoniasis. *Viruses and Parasites; Immunodiagnosis and Prevention of Infectious Diseases*, Plenum Publishing Corp., New York.

Aksoy, S., Williams, S., Chang, S. and Richards. F.F. (1990) SLACS Retrotransposon from *Trypanosoma brucei* gambiense is similar to mammalian LINEs. *Nucleic Acids Research*, 18, 785–92.

Alderete, J.F (1983a) Antigenic analysis of several pathogenic strains of *Trichomonas vaginalis*. *Infect. Immun.*, 39, 1041–7.

Alderete, J.F. (1983b) Identification of immunogenic and antibody-binding proteins on the membrane of pathogenic *Trichomonas vaginalis*. *Infect. Immun.*, 40, 284–91.

Alderete, J.F., and Pearlman, E. (1984) Pathogenic *Trichomonas vaginalis* cytotoxicity to cell culture monolayers. *Brit. J. Vener. Dis.*, 60, 99–105.

Alderete, J.F. and Garza, G.E. (1985) Specific nature of *Trichomonas vaginalis* parasitism of host cell surfaces. *Infect. Immun.*, 50, 701–8.

Alderete, J.F., Suprun–Brown, L., Kasmala, L. *et al.*, (1985) Heterogeneity of *Trichomonas vaginalis* and discrimination among trichomonal isolates and subpopulations by sera of patients and experimentally infected mice. *Infect. Immun.*, 49, 463–8.

Alderete, J.F., Garza, G.E., Smith, J. and Spence, M. (1986) *Trichomonas vaginalis*: Electrophoretic analysis reveals heterogeneity among isolates due to high molecular weight trichomonad proteins. *Exp. Parasitol.*, 61, 244–51.

Alderete, J.F. and Kasmala, L. (1986) Monoclonal antibody to a major glycoprotein immunogen mediates differential complement-independent lysis of *Trichomonas vaginalis*. *Infect. Immun.*, 53, 697–9.

Alderete, J.F., L. Kasmala, Metcalfe, E.C. and Garza, G.E. (1986b) Phenotypic variation and diversity among *Trichomonas vaginalis* and correlation of phenotype with contact-dependent host cell cytotoxicity. *Infect. Immun.*, 53, 285–93.

Alderete, J.F., Suprun–Brown, L. and Kasmala, L. (1986) Monoclonal antibody to a major surface immunogen differentiates isolates and subpopulations of *Trichomonas vaginalis*. *Infect. Immun.*, 52, 70–5.

Alderete, J.F. (1987) *Trichomonas vaginalis* phenotypic variation may be coordinated for a repertoire of trichomonad surface immunogens. *Infect. Immun.*, 55, 1957–62.

Alderete, J.F., Demeš, P., Gombošova, A. *et al.* (1987) Phenotype and protein/epitope phenotypic variation among fresh isolates of *Trichomonas vaginalis*. *Infect. Immun.*, 55, 1037–41.

Alderete, J.F. (1988) Alternating phenotypic expression of two classes of *Trichomonas vaginalis* surface markers. *Rev. Infect. Dis.*, 10, S408–12.

Alderete, J.F. and Garza, G.E. (1988) Identification and properties of *Trichomonas vaginalis* proteins involved in cytadherence. *Infect. Immun.*, 56, 28–33.

Alderete, J.F., Demeš, P., Gombošova, A., Valent, M., Fabušova, M., Janoška, A., Stefanovic, J. and Arroyo, R. (1988) Specific parasitism by *Trichomonas vaginalis* of purified vaginal epithelial cells. *Infect. Immun.*, 56, 2558–62.

Alderete, J.F. and Neale, K.A. (1989) Relatedness of major immunogen among all *Trichomonas vaginalis* isolates. *Infect. Immun.*, 56, 1849–53.

Alderete, J.F., Boothroyd, J.C., Dailey, D.C. and McKay, J.P. (1990) Identification of a differentially regulated gene from *Trichomonas vaginalis* isolates which undergo phenotypic variation. *Annual Meetings of the American Society for Microbiology*, Anaheim, CA.

Alderete, J.F., Newton, E., Dennis, C. and Neale, K.A. (1991a) Antibody in sera of patients infected with *Trichomonas vaginalis* is to trichomonad proteinases. *Genitourin. Med.*, 67, 331–4.

Alderete, J.F., Newton, E., Dennis, C. and Neale, K.A. (1991b) The vagina of women infected with *Trichomonas vaginalis* has numerous proteinases and antibody to trichomonad proteinases. *Genitourin. Med.*, 67, in press.

Arroyo, R. and Alderete, J.F. (1989) *Trichomonas vaginalis* proteinase activity is necessary for parasite cytadherence. *Infect. Immun.*, 57, 2991–7.

Arroyo, R., Engbring, J. and Alderete, J.F. (1992) Molecular basis of host epithelial cell recognition by *Trichomonas vaginalis*. *Molec. Microbiol.*, 6, 853–62.

Baseman, J.B. and Hayes, E.C. (1980) Molecular characterization of receptor binding proteins and immunogens of virulent *Treponema pallidum*. *J.Exp. Med.*, 151, 573–86.

Baseman, J.B., Cole, R.M., Krause, D.C. and Leith, D.K. (1982) Molecular basis for cytadsorption of *Mycoplasma pneumoniae*. *J. Bacteriol.*, 151, 1514–22.

Chulay, J.D., Lyon, J.A., Haynes, J.D. *et al.* (1987) Monoclonal antibody characterization of *Plasmodium falciparum* antigens in immune complexes formed when schizonts rupture in the presence of immune serum. *J. Immunol.*, 139, 2768–74.

Chirgwin, J.M., Przybyla, A.E., MacDonald, R. and Rutter, W.J. (1979) Isolation of biologically active ribonucleic acid from sources enriched in ribonuclease. *Biochem.*, 18, 5294–9.

Cohen, M.S., Britigan, B.E., French, M. and Bean, K. (1987) Preliminary observations on lactoferrin secretion in human vaginal mucus: variation during menstrual cycle, evidence of hormonal regulation and implications for infection with *Neisseria gonorrhoeae*. *Am. J. Obstet. Gynecol.*, 157, 1122–5.

Coombs, G.H. and North, M.J. (1983) An analysis of the proteinases of *Trichomonas vaginalis* by acrylamide gel electrophoresis. *Parasitol.*, 86, 1–6.

Dailey, D.C., Chang, T. and Alderete, J.F. (1990) Characterization of a hemolysin of *Trichomonas vaginalis*. *Parasitol.*, 101, 171–5.

Dailey, D.C. and J.F. Alderete. (1991) The phenotypically variable surface protein of *Trichomonas vaginalis* has a single, tandemly repeated immunodominant epitope. *Infect. Immun.*, 59, 2083–8.

Diamond, L.S. (1957) The establishment of various trichomonads of animals and man in axenic cultures. *J. Parasitol.*, **43**, 488–90.

Dintzis, H.M., Dintzis, R.Z. and Vogelstein, B. (1976) Molecular determinants of immunogenicity: the immunon model of immune response. *Proc. Natl Acad. Sci. U.S.A.*, **73**, 3671–5.

Dintzis, R.Z., Middleton, M.H. and Dintzis, H.M. (1983) Studies on the immunogenicity and tolerogenicity of T-independent antigens. *J. Immunol.*, **131**, 2196–203.

Flegr, J., Cerkasov, J. and Stokrova, J. (1988) Multiple population of double-stranded RNA in two virus-harbouring strains of *Trichomonas vaginalis*. *Folia Microbiol.*, **33**, 462–5.

Gorrell, T.E. (1985) Effect of culture medium iron content on the biochemical composition and metabolism of *Trichomonas vaginalis*. *J. Bacteriol.*, **161**, 1228–30.

Griffith, E. (1987) The iron-uptake systems of pathogenic bacteria, in *Iron and Infection: Molecular, Physiological and Clinical Aspects*, John Wiley and Sons, New York.

Helms, S.D., Oliver, J.D. and Travis, J.C. (1984) Role of heme compounds and haptoglobin in *Vibrio vulnificus* pathogenicity. *Infect. Immun.*, **45**, 345–9.

Holz, G.G., Lindmark, D.G., Beach, D.H. *et al.* (1987) Lipids and lipid metabolism of trichomonads. *Symposium on Trichomonads and Trichomoniasis*. Acta Universitatis Carolinae Biologica, Prague, Czechoslovakia.

Honigberg, B.M. (1978) Trichomonads of importance in human medicine. *Parasitic Protoza*, Academic Press, New York.

Honigberg, B.M., Livingston, M.C. and Frost, J.K. (1966) Pathogenicity of fresh isolates of *Trichomonas vaginalis*: the mouse assay versus clinical and pathologic findings. *Acta Cytol.*, **10**, 353–61.

Hopp, T.P. and Woods, K.R. (1981) Prediction of protein antigenic determinants from amino acid sequence. *Proc. Natl Acad. Sci. USA*, **78**, 3824–8.

Kemp, D.J., Conman, A.F. and Walliker, D. (1990) Genetic diversity in *Plasmodium falciparum*. *Adv. Parasitol.*, **29**, 75–128.

Kott, H. and Adler, S. (1961) The serological study of *Trichomonas* sp. parasitic in man. *Trans. R. Soc. Trop. Med. Hyg.*, **55**, 333–44.

Krieger, J.N. (1981) Urologic aspects of trichomoniasis. *Invest. Urol.*, **18**, 411–17.

Krieger, J.N., Poisson, M.A. and Rein, M.F. (1983) Beta hemolytic activity of *Trichomonas vaginalis* correlates with virulence. *Infect. Immun.*, **41**, 1291–5.

Krieger, J.N., Ravidin, J.I. and Rein, M.F. (1985a) Contact-dependent cytopathogenic mechanisms of *Trichomonas vaginalis*. *Infect. Immun.*, **50**, 778–86.

Krieger, J.N., Holmes, K.K., Spence, M.R. *et al.* (1985b) Geographic variation among isolates of *Trichomonas vaginalis*: demonstration of antigenic heterogeneity by using monoclonal antibodies and the indirect immunofluorescence technique. *J. Infect. Dis.*, **152**, 979–84.

Krieger, J.N., Wolner-Hanssen, P., Stevens, C. and Holmes, K.K. (1990) Characteristics of *Trichomonas vaginalis* isolates from women with and without colpitis macularis. *J. Infect. Dis.*, **161**, 307–11.

Laga, M., Nzila, H., Manoka, A.T. *et al.* (1989) High prevalence and incidence of HIV and other sexually transmitted diseases (STD) among 801 Kinshasa prostitutes. *Abstracts of the V International Conference on AIDS*, Montreal, Canada.

Laga, M., Nzila, N., Manoka, A.T. *et al.* (1990) Non ulcerative sexually transmitted diseases (STD) as risk factors for HIV infection. *Abstracts of the VI International Conference on AIDS*, San Francisco, CA.

Lanceley, F. (1958) Serological aspects of trichomoniasis. *Brit. J. Vener. Dis.*, 34, 3–8.

Lehker, M.L., Chang, T.H., Dailey, D.C. and Alderete, J.F. (1990) Specific erythrocyte binding is an additional nutrient acquisition system for *Trichomonas vaginalis*. *J. Exp. Med.*, 171, 2168–70.

Lehker, M. and Alderete, J.F. (1990a) Iron levels modulate adherence of *Trichomonas vaginalis* NYH 286 to HeLa cells. *Annual Meetings of the American Society for Microbiology*, Anaheim, CA.

Lehker, M.W. and Alderete, J.F. (1990c) Properties of *Trichomonas vaginalis* grown under chemostat controlled growth conditions. *Genitourin. Med.*, 66, 193–9.

Lehker, M.W., Arroyo, R. and Alderete, J.F. (1991) The regulation by iron of the synthesis of adhesins and cytoadherence levels in the protozoan *Trichomonas vaginalis*. *J. Exp. Med.*, 174, 311–18.

Lockwood, B.C., North, M.J., and Scott, K.I. (1987) The use of a highly sensitive electrophoretic method to compare the proteinases of trichomonads. *Mol. Biochem. Parasitol.*, 24, 89–95.

Lockwood, B.C., North, M.J. and Coombs, G.H. (1988) The release of hydrolases from *Trichomonas vaginalis* and *Tritrichomonas foetus*. *Molec. Biochem. Parasitol.*, 30, 135–42.

Lund, P.G. and Shorb, M.S. (1962) Steroid requirement of trichomonads. *J. Protozool.*, 9, 151–4.

Maniatis, T., Fritsch, E.F. and Sambrook, J. (1982) Molecular cloning, A laboratory manual, *Cold Spring Harbor Laboratory*, New York.

Masson, P.L., Heremans, J.F. and Dive, C.H. (1966) An iron-binding protein common to many external secretions. *Clin. Chim. Acta.*, 14, 729–34.

Müller, M., Meingassner, J.G., Miller, W.A. and Ledger, W.J. (1980) Three metronidazole-resistant strains of *Trichomonas vaginalis* from the USA. *Am. J. Obstet. Gynecol.*, 138, 808–12.

Müller, M. (1983) *Trichomonas vaginalis* and other sexually transmitted protozoan infections. *International Perspectives of Neglected Sexually Transmitted Diseases*, Hemisphere Publishing Corp., New York.

Muller-Eberhard, U. and Liem, H.H. (1974) Hemopexin, the heme-binding serum β-glycoprotein. *Structure and Function of Plasma Proteins*, Plenum Press, London.

Nagel, R.L. and Gibson, Q.H. (1967) Kinetics and mechanism of complex formation between hemoglobin and haptoglobin. *J. Biol. Chem.*, 242, 3428–34.

Neale, K.A. and Alderete, J.F. (1990) Analysis of the proteinases of representative *Trichomonas vaginalis* isolates. *Infect. Immun.*, 58, 157–62.

Nelson, G.J. (1967) Lipid composition of erythrocytes in various mammalian species. *Biochim. Biophys. Acta*, 144, 221–32.

Peterson, K.M. and Alderete, J.F. (1982) Host plasma proteins on the surface of pathogenic *Trichomonas vaginalis*. *Infect. Immun.*, 37, 755–62.

Peterson, K.M. and Alderete, J.F. (1983) Acquisition of α_1 antitrypsin by pathogenic *Trichomonas vaginalis*. *Infect. Immun.*, **40**, 640–6.

Peterson, K.M. and Alderete, J.F. (1984a) Selective acquisition of plasma proteins by *Trichomonas vaginalis* and human lipoproteins as a growth requirement by his species. *Mol. Biochem. Parasitol.*, **12**, 37–48.

Peterson, K.M. and Alderete, J.F. (1984b) Iron uptake and increased intracellular enzyme activity follow lactoferrin binding by *Trichomonas vaginalis* receptors. *J.Exp.Med.*, **160**, 398–410.

Peterson, K.M. and Alderete, J.F. (1984c) *Trichomonas vaginalis* is dependent on uptake and degradation of human low density lipoproteins. *J.Exp.Med.*, **160**, 1261–71.

Rein, M.F. and Chapel, T.A. (1975) Trichomoniasis, candidiasis, and the minor venereal diseases. *Clin.Obstet.Gynecol.*, **18**, 73–88.

Roitman, I., Heyworth, P.G. and Gutteridge, W.E. (1978) Lipid synthesis by *Trichomonas vaginalis*. *Ann. Trop. Med. Parasitol.*, **72**, 583–5.

Spence, M.R., Hollander, D.H., Smith, J. *et al.* (1980) The clinical and laboratory diagnosis of *Trichomonas vaginalis* infection. *Sex. Trans. Dis.*, **7**, 168–71.

Su-Lin, K.E. and Honigberg, B.M. (1983) Antigenic analysis of *Trichomonas vaginalis* strains by quantitative fluorescent antibody methods. *Z.Parasitenkd.*, **69**, 161–81.

Tarr, P.I., Aline, R.F. Jr, Smiley, B.L. *et al.* (1988) LRI: A candidate RNA virus of *Leishmania*. *Proc. Natl Acad. Sci. USA*, **85**, 9572–5.

Tempest, D.W. (1970) The continuous cultivation of microorganisms: 1. Theory of the chemostat. *Methods in Microbiology*, Academic Press, London.

Teras, J.K. (1966) Difference in the antigenic properties within strains of *Trichomonas vaginalis*. *Wiad. Parazytol.*, **12**, 357–63.

Wang, A.L. and Wang, C.C. (1985a) A linear double-stranded RNA in *Trichomonas vaginalis*. *J. Biol. Chem.*, **260**, 3697–702.

Wang, A.A. and Wang, C.C. (1985b) Isolation and characterization of DNA from *Trichomonas vaginalis* and *Tritrichomonas foetus*. *Molec. Biochem. Paratritol.*, **14**, 323–35.

Wang, A.L. and Wang C.C. (1986a) Discovery of a specific double-stranded RNA virus in *Giardia lamblia*. *Molec. Biochem. Parasitol.*, **21**, 269–76.

Wang, A.L. and Wang, C.C. (1986b) The double stranded RNA in *Trichomonas vaginalis* may originate from virus-like particles. *Proc. Natl Acad. Sci. USA*, **83**, 7956–60.

Wang, C.C., Wang, A. and Alderete, J.F. (1987) *Trichomonas vaginalis* phenotypic variation occurs only among trichomonads with double-stranded RNA virus. *J.Exp.Med.*, **166**, 142–50.

Weinberg, E.D. (1978) Iron and infection. *Microbiol.Rev.*, **42**, 45–66.

Widmer, G., Comeau, A.M., Furlong, D.B. *et al.* (1989) Characterization of a RNA virus from the parasite *Leishmania*. *Proc. Natl Acad. Sci. USA.*, **86**, 5979–82.

7

Hepatitis B virus and hepatitis delta virus

TIM J. HARRISON AND GEOFFREY M. DUSHEIKO

7.1 INTRODUCTION

Hepatitis B virus (HBV) may cause acute and chronic infection of the liver. Acute hepatitis B infection may cause serious icteric hepatitis or even fulminant hepatitis, though the infection may be anicteric and asymptomatic in a high proportion of cases. Chronic hepatitis B is defined by the presence of hepatitis B surface antigen (HBsAg) in serum for longer than six months. Several hundred million people worldwide are thus infected. The expression of the disease is complex: varying levels of viral replication occur and there is a spectrum of disease ranging from benign to severe forms of chronic hepatitis. Serum aminotransferases may or may not be elevated, and morphologic reactions in the liver range from minimal inflammatory change to destruction of hepatocytes and widespread inflammatory infiltration, cirrhosis or even hepatocellular carcinoma (HCC) (Beasley, 1988). The disease may remain clinically silent for decades but nonetheless progresses to cirrhosis and to HCC. Superinfection with hepatitis D may affect the expression of the disease. Variants of HBV infection with atypical serological markers add to the complexity of the infection. Although HBV is predominantly hepatotrophic, extrahepatic infection and complications can occur.

7.2 BIOLOGY OF HEPATITIS B VIRUS

7.2.1 Structure of the virus

The hepatitis B virion is a 42nm particle comprising an electron-dense core (nucleocapsid) 27nm in diameter surrounded by an outer envelope of the surface protein (HBsAg) embedded in membranous lipid derived from

Molecular and Cell Biology of Sexually Transmitted Diseases
Edited by D. Wright and L. Archard
Published in 1992 by Chapman and Hall, London ISBN 0 412 36510 3

the host cell. The surface antigen is produced in excess by the infected hepatocytes and is secreted in the form of 22nm particles and tubular structures of the same diameter. These subviral particles by far outnumber the virions in the circulation of infectious individuals and may prevent an effective humoral immune response to the surface protein in those who become chronically infected.

The 22nm particles are composed of the major surface protein in both non-glycosylated (p24) and glycosylated (gp27) form in roughly equimolar amounts along with a minority component of the so-called middle proteins (gp33 and gp36) which contain the pre-S2 domain, a glycosylated 55 amino acid N-terminal extension of the major surface protein. The surface of the virion has a similar composition but also contains the large surface proteins (p39 and gp42) which include both the pre-S1 and pre-S2 regions. These large surface proteins are not found in the 22nm spherical particles (but may be present in the tubular forms in highly viraemic individuals) and their detection in serum correlates with viraemia. The domain which binds to the specific HBV receptor on the hepatocyte appears to reside within the pre-S1 region (Neurath et al., 1986). Further details of the structure and synthesis of the pre-S proteins are given below (section 7.2.2).

The nucleocapsid of the virion consists of the viral genome surrounded by the core antigen (HBcAg). The carboxyl-terminus of the core protein is arginine rich and this highly basic domain is believed to interact with the nucleic acid. The genome, which is approximately 3.2 kilobases in length, has an unusual structure and is composed of two linear strands of DNA held in a circular configuration by base-pairing at the 5' ends (cohesive end region). One of the strands is less than genome length and the 3' end is associated with a DNA polymerase molecule which is able to complete this strand when supplied with deoxynucleoside triphosphates. In the past, this endogenous DNA polymerase reaction was used as a serological assay for the hepatitis B virion but this has now been superseded by more sensitive techniques such as DNA–DNA hybridization or the polymerase chain reaction. The 5' ends of both strands of the genome are modified: the 5' end of the complete strand is covalently linked to a protein and the 5' end of the incomplete strand is an oligoribonucleotide. In both cases, these moieties appear to be primers for the synthesis of the respective strands during the replication of the genome, as described below (section 7.2.3). A motif of 12 base pairs is directly repeated in the genome near to the 5' ends of the complete and incomplete strands (DR1 and DR2 respectively) and these sequences play an important role in the replication of the genome.

7.2.2 Organization of the HBV genome

To date, the genomes of more than a dozen isolates of hepatitis B virus have been cloned and the complete nucleotide sequences determined. Analysis of the coding potential of the genome reveals four open reading frames (ORFs) which are conserved amongst all of these isolates (Figure 7.1). These have the same polarity as the incomplete strand of genomic DNA which has accordingly been designated the plus strand.

The first ORF encodes the various forms of the surface protein and contains three in-frame methionine codons which are used for initiation of translation. Both the middle (gp33 and gp36) and major (p24 and p27) proteins are translated from a family of 2.1kb mRNAs which are transcribed from a promoter located in the pre-S1 region and are polyadenylated in response to a signal sequence located just downstream from the start of the core ORF. This promoter has homology to the late promoter of SV40 and, like that promoter, lacks the TATA box which normally determines the cap

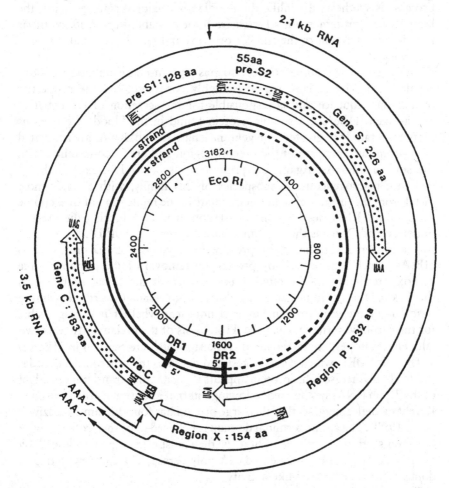

Figure 7.1 The structural organization of the Hepatitis B virus genome. See text for details. Modified from Tiollais *et al.* (1985).

205

site of eukaryotic mRNAs. In consequence, the 5' ends of this family of mRNAs are heterogeneous and span the pre-S2 initiation codon and the absence of this signal from many of the transcripts, perhaps coupled with the fact that it appears to be in a poor context for recognition by the scanning ribosome, leads to the synthesis of a vast excess of the major proteins (p24 and gp27) over the middle proteins.

A second promoter, which contains an extended TATA box (TATATAA), is located upstream of the pre-S1 initiation codon and directs the synthesis of a 2.4kb mRNA which is co-terminal with the other surface protein messengers and is translated to yield the large (pre-S1) surface proteins. This promoter appears to be weak (or may be down-regulated) so that the message is of low abundance and relatively little of the large surface proteins is synthesized. Unlike the middle and major surface proteins, the large surface protein is not secreted from the cell and, in fact, its synthesis inhibits the secretion of the smaller proteins and may be a signal for virus assembly.

The core open reading frame also has two in-phase initiation codons (Figure 7.1). The precore region is highly conserved, has the properties of a signal sequence and is responsible for the secretion of the hepatitis B e antigen (HBeAg), a soluble protein found in the blood of infectious carriers. A family of greater than genome length (3.5kb) RNAs are generated by a promoter upstream of the precore region and are 3' co-terminal with the surface mRNAs. Again, the promoter has no TATA-box and the 5' ends are heterogeneous. A subspecies of this family, with 5' ends most distal from the promoter, is an intermediate in the replication of the genome (pregenomic RNA, see below). Translation of mRNA lacking the precore initiation codon yields the structural core antigen (M_r=21kD). Translation of the longer mRNA from the precore AUG yields a 25kD precursor to HBeAg which is processed by proteolytic removal of the protamine-like carboxyl-terminus and secreted after cleavage of the signal sequence as the 17kD HBeAg. A protease-like domain is present towards the amino terminus of the core protein but it is not clear whether it is responsible for the carboxyl-terminal cleavage. Proteolysis of the precursor reveals two antigenic epitopes, characteristic of HBeAg, which are masked in HBcAg.

The third ORF, which is the largest and overlaps the other three, encodes the viral polymerase. This protein appears to be another translation product of the 3.5kb RNA, synthesized following internal initiation by the ribosome. Recent evidence suggests that this large protein has four domains (Radziwill et al., 1990). The amino terminal domain is believed to be the protein primer for minus strand synthesis. There is then a spacer region followed by the (RNA and DNA-dependent) DNA polymerase. The carboxyl-terminal domain has an RNase H-like activity.

The fourth ORF was designated X for the unknown function of its small gene product. However, this has now been shown to be a transcriptional transactivator and it may effectively be an early gene product which functions to up-regulate the core promoter. The X protein is active on a wide range of promoters and appears to act not by binding directly to DNA but by binding to cellular factors which themselves are DNA-binding. The X gene has its own promoter, which is located adjacent to a transcriptional enhancer and transcribes an RNA of around 0.8kb.

7.2.3 Replication of the HBV genome

Following infection of the hepatocyte, the single-stranded region of the virion DNA is repaired (presumably by the endogenous polymerase) and the genome appears in a covalently closed, circular form in the nucleus. This DNA is the template for the transcription of all of the viral RNAs which are 3' co-terminal, being polyadenylated in response to a signal (TATAAA) near to the start of the core ORF. During transcription of the 3.5 kb RNAs, this signal is read through at the first pass, and is perhaps not recognized because of its close proximity to the promoter, so that greater than genome length transcripts are generated. As mentioned above, a subset of the 3.5kb RNA is destined to become pregenomes and these are encapsidated along with the polymerase in immature core particles in the cytoplasm. Synthesis of minus strand DNA is primed by a protein, now believed to be the amino-terminal domain of the polymerase, and proceeds with the concomitant degradation of the RNA template by the RNase H activity. It is not clear whether one or two copies of the polymerase molecule are involved nor whether the primer domain is cleaved from the remainder of the polymerase.

The pregenomic RNA contains two copies of DR1 and minus strand synthesis is primed at the 3'-most copy so that the completed strand has a short terminal redundancy (Figure 7.2). Degradation of the pregenome by the RNase H activity leaves a short, capped oligoribonucleotide which contains a copy of DR1 and this is now translocated to the copy of DR2 on the minus strand where it primes plus strand synthesis. If the primer for minus strand synthesis and the polymerase are indeed the same molecule, the two ends of the minus strand will be held in close proximity and this may facilitate the translocation of the primer for the plus strand. Plus strand synthesis proceeds to the 5' end of the minus strand and a second template switch is required as the molecule is circularized via the redundancy in the minus strand. Completion of the core during plus strand synthesis starves the polymerase of deoxynucleoside triphosphate precursors and leaves the single-stranded region characteristic of genomic DNA.

There is no semi-conservative replication of the covalently-closed circular DNA in the nucleus and the pool of template DNA is initially built up by transfer of some of the progeny DNA from the cytoplasm to the nucleus. The mode of replication of the viral genome resembles that of the retroviruses

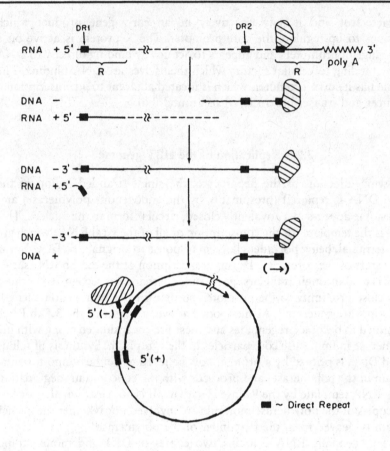

Figure 7.2 Replication of the Hepadnavirus Genome. See text for details. Modified from Lien *et al.* (1986).

and, more closely, a family of plant viruses (the caulimoviruses) and these viruses appear to be related phylogenetically. The HBV polymerase has homology with the polymerases of these viruses at the amino acid level.

7.2.4 Detection of HBV DNA in the liver

Southern hybridization to DNA extracted from the liver of patients who are seropositive for HBeAg reveals intermediates in the replication of the genome (Figure 7.3, lane A6). There is a bimodal smear of hybridization in the low molecular weight region of the gel. The more intense, faster migrating species are minus strand DNA molecules which were synthesized on the RNA template. The higher molecular weight species, up to the genome length of 3.2kb, are partially double-stranded molecules and represent plus strand synthesis. This apparent asynchrony of viral DNA synthesis was one of the clues that led to the discovery that the HBV genome is not replicated semiconservatively but through an RNA intermediate.

Occasionally, in restriction endonuclease digested liver DNA, higher molecular weight species may be seen migrating in the upper portion of the gel. These are an indication that the viral genome has integrated into cellular DNA (Figure 7.3, lanes B and C). Their detection implies either that identical integration events have occurred in a number of cells or that, following an initial integration event, there has been a clonal expansion of that cell. Since there is no evidence for specific sites of integration of HBV DNA in the human genome, the latter appears to be the case and this integration and clonal expansion may be the first step in a process that may

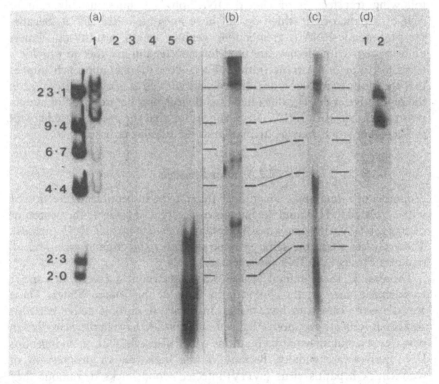

Figure 7.3 Hepatitis B virus DNA in the liver. Lane A6; Biopsy from an HBeAg-positive patient showing replicative forms. Lane C; Biopsy from an HBeAg-positive patient showing both replicative and integrated viral DNA. Lane B; Integrated viral DNA in an anti-HBe-positive patient. Lanes D; Liver DNA from an anti-HBe-positive patient with hepatocellular carcinoma; D1: DNA from non-tumourous liver, D2: tumour DNA. In all cases extracted DNA was digested with the restriction enzyme Hind III. Left lane; radiolabelled Hind III fragments of bacteriophage lambda DNA as size markers (sizes given in kilobase pairs to the left of the figure). Modified from Harrison *et al.* (1986).

eventually lead to tumour formation, as discussed below. Such integrations may sometimes also be detected in the livers of HBsAg carriers who have cleared virus replication and their expression may lead to the production of HBsAg in the absence of virus replication.

7.2.5 Expression of HBV *in vitro*

Over the years, many unsuccessful attempts have been made to propagate HBV in cells in culture both using primary hepatocytes and using established cell lines. Following the molecular cloning of the viral DNA, the various gene products have been expressed using both prokaryotic and eukaryotic systems. The production of complete virions has also been achieved following transfection of cloned DNA into cell lines, usually tumour cells of hepatic origin which do not have integrated HBV DNA. Because the pregenomic RNA is greater than genome length, head-to-tail dimers or recircularized molecules are used for transfection in order to provide a suitable template for transcription. The success of this approach implies that the block to infection of cultured cells with virus is at an early stage in the replicative cycle. The duck hepatitis B virus may be produced in human cells transfected with DHBV DNA suggesting that the species specificity of the hepadnaviruses may be attributed to the cellular receptor.

7.2.6 Animal models

A number of animal species have been found to be susceptible to infection by viruses related to HBV and the human virus is now considered the prototype of a group of viruses known as hepadnaviruses (*hepa*tropic DNA viruses). These viruses share the same genome structure and organization and all replicate through an RNA intermediate.

The first of these animal viruses was detected in a colony of captive woodchucks held at the Philadelphia Zoo in the United States. These animals were known to have a high incidence of chronic active hepatitis and hepatocellular carcinoma and were observed to have in their circulation particles resembling hepatitis B virions which also exhibited an endogenous DNA polymerase activity. Because of the high rate of progression of woodchuck hepatitis virus (WHV)-infected woodchucks to tumour they have become an attractive model for the study of hepadnavirus-associated hepatocellular carcinoma despite the difficulties in working with an animal which is both wild and given to hibernation. WHV-infected woodchucks have been successfully superinfected with hepatitis delta virus (HDV) and so provide a model for the study of that agent.

Beechey ground squirrels may also be infected by a hepadnavirus, ground squirrel hepatitis virus (GSHV), though in this case the virus appears to be less pathogenic with a low rate of progression to primary liver cancer.

The viral genome has approximately 90% nucleotide homology with WHV, while both rodent viruses are approximately 70% homologous with HBV.

Duck hepatitis B virus (DHBV) naturally infects Pekin ducks and the virus is endemic in many domestic flocks. Transmission is vertical and hatchlings have a high level of viraemia which later declines. Similar viruses have been described in wild ducks and herons. Not surprisingly, the avian viruses have diverged from those infecting mammals; there is a slightly shorter genome (around 3.0kb) and the X gene is absent. The Pekin duck has also proved useful to the study of virology and it was with this model that Summers and Mason (1982) elucidated the mechanism of replication of hepadnavirus genomes. DHBV-infected ducks are used in many laboratories for the evaluation of potential anti-viral drugs.

7.3 EPIDEMIOLOGY AND TRANSMISSION OF HEPATITIS B

There are important geographical differences in the prevalence of hepatitis B and in the mode and age of exposure. In Western Europe, North America and other developed countries, hepatitis B occurs sporadically; within these countries, the major risk factors are male homosexuality, low socio-economic status, drug abuse, ethnic group, sexual promiscuity, residence in institutions, mental handicap and employment in health professions (Baddour et al., 1988; Van Ditzhuijsen et al., 1988). The disease is spread by percutaneous and permucosal routes. Sporadic cases are often transmitted from carriers. Familial clustering may occur. Infusion of pooled blood products also carries a high risk for the transmission of hepatitis B. In areas of the world where hepatitis B is endemic, the acquisition of infection in childhood is an important mode of transmission.

Anicteric and asymptomatic infection and chronicity are hallmarks of childhood and perinatal infection, as the likelihood of chronic infection is inversely related to the age of infection. This pattern establishes a prevalence of endemic disease in many parts of the world. Neonatal (maternal/infant) transmission is relatively common in Chinese compared with Africans, reflecting the higher degree of infectivity in pregnant women among the former. The carrier rate is usually 2–3 times higher in males.

7.3.1 Subtypes of hepatitis B virus

The discovery of variation in the epitopes presented on the surface of the virions and subviral particles enabled the definition of subtypes of HBV which differ in their geographical distribution. All isolates of the virus share a common epitope, a, which is a domain of the major surface protein. Anti-a antibodies appear to be protective against infection and constitute the major antibody response in vaccinated individuals. Two other pairs of mutually exclusive antigenic determinants, d or y and w or r, are also present on

211

the major surface protein. These variations have been correlated with single nucleotide changes in the ORF which lead to variation in single amino acids in the surface protein (Okamoto *et al.*, 1987).

Thus, four principal subtypes of HBV are recognized: *adw*, *adr*, *ayw* and *ayr*. Subtype *adw* predominates in northern Europe, the Americas and Australia and is also found in Africa and Asia whilst *ayw* is found in the Mediterranean region, eastern Europe, northern and western Africa, the Near East and the Indian subcontinent. In the Far East, *adr* predominates but the rarer *ayr* may occasionally be found in Japan and Papua New Guinea. These geographic distinctions are breaking down as a result of mass migration but more sophisticated methods, such as a combination of PCR and DNA sequencing, are now available for studying the routes of transmission of HBV.

7.3.2 Variants of Hepatitis B virus

The description of cases of HBV infection in Senegal which were characterized serologically by a lack of anti-HBc and low levels of HBsAg led to the hypothesis that there was a second form of HBV (HBV 2). However, it now appears likely that these cases arose from a lack of host responsiveness rather than a true variant of the virus.

There have been a number of descriptions of patients in the Mediterranean region whose infections are characterized by high levels of circulating virus in the absence of HBeAg. Sequence analysis of viral DNA from these infections reveals that there are frequently mutations in the precore region (often a point mutation leading to a termination codon or, more rarely, the loss of the precore initiation codon) such that HBeAg cannot be synthesized. It has been suggested that these mutations may lead to increased virulence.

7.3.3 Sexual transmission

Hepatitis B is an important cause of hepatitis amongst homosexuals, and indeed sexual spread may be the most common mode of spread of hepatitis B in some Western countries (Alter *et al.*, 1989). This route of transmission may be important in the developed countries in which HBsAg carrier rates are low and where hepatitis B is acquired by susceptible sexually active young adults and homosexual men. This difference in prevalence of sexually acquired hepatitis B between high exposure and low exposure groups has been demonstrated in South Africa. A study conducted in 1977 showed that the prevalence of HBsAg was significantly greater in Caucasian patients attending venereal disease clinics than in healthy white blood donors. However, no such difference was found amongst blacks with venereal disease (Schneider, 1977). This evidence suggests that HBV transmission by sexual contact between adults is less likely in developing countries in whom HBV infection is acquired before sexual maturity. However, with a shift in living standards and socioeconomic status, it is likely that increasing numbers of patients in developing countries do acquire HBV infection through sexual spread.

Secondary cases of type B hepatitis are not infrequent amongst susceptible spouses, and the sexual partners of patients with acute or chronic type B hepatitis. Prospective studies of such persons have established that the secondary attack rate is of the order of 20%. Infection may be transmitted more frequently from males to females, but spread in the reverse direction can occur. There is a high incidence of hepatitis B amongst the sexually promiscuous including prostitutes, and patients with other sexually transmitted diseases but not in lesbians. Although all sexual partners of infected persons are at risk, the prevalence of HBV markers is highest amongst promiscuous homosexual men. Point prevalence studies, such as the Belle Glade, Florida community study, clearly show that the risk of heterosexual transmission of HBV correlates with the number of lifetime sexual partners, and with a positive serologic test for syphilis (Rosenblum, 1988).

Hepatitis B is also not infrequent amongst travellers to Far Eastern countries who consort with prostitutes. The mode of sexual spread depends on several factors. HBsAg can be found in saliva of chronic carriers and kissing or oral contact alone may be a vehicle. Oral inoculation of hepatitis B has been shown to occur experimentally, although transmission probably occurs more frequently when associated with traumatic procedures or when associated with large open wounds in the mouth. The mucus membranes of the genital tract and possibly the lower gastrointestinal tract are equally susceptible. Animal experiments have shown that semen from HBsAg-positive carriers is infectious, and semen introduced intra-vaginally has resulted in infection.

The extremely high incidence of HBV infection amongst male homosexuals is accounted for by several factors. Infectivity in this group no doubt relates to the high frequency of HBeAg in blood of young male hepatitis B carriers who have recently acquired the carrier state. Moreover, male homosexuals, at least prior to the HIV epidemic, were known to have a large number of sexual partners, as many as a 1000 in a lifetime. Their promiscuity accounts for the fact that hepatitis B markers are far more common in them than in the general population of the United States (60% v. 5–7%). Patterns of sexual behaviour amongst homosexual men are of major importance in the acquisition of infection. HBV seropositivity in homosexuals has been significantly correlated with the duration of regular homosexual activity and with the number of male sexual contacts. Anal intercourse, oral–anal intercourse and rectal trauma dispose to HBV infection. The highest relative risk has been associated with passive anal-genital intercourse with casual partners. Asymptomatic rectal bleeding occurs in a high proportion (59% in one study) of homosexuals, and the friable rectal mucosa appears to be the portal of entry for infected semen. Surprisingly, active oral–anal intercourse has also been related to HBV infection, perhaps because of buccal inoculation of blood-contaminated rectal mucus. With the advent of the HIV epidemic, sexual patterns in homosexuals have changed and the frequency of cases of HBV infection is declining, in contrast to that seen in the general population and amongst intravenous drug abusers.

7.4 ACUTE HEPATITIS B

The incubation period of acute hepatitis B is 28–225 days (mean 75). The clinical features of acute hepatitis B are frequently so similar as to be indistinguishable from the other viral hepatitides; only minor features of the clinical disease, taken together with the incubation period and the epidemiologic history, serve to distinguish the various acute hepatitides, and specific diagnosis requires serologic testing. The initial symptoms in the pre-icteric phase are malaise, fatigue, listlessness, lack of energy and weakness. Anorexia and nausea are often present, accompanied by vomiting which may be induced by fatty food. Patients may also have a distaste for cigarettes and complain of right upper quadrant pain. Diarrhoea or constipation can occur. Non-specific symptoms such as diffuse aches and pains or myalgias are also common. Acute hepatitis B may be accompanied by serum sickness-like syndrome in 5–15% of patients, manifested by a low grade fever, urticarial rash and arthralgias of the wrist, elbow, knee and ankle.

The initial symptoms may then be followed by the appearance of dark urine and jaundice in icteric cases. Some patients will have jaundice without any preceding symptoms. Typically, once jaundice appears some of the prodromal symptoms improve. The jaundice may deepen within 5–10 days, and anorexia and fatigue may worsen during this period. A degree of weight loss can occur.

Towards recovery, nausea disappears and appetite and well-being return. If the jaundice is prolonged and deep, some pruritus may be noted. Malaise and fatigue may persist for some time and mild relapses, marked by an elevation of the serum transaminases or serum bilirubin, can occur in 1–5% of patients. The relationship of such relapses to a return to work or vigorous exercise has been noted, but is unproven. However, exercise tolerance is generally decreased in the early convalescent phase and depression may be a prominent symptom.

7.4.1 Physical signs

The physical signs are usually minimal. The most common abnormal physical findings are jaundice, hepatic tenderness, hepatomegaly, splenomegaly (5–10%) and, in some patients, lymphadenopathy. Jaundice may affect only the sclera in mild cases. Jaundice usually decreases within a few days to two weeks, but can persist for as long as six weeks. The liver is not usually markedly increased in size, but some hepatic tenderness will be elicited in most patients. A soft non-tender splenic tip can be palpated when enlarged. The rest of the physical examination is usually unremarkable. If fulminant hepatitis supervenes, the symptoms and signs are correspondingly more severe. Skin rashes may be noted. In children with acute type B hepatitis, papular acrodermatitis (Gianotti's syndrome), characterized by erythematous papules on the arms, legs and face may occur. Signs of portal hypertension such as ascites, prominent abdominal venous patterns, oesophageal variceal bleeding, gynaecomastia, palmar erythema, wasting and peripheral oedema are not present in uncomplicated acute hepatitis B. In the majority of young children and those adults with anicteric disease, the disease may pass completely unnoticed or only a mild fatigue may be noticed. There are no abnormal physical findings, and the only indications of an acute hepatitis are raised serum aminotransferases.

7.4.2 Laboratory tests

Acute hepatitis B is marked by an elevation in the serum aminotransferase activities (ALT and AST). Both are usually elevated in acute icteric hepatitis to more than eight times the upper limit of normal. ALT is usually as high or higher than the AST. The serum bilirubin concentrations may increase commensurate with the severity of the illness. In anicteric viral hepatitis the ALT and AST are raised in the absence of major symptoms. With recovery, the serum aminotransferase concentrations decline rapidly and usually become normal within 10–12 weeks of onset. The persistence of abnormal ALT activity or the persistence of HBsAg in serum for more than six months after the onset of acute hepatitis is usually indicative of the progression of the disease. Haemoglobin, white cells and platelets usually do not change unless the disease is severe, or is associated with fulminant hepatic failure. The prothrombin time is generally normal in uncomplicated acute viral hepatitis, but may be prolonged in fulminant hepatitis. Serum albumin and globulin are usually normal at the onset, but serum albumin concentrations may fall slightly during the course of hepatitis, and serum globulins may rise especially if the illness is severe and prolonged. Serum glucose is usually normal in uncomplicated viral hepatitis, but the majority of patients with fulminant disease will have hypoglycaemia. The α-foetoprotein value is increased transiently in patients with acute viral hepatitis. Liver biopsy is not usually performed in acute viral hepatitis. The typical outcome of acute hepatitis B is indicated in Figure 7.4. Following exposure, HBsAg can be detected in serum for several weeks before increases in serum aminotransferases occur. HBsAg persists during the prodromal phase and is not usually cleared from the

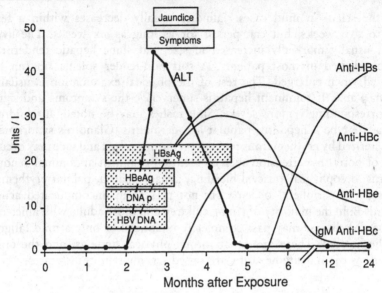

Figure 7.4 Typical serological course in a patient with acute hepatitis B.

serum until convalescence. Shortly after HBsAg is detectable, HBV DNA and DNA polymerase, anti-HBc and HBeAg appear in the circulation and are evidence of viral replication occuring before the advent of biochemical or clinical hepatitis. A positive IgM anti-HBc test typically distinguishes acute from chronic hepatitis B, as commercial assays are artificially designed to detect higher levels of IgM anti-HBc in serum.

By the time the patient consults the physician, HBV DNA and HBeAg are often no longer detectable in serum, indicative of cessation of viral replication. The loss of HBeAg is a good prognostic sign, and indicates that the patient will clear HBsAg and will not develop chronic infection. Disappearance of HBeAg is usually followed by appearance of serum anti-HBe. HBsAg usually persists throughout the illness and disappears with recovery. Anti-HBs is the last marker to appear in serum and is only detectable as antibody excess once HBsAg has been cleared from serum. Many patients will have a period after the clearance of HBsAg during which anti-HBs is not detected and during this period anti-HBc may be the only serologic marker. Ten percent of patients never produce anti-HBs despite a complete clinical recovery. Pre-S proteins can be detected in patients with acute HBV infection and show a good correlation with the detection of HBV DNA.

Following a minor infection, HBsAg is present only transiently in serum. The only evidence of infection may be the sudden development of anti-HBc and anti-HBs. This pattern of infection is probably the most common outcome of an infection with HBV. A patient with fulminant hepatitis

216

may become HBeAg-negative and HBsAg-negative, even as his condition deteriorates. A few patients become HBsAg-negative before they develop significant symptoms, and will test negative for HBsAg though they are suffering from acute type B hepatitis. IgM anti-HBc is positive in these patients. Some patients with acute hepatitis who are HBsAg-positive do not have acute type B hepatitis. This situation occurs when a chronic carrier develops a second type of hepatitis such as A, NANB or delta hepatitis. IgM anti-HBc will distinguish acute from chronic hepatitis B. A superimposed hepatitis A can be diagnosed by means of IgM anti-HAV.

7.4.3 Treatment of acute hepatitis

Most patients can be cared for at home. Admission to hospital may be required for social reasons, for diagnosis, if persistent vomiting ensues, if the prothrombin time is more than two seconds prolonged, or if changes in personality or sleep behaviour are observed. In hospital patients require barrier nursing. There is no specific treatment for acute viral hepatitis. Rest is recommended and the patient should avoid alcohol and exhausting exercise for at least three months. Most people require at least three weeks rest before returning to work. There is no evidence to suggest that antiviral treatment, particularly alpha interferon, will reduce the morbidity of acute hepatitis B; Serum bilirubin and ALT levels should be measured weekly. HBsAg should be tested until no longer detectable.

7.4.4 Complications

The hepatic complications of acute hepatitis B include relapse, fulminant hepatitis, prolonged cholestasis, and chronic viral hepatitis. Case fatality rates for hepatitis B, C and D are higher than those for hepatitis A. The more common extrahepatic complications include polyarteritis nodosa, glomerulonephritis, Guillain Barre syndrome, meningoencephalitis, peripheral neuropathy, aplastic anaemia, agranulocytosis haemolytic anaemia, myocarditis, adult respiratory distress syndrome, pleural effusions, diarrhoea, acute pancreatitis, urticaria, papular acrodermatitis and polyarthritis.

7.5 CHRONIC HEPATITIS B

7.5.1 The carrier state

The carrier state is defined as the continued presence of HBsAg in serum for six months or more. Following some acute HBV infections, often those which are asymptomatic, the immune system fails to clear the virus from the liver and a persistent infection ensues. The probability of an acute infection of a healthy adult becoming chronic is 5–10% but this rises as high as 90% in immunologically immature infants who are infected perinatally.

In the first phase of chronicity, virus replication is ongoing in the liver and replicative intermediates of the viral genome may be detected in DNA extracted from biopsies (Figure 7.3, lane A6). Markers of virus replication such as HBV DNA, HBeAg and the pre-S1 proteins may be detected in the serum. In those infected at a very young age, this phase may persist for life but more usually virus levels decline over time and there is eventually immune clearance of infected hepatocytes associated with seroconversion from HBeAg- to anti-HBe-positivity (Figure 7.5). Antiviral therapy seeks to facilitate this process. Rarely, seroconversion to anti-HBs also occurs but, more frequently, HBsAg persists as a result of the expression of integrated viral DNA.

7.5.2 Hepatitis B e antigen and antibody

Detection of HBeAg has proven useful as a readily detectable and reasonably sensitive index of the presence of virions in serum, and therefore of higher levels of infectivity. Commercial RIA and ELISA techniques are now widely available for the detection of HBeAg and anti-HBe. The persistence of HBe in a patient's serum for greater than ten weeks during acute hepatitis B infection indicates progression of that patient's disease to persistent infection, and the presence of HBeAg in the serum of a chronic carrier indicates a greater potential for transmission of HBV and of active disease. In contrast anti-HBe has been associated with the asymptomatic chronic carrier state, and with less risk of parenteral and maternal–infant transmission. However, HBsAg positive blood containing anti-HBe may still contain detectable HBV DNA and be infectious (Seeff, 1988).

7.5.3 Pathogenesis

The mechanism of chronicity, and the range of disease spectrums is not understood; continuing viral replication in the liver reflects an inadequate immune response to the virus, which may result from an inadequately developed immune response in those infected at an early age, or to immune modulation by circulating maternal anti-HBc, or a relative deficiency or suppression of the interferon system (Onji et al., 1989; Twu et al., 1988). However, under certain circumstances, clearance of replicating virus after prolonged chronic hepatitis, can be achieved, and demonstrable induction and activation of the interferon system has been documented in such longstanding carriers (Shindo et al., 1988; Heathcote et al., 1989). The virus itself may not be cytopathic, and hepatic injury may require a targeted host response, abetted by antigen presentation in association with HLA molecules. HBcAg may be the target antigen to which immune mediated cytotoxic T cell attack is directed, whereas envelope protein epitopes are recognized by B cells (Milich *et al.*, 1988).

7.5.4 Clinical features of chronic hepatitis B

The clinical features depend upon the stage and activity of the disease. Like other forms of chronic hepatitis, the disease can be classified as chronic persistent or chronic active hepatitis (CPH or CAH), but these pathological definitions are an oversimplification. Many patients may be asymptomatic; in patients with more severe hepatitic injury, fatigue may be prominent, or the patient may present with established chronic active hepatitis, or even cirrhosis with jaundice, ascites, portal hypertension and encephalopathy. Clinical examination may be normal, or patients may have hepatomegaly, spider naevi, palmar erythema and splenomegaly, ascites, bleeding varices or even hepatocellular carcinoma.

7.5.5 Natural history

The natural history is variable. The prognosis of histological CAH is worse than CPH. Prospective follow-up of patients with chronic hepatitis B indicate that approximately 5–15% of HBeAg-positive carriers spontaneously remit

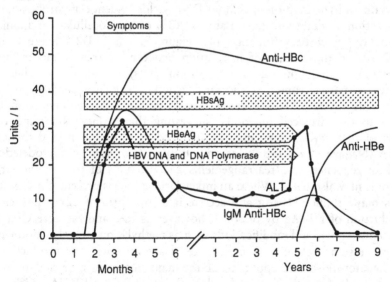

Figure 7.5 Typical serological course in a patient with chronic hepatitis B. In the first phase of the infection, the disease is predominantly replicative, and markers of HBV replication including HBeAg and HBV DNA are found in serum. After a varying interval, HBV DNA is no longer detectable in serum by dot-blot hybridization, and the patient seroconverts to anti-HBe.

and seroconvert to anti-HBe each year. Spontaneous clearance of HBeAg may be associated with an abrupt elevation of serum ALT concentrations in many patients. Loss of HBsAg is, however, rare. Some anti-HBe patients remain HBV DNA-positive and have raised serum aminotransferases. This may reflect either infection with a HBV variant, or very low levels of HBV replication where HBeAg is mopped up by excess antibody. Reactivation of HBV replication may occur after chemotherapy. Superinfection with HIV may lead to accelerated disease. Alcohol abuse may also adversely affect the outcome.

7.5.6 HBV and hepatocellular carcinoma

When tests for HBsAg became widely available, it became clear that the regions of the world where the chronic carrier state is common are coincident with those where there is a high prevalence of primary liver cancer. Furthermore, in these areas patients with tumour are almost invariably seropositive for HBsAg. In an elegant prospective study, Beasley and colleagues (Beasley et al., 1981) investigated the incidence of this tumour in male HBsAg carriers and controls in Taiwan. After more than 11 years of follow up, 184 cases of hepatocellular carcinoma had appeared in the 3454 individuals who were HBsAg-positive at the start of the study but only ten such tumours arose in the 19 253 individuals who were HBsAg-negative.

Southern hybridization of tumour DNA yields evidence of chromosomal integration of viral sequences in at least 80% of hepatocellular carcinomas from HBsAg carriers (for example, Figure 7.3, lane D2). There is no similarity in the pattern of integration between different tumours, and variation is seen both in the integration site(s) and in the number of copies or partial copies of the viral genome. Sequence analysis of the integrants reveals that the direct repeats in the viral genome often lie close to the virus/cell junctions, suggesting that sequences around the ends of the viral genome may be involved in recombination with host DNA (Nagaya et al., 1987). Integration appears to involve microdeletion of host sequences and rearrangements and deletions of part of the viral genome may also occur. Where an intact surface gene is present, the tumour cells may produce HBsAg and secrete it in the form of 22nm particles. Production of HBcAg by tumours, however, is rare and the core ORF is often incomplete and modifications such as methylation may also modulate its expression. Cytotoxic T cells targeted against core gene products on the hepatocyte surface appear to be the major mechanism of clearance of infected cells from the liver and cells with integrated viral DNA which are capable of expressing these proteins may also be lysed. Thus, there may effectively be immune selection of cells with integrated viral DNA which are incapable of HBcAg expression.

Despite considerable effort by many research groups, the mechanism of oncogenesis by HBV remains obscure. It has been suggested that HBV may act non-specifically through the active regeneration and cirrhosis which may be associated with long-term chronicity. However, HBV-associated tumours often arise in the absence of cirrhosis and such theories do not explain the frequent finding of integrated viral DNA in tumours. In rare instances the viral genome has been found to be integrated into cellular genes such as cyclin A (Wang *et al.*, 1990) or a retinoic acid receptor (Dejean *et al.*, 1986). Translocations and other chromosomal rearrangements have also been observed. Although insertional mutagenesis remains an attractive hypothesis to explain the oncogenicity of this virus, supportive evidence has generally not been forthcoming.

An alternative possibility is that tumour formation is associated with the action of a viral gene product. The product of the X gene is known to be a fairly non-specific transactivator of transcription and so may inappropriately up-regulate cellular genes. It has recently been shown that truncated forms of HBsAg, which may be produced from incomplete surface ORFs integrated in tumour cells, may also have transactivating activity (Kekule *et al.*, 1990), presumably through interaction with receptors in the cell membrane.

As is the case with most cancers, development of hepatocellular carcinoma is likely to be a multistep process involving a number of factors. The clonal expansion of cells with integrated viral DNA appears to be an early stage in this process and such clones may accumulate in the liver throughout the period of active virus replication. In areas where the prevalence of primary liver cancer is high, virus infection usually occurs at an early age and chronic virus replication may be prolonged, yet the peak incidence of tumour is some decades after the initial infection. Hepatitis B acquired by sexual transmission generally occurs later in life so that the probability of progression to chronicity is lower and the period of active virus replication likely to be shorter.

7.6 TREATMENT

Although effective vaccines against HBV have been developed, the need for effective treatment remains to prevent morbidity from the disease and to reduce the infectivity of individual carriers remains. There has been a long and intensive search for agents that inhibit or eradicate HBV and are not injurious to the host. Few agents have achieved clinical applicability.

7.6.1 Corticosteroids

There is no sound foundation for advocating corticosteroid therapy in patients with persistent HBV infection, and recent evidence suggests that long-term treatment with steroids is not beneficial and may be detrimental.

7.6.2 Antiviral agents

Acyclovir, a synthetic purine nucleoside analogue is effective against herpes viruses, particularly because herpes virus thymidine kinase accelerates the phosphorylation of ACV, adding to the drug's action as a viral DNA chain terminator. ACV is a modest inhibitor of HBV replication. In two controlled trials ACV, given at high doses, intravenously, for 28 days had little therapeutic benefit. ACV and ACV analogues have also been used in combination with interferon, but their efficacy compared with IFN alone is uncertain (De Man *et al.*, 1988; Dusheiko, 1988).

Foscarnet (Trisodium phosphonoformate) treatment decreases serum levels of HBV DNA. Patients with fulminant hepatitis may be benefited by this drug, but further research is required.

3-Azido-3'-deoxythymidine triphosphate, a reverse transcriptase inhibitor, has an inhibitory effect upon HBV replication. The drug has been used to treat patients with HIV and HBV infection, but would probably only merit consideration in this group.

Adenine arabinoside is an analogue of adenine and a potent inhibitor of HBV DNA polymerase. Adenine arabinoside 5'-monophosphate, a water soluble congener, allows intramuscular administration. A large number of hepatitis B carriers have been treated with Ara-AMP, and in these studies, suppression of HBV DNA and DNA polymerase has been demonstrated (Marcellin *et al.*, 1989). Usually eight weeks of treatment is required, but serious neurotoxicity develops in many after four weeks. The drug may prove to be of use in combination therapy.

7.6.3 Interferons

These complex antiviral and immunomodulatory agents are the most widely used antiviral compounds for the treatment of chronic HBV infection. Both alpha, beta and gamma interferons have been used. Several controlled trials using alpha interferon have been completed, and the overall results suggest that a proportion of patients may respond to treatment with a permanent inhibition of HBV replication and amelioration of disease (Hoofnagle *et al.*, 1988). Typically, patients who clear HBeAg or HBsAg develop a transient flare, or exacerbation of the disease in the latter weeks of treatment. Expected response rates are 35–45%. Improved responses have been observed in Caucasians, in those with recent onset of disease, those with more active disease, and in anti-HIV negative patients (Lok *et al.*, 1988). This characterization is not absolute, adding to the difficulty in treating patients. The appropriate dose and duration of treatment are still under study, but a dose of 5–10 million units three times a week for 12–16 weeks is currently used.

Combinations of alpha, beta and gamma interferon are still being studied.

Their efficacy should be compared with that of alpha interferon alone (Bissett *et al.*, 1988; Caselmann *et al.*, 1989).

7.6.4 Pulsed corticosteroid treatment and interferon

It has been observed that patients with chronic hepatitis B who are withdrawn from corticosteroid or cytotoxic therapy not infrequently have a rebound exacerbation of the disease associated with an increase in serum aminotransferases and occasionally loss of HBsAg. The first trials of pulsed corticosteroid withdrawal and subsequent interferon proved encouraging (Perrillo *et al.*, 1988). A subsequent multicentre controlled trial comparing the efficacy of this combined regimen with α-interferon alone did not show an added benefit of corticosteroid plus interferon. However, in those patients with low pretreatment ALT, the regimen may be beneficial. The treatment should be used with caution in those patients with decompensated hepatitis B because of the risk of inducing severe hepatic necrosis.

7.6.5 Other treatments

Levamizole, an anti-helminthic, has been claimed to inhibit HBV replication in up to 60% of patients. However, it is not yet established whether the initial results have proved reproducible.

Phyllanthrus amarus: In a preliminary study, hepatitis B carriers were treated with this plant extract; a high proportion of carriers (59%) lost HBsAg (Thyagarajan *et al.*, 1988). The striking results in this study have also not been reproduced at the time of writing.

Interleukin-2. Patients have been treated for 28 days with 15 μg recombinant IL-2 intravenously. Some transient inhibition of HBV DNA polymerase was observed. The usefulness of IL-2 will no doubt be further explored, as will thymostimulin (Minuk and LaFreniere, 1988).

7.6.6 Liver transplantation

End-stage cirrhosis due to hepatitis B can be treated by liver transplantation. The major limiting factor at the present time is the very high rate of recurrence of infection in HBeAg, HBV DNA positive patients, despite the administration of hepatitis B immunoglobulin prophylaxis (Ferla *et al.*, 1988).

7.7 PREVENTION

Recovery from a natural infection with hepatitis B virus, or vaccination with the purified envelope proteins, leads to production of antibodies against HBsAg. The humoral immune response in man is generally directed against the A determinant of HBsAg which is common to all other serotypes of the virus, and therefore immunization with one serotype of HBsAg is generally protective against all serotypes of HBsAg. Three hydrophobic and two hydrophilic areas can be discerned in the HBsAg molecule; the central portion of the S domain (aa 120–160) is hydrophilic, and in the small HBsAg particles, it is exposed to the surface, where it forms an antigenic loop, which acquires a complicated conformation involving several disulphide bonds. The significance of conformation for HBsAg determinants has been well established. Most antibodies to the small HBsAg cannot usually therefore be detected by immune blotting. Minor determinants which are responsible for subtype variation have been discussed in section 7.3.1. Responses to these do not play a major role in the immune response to vaccine. Studies with peptide analogues of HBsAg have shown that the nonapeptide sequence 139–147 represents the total or an essential part of the a determinant of HBsAg (Bhatnagar et al., 1982).

Current vaccines are prepared from 22 nm viral coat particles obtained from plasma of chronic carriers, or by recombinant DNA technology (Andre, 1988; Murray, 1988). Both are satisfactorily inactivated and purified, and effective (Odaka et al., 1988), but recombinant yeast vaccines are gaining wider acceptance in the Western World.

The first vaccines produced for commercial use were derived from persons who were chronic carriers of HBV. Such an approach was possible because of the disproportionate production of HBsAg as compared with whole virus particles in carriers. The number of incomplete virus particles in serum of a typical carrier can range up to a trillion particles per ml of blood. Large volumes of plasma were obtained by plasmapheresis from carriers with high titres of HBsAg. By rate zonal and density gradient centrifugation, the 22 nm particles were isolated from Dane particles and tubular forms, and then inactivated by treatment with formaldehyde, urea and pepsin digestion.

Extensive safety tests were undertaken in chimpanzees and selected human populations. The aqueous vaccine preparations were more immunogenic when changed to include an immunoadjuvant, aluminium hydroxide (alum). The currently licensed vaccines has been shown to induce anti-HBs in 80–90% of recipients after immunization. The anti-HBs that develops has specificity for the common HBsAg determinant a, and is hence protective against both the ad and ay subtypes of HBV (Alter, 1982).

The construction of yeast vaccines has been described in detail; after cloning of HBV DNA (subtype adw) from human serum, HBV DNA was cloned as a 3200 base pair fragment in an E. coli vector. Expression plasmids were constructed, containing genetic elements typical of E. coli/S. cerevisiae shuttle vectors, employing a general strategy typical of such expression cassettes. The S gene mRNA transcript in these recombinant yeast strain is 1130 nucleotides (Harford et al., 1987). HBsAg protein is obtained by disruption of the cells, and removal of yeast contaminants by ultrafiltration, followed by anion exchange and size exclusion chromatography. Under standard conditions, contaminants amounting to 1% of the protein in the final product are detected; yeast DNA content is usually less than 5pg/20 ug protein dose.

A satisfactory expression level of p24 polypeptide in this system, combined with a high biomass in fermentors, and the recovery of antigen in the form of particles similar to those found in human plasma (but nonglycosylated) has enabled a satisfactory yeast recombinant vaccine to be produced.

More than 90% of healthy young adults vaccinated with the latter vaccine develop anti-HBs. The immunogenicity is comparable with that of the plasma derived vaccine. The antibody has the same region specificity as anti-HBs induced by plasma derived vaccine and HBV infection itself.

The recommended schedule of vaccination for at-risk adults is 20μg (1ml) given intramuscularly in the deltoid muscle at 0, 1 and six months. Recently schedules of 0, 1, 2 and 12 months were found to improve anti-HBs titres and to result in more rapid past-exposure seroconversions. Post exposure prophylaxis for sexual or parenteral exposure, and perinatal transmission of hepatitis B requires simultaneous contra-lateral administration of hepatitis B immunoglobulin (0.05ml/kg) and HBV vaccine which should be given as soon as possible after exposure.

Persistence of anti-HBs is related to the peak titres achieved after vaccination; preliminary evidence suggests that 100% of those responding to both yeast and plasma vaccine are anti-HBs positive after 36 months. Considerable decline in geometric titres of anti-HBs occurs with all schedules, however. It has been arbitrarily assumed that 10 iu/l is the protective antibody titre, and that 29–35% of initially responsive healthy young adults would have levels of antibody below 10 iu/l at five years. Lack of antibody may not be synonymous with loss of immunity, however. It was found that 15% of homosexuals with low maximal levels develope HBV infection two and a half years after anti-HBs titres declined to less than 10 iu/l. Virtually all cases of infection in this group are subclinical. Firm recommendations regarding booster doses have not been formulated at this stage. It may be reasonable to recommend that booster doses be administered after one to two years for those individuals with a low response (titres of 10–100 iu/l) The role of natural boosting in endemic areas is being explored at present

Response rates are significantly lower in anti-HIV positive homosexuals and 47–60% of such patients do not respond, compared with 5.6 to 13% of those anti-HIV negative (Groval, 1989; Laukamm-Josten, 1988). Other poor responders have been identified. These include those with genetically determined immune hypo-responsiveness, the elderly, obese and cigarette smokers (Chiou et al., 1988; Collier et al., 1988; Watanabe et al., 1988). A monocyte defect may correlate with non-responsiveness to vaccine in haemodialysed patients, but this defect may be overcome with IL-2 (Meuer *et al.*, 1989). A higher dose (40μg) is required in these patients.

A recent analysis of MHC HLA in subjects with low antibody responses has indicated that homozygotes for the extended MHC haplotype (HLA B8, SCO1, DR3) exhibit a poor immune response to HBV vaccine, which may be due to an absence of a dominant immune response gene (Alper *et al.*, 1989)

A HBV variant with a point mutation from guanosine to adenosine at nucleotide position 587 has been shown to be responsible for hepatitis B infection in a series of successfully vaccinated patients in Italy (Carman *et al.*, 1990). In 2.8% of a cohort of vaccinated persons in that region, HBsAg was detectable after immunization with HBV vaccine and seroconversion to anti-HBs. The effect of the nucleotide substitution in the mutant HBV was to cause an amino acid substitution from glycine to arginine at amino acid 145 of HBsAg, thus causing a (conformational) change affecting several epitopes in the region of the a determinant. Since mutant virus putatively lacks this epitope, it is not neutralized by antibody of this specificity. The public health implications of this finding remain to be determined.

Several countries have initiated state-funded vaccination programmes to integrate HBV vaccination into existing EPI programmes. Intradermal vaccination using a 2μg dose will cut the costs of vaccination, but imposes inherent problems.

7.8 NEW HEPATITIS B VACCINES

The coding sequence for HBsAg has been inserted into the vaccinia virus genome under control of vaccinia virus early promoters. HBsAg made by vaccinia virus recombinants are similar in physical properties to particles circulating in plasma of HBV carriers. Cells infected with these vaccinia virus recombinants synthesize and secrete HBsAg, and vaccinated rabbits produce antibodies to HBsAg (Smith, 1983). Preliminary studies in chimpanzees have also demonstrated the feasibility of this vaccine; however, to date large scale clinical trials of chimaeric vaccinia vaccines have not been undertaken. Synthetic peptide analogue vaccines have also been shown to be immunogenic, but whether these vaccines will replace current subunit recombinant vaccines remains uncertain. Other novel vaccines are the subject of some research. These include anti-idiotype vaccines, whereby antibodies act as antigens—the antibodies induced against them, i.e. anti-idiotype antibodies, recognize the idiotype of an antibody in the shape of the original antigen (Reviewed by Zuckerman, 1987). Studies on the genetic restriction of the immune response in inbred mouse strains to S and pre-S proteins indicate that the immune response to these respective determinants are regulated by distinct genes. Non-responsiveness to the S protein may be circumvented by immunization with pre-S determinants, and induce antibody in refractory persons. Clinical trials using pre-S vaccines are in progress.

Hepatitis B core particles of HBV are potent immunogens. They induce T cell independent production of anti-HBc. Chimpanzees immunized with genetically cloned hepatitis B core antigen have been protected against HBV infection when challenged (Iwarson *et al.*, 1985). Francis *et al.* have reported that rhinovirus peptide-hepatitis B core antigen fusion proteins are 10-fold more immunogenic than peptides coupled to other epitopes. The use of the core antigen to present foreign epitopes may offer the advantage of an enhanced helper T cell signal, and overcome the poor immunogenicity of B cell epitope peptides for vaccination (Francis *et al.*, 1990).

7.9 HEPATITIS DELTA VIRUS

7.9.1 Structure and replication of the Hepatitis Delta Virus (HDV)

The HDV particle, which exists in the serum of infected individuals in the presence of excess HBV subviral particles, is approximately 36nm in diameter and is composed of an RNA genome associated with the delta antigen and surrounded by hepatitis B surface antigen. The virus reaches higher concentrations in the circulation than HBV and titres of up to 10^{10} particles per ml are not uncommon. The helper function required from HBV is the synthesis of the viral envelope, which is presumably needed for assembly and release of the virus particle and for entry to the cell for the next round of replication. The precise constitution of the envelope depends on the level of HBV replication. Thus, in patients who are highly viraemic for HBV, pre-S1 and pre-S2 proteins will be present on the surface of the HDV particle, whilst in HBsAg-carriers without HBV replication, there will be no pre-S1 and relatively little of the pre-S2 proteins. The fact that HDV can infect HBsAg-carriers who are no longer replicating HBV suggests that further helper functions are not required.

The HDV genome is a closed circular RNA molecule of 1679 nucleotides with extensive sequence complementarity that permits pairing of approximately 70% of the bases to form an unbranched rod structure. The genome thus resembles those of the viroids and virusoids of plants and, in common with these, appears to be replicated via a rolling circle mechanism with autocatalytic cleavage and circularization of the progeny genomes via *trans*-esterification reactions. Consensus sequences of viroids which are believed to be involved in these processes are also conserved in HDV. Further details of the replication of the hepatitis delta virus genome can be found in a review by Taylor (1990).

Unlike the plant viroids, HDV codes for a protein, the nucleocapsid protein (delta antigen, HDAg). The protein is encoded in an open reading frame in the antigenomic RNA but four other open reading frames which are also present in the genome do not appear to be utilized. A polyadenylated mRNA, approximately 800nt in length and of anti-genomic polarity, may be detected in the cytoplasm of infected hepatocytes. The antigen was originally detected in the nuclei of infected hepatocytes and may be detected in serum only after stripping off the outer envelope of the virus with detergent. It is phosphorylated at serine residues, has RNA binding properties and appears to play some role in the replication of the genome in addition to its structural capacity. Two species of the delta antigen, 22–24 and 24–27 kDa, have been detected in the infected liver by polyacrylamide gel electrophoresis but the precise mechanism of their synthesis remains to be determined.

7.9.1 Hepatitis D virus infection

Transmission of HDV, as with HBV, may occur by parenteral or inapparent parenteral means. The incubation period is 35 days. High incidence areas include the Amazon basin, Equatorial Africa, Middle East, Asiatic Russia and the Mediterranean basin. Epidemics of severe hepatitis related to HDV infection have been documented. In developed countries, infection occurs mainly in drug addicts, haemophiliacs and institutionalized persons; perinatal transmission is relatively rare. Although not previously believed to be common in male homosexuals, recent studies have shown that 9–14% of homosexuals have evidence of HDV infection, and that sexual transmission of HDV may indeed occur through either homosexual or heterosexual activities (Pol, 1989; Leon, 1988; Solomon, 1988).

Infection by HDV can occur in two situations: either simultaneous infection with HBV and HDV may occur and this is termed coinfection, or superadded infection of a chronically infected HBV carrier may take place (superinfection). Coinfection usually results in short-lived limited expression of HDV because of the usual transient nature of acute HBV infection, and the disease is not usually progressive. A biphasic pattern can be observed with two distinct aminotransferase peaks. Outbreaks of fulminant hepatitis due to hepatitis B plus D are described, however. With superinfection of an already HBV infected carrier, a milieu exists for efficient replication and propagation of HDV and prolonged replication of HDV is therefore possible; thus chronic infection is the more frequent outcome in such cases.

The diagnosis of HDV coinfection is made by the presence of HBsAg, IgM anti-HBc and IgM anti-HD in a patient with acute hepatitis. Superinfection of a hepatitis B carrier by HDV is diagnosed by the presence of HBsAg in the absence of IgM anti-HBc, and high titres of IgG anti-HD or IgM anti-HD. HDV RNA in serum, or HDSAg (detected by Western blotting), correlate closely with the detection of HDAg by histochemical staining in hepatocytes in chronic HDV infection.

Carriers of HBV and HDV are more likely to have evidence of chronic active hepatitis and cirrhosis.

7.9.2 Treatment and prevention of hepatitis D

Approximately 50% of patients treated with 9 mega units of α interferon, three times a week for six months, show a decline in serum aminotransferase and HDV RNA while on treatment. Unfortunately most relapse when treatment is stopped. Data is still being collected to decide on the optimal regimen. Hepatitis D is prevented by vaccination against HBV. A recombinant purified HDAg vaccine has also been produced, for prophylaxis of HBsAg carriers.

REFERENCES

Alper, C.A., Kruskall, M.S., Marcus-Bagley, D. et al. (1989) Genetic prediction of non-response to hepatitis B vaccine. New Engl. J. Med., 321, 708–12.

Alter, M. J., Coleman, P. J., Alexander, W. J. et al. (1989) Importance of heterosexual activity in the transmission of hepatitis B and non-A, non-B hepatitis. JAMA, 262, 1201–5.

Alter, H.J. (1982) The evolution, implications and applications of the hepatitis B vaccine. JAMA, 247, 2272–5.

Andre, F.E. (1988) Clinical experience with a recombinant DNA hepatitis B vaccine. South East Asian J. Trop. Med. Public Health, 19, 501–10.

Baddour, L.M., Bucak, V.A., Somes, G. and Hudson, R. (1988) Risk factors for hepatitis B virus infection in black female attendees of a sexually transmitted disease clinic. Sex. Transm. Dis., 15, 174–6.

Beasley, R.P. (1988) Hepatitis B virus. The major etiology of hepatocellular carcinoma. Cancer, 61, 1942–56.

Beasley, R.P., Lin, C.C., Hwang, L.-Y. and Chien, C.S. (1981) Hepatocellular carcinoma and hepatitis B virus: a prospective study of 22,707 men in Taiwan. Lancet, ii, 1129–33.

Bhatnagar, I.K., Papas, E., Blum, H.E. et al. (1982) Immune response to synthetic peptide analogues of hepatitis B surface antigen specific for the a determinant. Proc. Natl Acad. Sci. USA, 79, 4400–4.

Bissett, J., Eisenberg, M., Gregory, P. et al. (1988) Recombinant fibroblast interferon and immune interferon for treating chronic hepatitis B virus infection: patients' tolerance and the effect on viral markers. J.Infect.Dis., 157, 1076–80.

Boilav, C. and Piot, P. (1989) Vaccination against hepatitis B in homosexual men. A review. Am. J. Med., 87, 215–55.

Carman, W.F., Zanetti, A.R., Karayiannis, P. et al. (1990) Vaccine-induced escape mutant of hepatitis B virus. Lancet, 336, 325–9.

Caselmann, W.H., Eisenburg, J., Hofschneider, P.H. and Koshy, R. (1989) Beta and gamma interferon in chronic active hepatitis B. A pilot trial of short-term combination therapy. Gastroenterology, 96, 449–55.

Chiou, S.S., Yamuchi, K., Nakanishi, T. and Obata, H. (1988) Nature of immunological non-responsiveness to hepatitis B vaccine in healthy individuals. *Immunology*, **64**, 545–50.

Collier, A.C., Core, L., Murphy, V.L. and Hasterfield, H.H. (1988) Antibody to human immunodeficiency virus (HIV) and suboptimal response to hepatitis B vaccination. *Ann. Intern. Med.*, **109**, 101–5.

De Man, R.A., Schalm, S.W., Heijtink, R.A. *et al.* (1988) Long-term follow-up of antiviral combination therapy in chronic hepatitis B. *Am.J.Med*, Suppl.2A, **85**, 150–4.

Dejean, A., Bougueleret, L., Grzeschic K.-H. and Tiollais, P. (1986) Hepatitis B virus DNA integration in a sequence homologous to *v-erb-A* and steroid receptor genes in a hepatocellular carcinoma. *Nature*, **322**, 70–2.

Dusheiko, G. M. and Zuckerman, A. J. (1991) Therapy for Hepatitis B. *Current Opinion Infectious Diseases*, **4**, 785–94.

Ferla, G., Colledan, M., Doglia, M. *et al.* (1988) B hepatitis and liver transplantation. *Transplant Proc.*, **20**, Suppl.1, 566–9.

Francis, M. J., Hastings, G. Z., Brown, A. L. *et al.* (1990) Immunological properties of hepatitis B core antigen fusion proteins. *Proc. Natl Acad. Sci. USA*, **87**, 2545–9.

Harford, N., Cabezou, T., Colau, B. *et al.* (1987) Construction and characterization of a saccharonyces cerevisiae strain (RIT4376) expressing hepatitis B surface antigen. *Postgraduate Med. J.*, **63**, suppl 2, 65–70.

Harrison, T. J., Anderson, M. G., Murray-Lyon, I. M. and Zuckerman, A. J. (1986) Hepatitis B virus DNA in the hepatocyte – A series of 160 biopsies. *J. Hepatol.*, **2**, 1–10.

Heathcote, J., Kim, Y. I., Yim, C. K. *et al.* (1989) Interferon-associated lymphocyte 2'5'-oligoadenylate synthetase in acute and chronic viral hepatitis. *Hepatology*, **9**, 105–9.

Hoofnagle, J. H., Peters, M., Mullen, K. D. *et al.* (1988) Randomized, controlled trial of recombinant human alpha-interferon in patients with chronic hepatitis B. *Gastroenterol*, **95**, 1318–25.

Iwarson, S., Tabor, E., Thomas, H. C. *et al.* (1985) Protection against hepatitis B virus infection by immunization with hepatitis B core antigen. *Gastroenterology*, **88**, 763–7.

Kekule, A. S., Lauer, U., Meyer, M. *et al.* (1990) The *preS2/S* region of integrated hepatitis B virus DNA encodes a transcriptional transactivator. *Nature*, **343**, 457–61.

Laukamm-Josten, U., Muller, O., Bienzle, U. *et al.* (1988) Decline of naturally acquired antibodies to hepatitis B surface antigen in HIV-1 infected homosexual men (letter). *AIDS*, **2**, 400–1.

Leon, P., Lopez, J. A., Contreras, G. and Echevarria, J. M. (1988) Antibodies to hepatitis delta virus in intravenous drug addicts and male homosexuals in Spain. *Eur. J. Clin. Microbiol. Infect. Dis.*, **7**, 533–5.

Lien, J.-M., Aldrich, C. E. and Mason, W. S. (1986) Evidence that a capped oligoribonucleotide is the primer for duck hepatitis B virus plus-strand synthesis. *J. Virol.*, **57**, 229–36.

Lok, A. S., Lai, C. L., Wu, P. C. and Leung, E.K. (1988) Long-term follow-up in a randomised controlled trial of recombinant alpha 2-interferon in Chinese patients with chronic hepatitis B infection. *Lancet*, i (8606), 298–302.

Marcellin, P., Ouzan, D., Degos, F. *et al.* (1989) Randomized controlled trial of adenine arabinoside 5′-monophosphate in chronic active hepatitis B: comparison of the efficacy in heterosexual and homosexual patients. *Hepatology*, 10, 328–31.

Meuer, S. C., Dumann, H., Meyer zum Buschenfelde, K. H. and Kohler, H. (1989) Low-dose interleukin-2 induces systemic immune responses against HBsAg in immunodeficient non-responders to hepatitis B vaccination. *Lancet*, i (8628), 15–18.

Milich, D. R., Hughes, J. L., McLachlan, A. *et al.* (1988) Hepatitis B synthetic immunogen comprised of nucleocapsid T-cell sites and an envelope B-cell epitope. *Proc. Natl Acad. Sci. USA*, 85, 1610–14.

Minuk, G. Y. and La Freniere, R. (1988) Interleukin-1 and interleukin-2 in chronic type B hepatitis. *Gastroenterol.*, 94, 1094–6.

Murray, K. (1988) Application of recombinant DNA techniques in the development of viral vaccines. *Vaccine*, 6, 164–74.

Nagaya, T., Nakamura, T., Tokino, T. *et al.* (1987) The mode of hepatitis B virus DNA integration in chromosomes of human hepatocellular carcinoma *Gen. Develop.*, 1, 773–82.

Neurath, A.R., Kent, S.B.H., Strick, N. and Parker, K. (1986) Identification and chemical synthesis of a host cell receptor binding site on hepatitis B virus. *Cell*, 46, 429–36.

Odaka, N., Eldred, L., Cohn, S. *et al.* (1988) Comparative immunogenicity of plasma and recombinant hepatitis B virus vaccines in homosexual men. *JAMA*, 260, 3635–7.

Okamoto, H., Imai, M., Tsuda, F. *et al.* (1987) Point mutation in the S gene of hepatitis B virus for a d/y or w/r subtypic change in two blood donors carrying a surface antigen of compound subtype adyr or adwr. *J. Virol.*, 61, 3030–4.

Onji, M., Lever, A. M., Saito, I. and Thomas, H.C. (1989) Defective response to interferons in cells transfected with the hepatitis B virus genome. *Hepatology*, 9, 92–6.

Perrillo, R. P., Regenstein, F. G., Peters, M. G. *et al.* (1988) Prednisone withdrawal followed by recombinant alpha interferon in the treatment of chronic type B hepatitis. A randomized, controlled trial. *Ann. Intern. Med.*, 109, 95–100.

Pol, S., Dubois, F., Roingeard, P., Zignego, L., Housset, C. *et al.* (1989) Hepatitis delta virus infection in French male HBsAg-positive homosexuals. *Hepatology*, 10, 342–5.

Radziwill, G., Tucker, W. and Schaller, H. (1990) Mutational analysis of the hepatitis B virus P gene product: domain structure and RNase H activity. *J. Virol.*, 64, 613–20.

Rosenblum, L. S., Hadler, S. C., Castro, K. G. *et al.* (1990) Heterosexual transmission of hepatitis B virus in Belle Glade, Florida. *J. Infect. Dis.*, 161, 407–11.

Saracco, G., Mazzella, G., Rosina, F. *et al.* (1989) A controlled trial of human lymphoblastoid interferon in chronic hepatitis B in Italy. *Hepatology*, 10, 336–41.

Schneider, J., King, L., Macnab, G. M. and Kew, M. C. (1977) Hepatitis B surface antigen and antibody in black and white patients with venereal diseases. *Br. J. Vener. Dis.*, 53, 372–4.

Seeff, L. B. (1988) The gold standard serologic marker for hepatitis B virus infectivity. *Hepatology*, 8, 1711–13.

Shindo, M., Okuno, T., Matsumoto, M. *et al.* (1988) Serum 2′, 5′-oligoadenylate synthetase activity during interferon treatment of chronic hepatitis B. *Hepatology*, 8, 366–70.

Solomon, R. E., Kaslow, R. A., Phair, J. P. *et al.* (1988) Human immunodeficiency virus and hepatitis delta virus in homosexual men. A study of four cohorts. *Ann. Intern. Med.*, 108, 51–4.

Summers, J. and Mason, W.S. (1982) Replication of the genome of a hepatitis B-like virus by reverse transcription of an RNA intermediate. *Cell*, 29, 403–15.

Taylor, J. M. (1990) Hepatitis delta virus: *cis* and *trans* functions required for replication. *Cell*, 61, 371–3.

Thyagarajan, S. P., Subramanian, S., Thirunalasundari, T. *et al.* (1988) Effect of *Phyllanthus amarus* on chronic carriers of hepatitis B virus. *Lancet*, ii (8614), 764–6.

Tiollais, P., Pourcel, C. and Dejean, D. (1985) The hepatitis B virus. *Nature*, 317, 489–95.

Twu, J. S., Lee, C. H., Lin. P. M. and Schloemer, R.H. (1988) Hepatitis B virus suppresses expression of human beta-interferon. *Proc. Natl Acad. Sci USA.*, 85, 252–6.

van Ditzhuijsen, T. J., de Witte van der Schoot, E., van Loon, A. M. *et al.* (1988) Hapatitis B virus infection in an institution for the mentally retarded. *Am. J. Epidemiol.*, 128, 629–38.

Wang, J., Chenivesse, X., Henglein, B. and Brechot, C. (1990) Hepatitis B virus integration in a cyclin A gene in a hepatocellular carcinoma. *Nature*, 343, 555–7.

Watanabe, H., Matsushita, S., Kamikawaji, N. *et al.* (1988) Immune suppression gene on HLA-Bw54-DR4-DRw53 haplotype controls nonresponsiveness in humans to hepatitis B surface antigen via CD8+ suppressor T cells. *Hum. Immunol.*, 22, 9–17.

Zuckerman, A. J. (1991) Present and future hepatitis B vaccines. *Oxford Textbook of Clinical Hepatology* (eds N. McIntyre, J. P. Benhamou, J. Bircher *et al.*), Vol. I, 620–6.

8

Molluscum contagiosum virus

COLIN D. PORTER, NEIL W. BLAKE, JEFFREY J. CREAM and
LEONARD C. ARCHARD

8.1 INTRODUCTION

Molluscum contagiosum virus (MCV) is a poxvirus that causes benign
skin tumours in man, consisting of a localized mass of hypertrophied and
hyperplastic epidermis due to enhanced basal cell division. The disease was
first described in 1814 (Bateman, 1814). The intracytoplasmic inclusion
bodies (molluscum or Henderson-Paterson bodies) were described in 1841
(Henderson, 1841; Paterson, 1841) and the viral nature of the disease was
eventually established in 1905 (Juliusberg, 1905). MCV is one of two
poxviruses regarded as having specificity for the human host, the other
being variola, the agent of smallpox. There have been occasional reports
of *molluscum contagiosum* in other species (Brown *et al.*, 1981) although
these have not been substantiated by molecular analysis. MCV is also one
of a small number of poxviruses which induce tumour formation in their
natural hosts.

MCV is antigenically unrelated to any of the established genera of
poxviruses, within which species are serologically related, and is thus
described as unclassified. Attempts at propagating the virus *in vitro* have
been unsuccessful. Molecular studies have identified three subtypes of the
virus that show a greater degree of diversity with respect to physical
maps of their genomes than that seen amongst species of other poxvirus
genera for which genomic mapping data are available. The MCV subtypes
are independent but closely related and cause clinically indistinguishable
lesions: as such, they could be regarded as forming a new genus of
poxviruses. The structural features of the MCV genome and the recognition
of control elements by the transcriptional machinery of vaccinia virus
suggest a replication strategy similar to that of the latter. The mechanism
of tumourigenesis remains unknown.

Molecular and Cell Biology of Sexually Transmitted Diseases
Edited by D. Wright and L. Archard
Published in 1992 by Chapman and Hall, London ISBN 0 412 36510 3

8.2 CLINICAL FEATURES

Molluscum contagiosum lesions consist of discrete, flesh-coloured, dome-shaped protruberances or papules usually 2 to 4 mm across. An umbilicated centre, from which a white curdy core can be expressed, is a diagnostic feature (Brown *et al.*, 1981). Giant, often solitary lesions, with diameters of 15 mm or more, may occur. Intrafollicular lesions – small, pearly, semi-translucent papules in the upper dermis – are a rare variant and may be difficult to identify as *molluscum contagiosum* because they lack the characteristic central umbilication (Ive, 1985). The lesions can occur anywhere on the skin and vary in number from solitary to numerous but are rare on the palms and soles (Baxter and Carson, 1966): lesions on the buccal mucosa have been described (Laskaris and Sklavounou, 1984). At any site they are often barely apparent, and may go unnoticed by the patient, for many of whom they are no more than a minor cosmetic problem. However, eczema can develop around the lesions and, usually as a prelude to resolution, they can become inflamed and infected secondarily. *Molluscum contagiosum* on the eyelids may be associated with conjunctivitis. Individual lesions usually resolve spontaneously within a few months. However, complete resolution of the disease takes longer due to autoinoculation with virus from lesion cores. Persistence over very long periods implies impaired immunity. Treatment, when given, involves lesion disruption by expression of the core, or removal by curettage.

8.3 PATHOLOGY

In the *molluscum contagiosum* lesion, the epidermis grows down into the dermis as pear-shaped lobules and projects above the surface as a papule. The basement membrane remains intact. The basal cells are larger and more columnar than usual and intracytoplasmic inclusion bodies (Henderson-Paterson or molluscum bodies), the site of virus assembly, appear first in the cells of the *stratum malpighii* at a level of one or two cells above (Sutton and Burnett, 1969; Epstein and Fukuyama, 1973). As the cells become more superficial, the molluscum bodies expand and displace the nucleus and remaining cytoplasm into a thin crescent at the periphery of the cell. Virus particles are present exclusively in the cytoplasm of infected cells and do not occur in basal cells or cells adjacent to the lesion (Shirodaria and Matthews, 1977). The latter continue to develop normally. Within each infected keratinocyte, the virus inclusion body is enclosed by a well-defined sac which can be identified by scanning electron microscopy (Shelly and Burmeister, 1986) and microdissection (Van Rooyen, 1938). This sac may provide a site that is anatomically and immunologically privileged, thus permitting unimpeded growth of the virus. In the *stratum corneum*, the degenerating keratinocytes with their inclusion bodies become enmeshed in the keratin network. The central core of the lesion is an expressable material consisting of husks of epidermal cells packed with the elementary bodies of the virus (Sutton and Burnett, 1969).

The *molluscum contagiosum* tumour results both from hypertrophy of epidermal cells, due to the presence of the virus-packed inclusion bodies, and from hyperplasia of basal cells. A reduction in the basal cell division time, from 6.1 to 3.4 days, was observed following autoradiography of lesion sections after injection of tritiated thymidine into skin lesions *in situ* (Epstein *et al.*, 1966). The use of colcemid (deacetyl–methyl-colchicine) to arrest cell division and count mitotic bodies confirmed the reduced renewal time. Cytoplasmic labelling with radioactive thymidine was observed in the cells containing inclusion bodies due to viral DNA replication at these sites. Infected cells migrate through the epidermis to reach the *stratum granulosum* at a more uniform rate of 9–15 days compared with 9–30 days for uninfected cells. Viral DNA replication occurs within the first 4–5 days and studies using tritiated uridine demonstrated shut-off of host transcription and the presence of viral transcription within the inclusion body (Epstein and Fukuyama, 1973). Some hyperplasia of the dermal fibroblasts subjacent to the lesion has been reported although these cells, like basal cells, do not harbour virus (Sutton and Burnett, 1969).

8.4 IMMUNOLOGY

The lesions can persist in normal individuals for months or years without any sign of inflammation. Even when some lesions are removed, others may continue to appear and the virus can be difficult to eradicate. T-lymphocytes and natural killer cells are not present in the base of the lesions and there is no inflammatory infiltrate in the early eruptive phase (Heng *et al.*, 1989). However, resolution may be accompanied by erythema initially at the base of the lesion and a predominantly mononuclear infiltrate has been taken to imply a cell-mediated immune response. In one study, only 46 of 67 patients had an antibody response, predominantly of the IgG class (Shirodaria and Matthews, 1977). The lack of a response in the remainder may be due to the superficial site of infection and the outward migration of infected cells within the epidermis. Most patients develop virus-specific antibody after treatment.

The role of immune responses to *Molluscum contagiosum* virus in normal subjects is uncertain but impaired immunity results in widespread multiple and recurrent *Molluscum contagiosum*, as may be seen with immunosuppressive therapy (Cotton *et al.*, 1987), in sarcoidosis, leukaemia, and the acquired immunodeficiency syndrome (AIDS) (Brown *et al.*, 1981). In one study, *Molluscum contagiosum* was found in 9% of 117 adults with AIDS and AIDS-related complex (Goodman *et al.*, 1987). Children with atopic dermatitis commonly have *Molluscum contagiosum*, predominantly at the sites of the dermatitis, which may simply reflect ease of entry of the virus at these sites. However, patients with severe atopic dermatitis can have numerous and persistent lesions and some of these patients do have impaired immunity.

8.5 TRANSMISSION AND EPIDEMIOLOGY

Juliusberg established the viral nature of the disease in 1905 when he was able to induce new lesions in one of three individuals who had been inoculated with a filtered extract of *Molluscum contagiosum* lesions (Juliusberg, 1905). Wiles and Kingery repeated the experiment and noted papules at two to three weeks and typical lesions at eight weeks (Wiles and Kingery, 1919). However, others have not been able to induce experimental lesions, although papular reactions were observed (Pinkus and Frisch, 1949; Blank and Rake, 1955; Goldschmidt and Kligman, 1958). Accidental inoculation has also been reported after tatooing (Foulds, 1982).

Virus reaches the surface of the lesion in quantity by extrusion from the central core. The mode of spread is uncertain but it seems likely that direct contact is responsible. Lesions are commonly seen on opposing surfaces and transmission after close physical contact from baby's face to mothers's breast, sibling to sibling, surgeon to patients, husband to wife, prostitutes to soldiers and masseuse to client has been reported (Editorial, British Medical Journal, 1968). However, the conditions favouring the spread of infection are incompletely understood. The incidence of *Molluscum contagiosum* varies in different communities, from 12 of 1000 out-patients in Aberdeen between 1956 and 1963, to as high as 4.5% of the entire population of a village surveyed in Fiji in 1966 (Postlethwaite, 1970; Postlethwaite *et al.*, 1967). Infectivity is normally low in a European environment, as shown by the Aberdeen study, where spread within households and in schools was rare. The peak incidence was between 10 and 12 years of age. In contrast, the Fiji study showed that 25% of households harboured more than one case and the peak incidence was between two and three years.

Molluscum contagiosum cannot be regarded as a disease solely transmitted by sexual intercourse. However, as is the case with herpes simplex virus and human papilloma virus, sexual transmission of *Molluscum contagiosum* virus is undoubted on the basis of a distribution that can be confined to the genital area, the prevalence of *Molluscum contagiosum* in the sexually active (Lynch and Minki, 1968; Brown and Weinberger, 1974; Cobbold and Macdonald, 1970; Gudgel, 1954; Wilkin, 1977), its occurrence in sexual partners (Jacob, 1970), an association with sexually transmitted diseases – 65% of cases in one study (Brown and Weinberger, 1974) – and its rising incidence in common with other sexually transmitted diseases (Oriel, 1987;

Becker *et al.*, 1986). In England, the number of cases of genital *Molluscum contagiosum* quadrupled between 1971 and 1985 (Department of Health and Social Security, Extract from the annual report of the Chief Medical Officer; 1971, 1983). Cases of *Molluscum contagiosum* infection account for up to 1.0% of patients attending Sexually Transmitted Disease Clinics and the disorder is most commonly seen in the decade 15 to 24. Affected women tend to be younger than affected men, a pattern that has been observed with genital herpes infection (Becker *et al.*, 1986). In itself innocent and self-limiting in most patients, genital *Molluscum contagiosum* may, nevertheless, be a marker of other more serious sexually transmitted infections.

8.6 VIRION STRUCTURE AND LIFE CYCLE

The *Molluscum contagiosum* virus (MCV) particle is approximately 100 × 200 × 300 nm in size. On the strength of morphological studies by electron microscopy and histopathological similarities to infection with fowlpox virus, MCV was identified as a poxvirus (Goodpasture and Woodruff, 1931; Goodpasture, 1933; Postlethwaite, 1970). The mature virion consists of inner and outer membranes enclosing a dumb-bell shaped core, with two lateral bodies lying between the membranes. The virion core contains the virus genome and virus-encoded enzymes necessary to establish the cytoplasmic site of replication in the newly infected cell, including those for viral gene transcription and mRNA modification.

Virions purified from clinical lesions by sucrose density gradient centrifugation have similar physical and chemical properties to those of vaccinia virus with respect to sedimentation characteristics, DNA content and absence of RNA (Pirie *et al.*, 1971). Molecular biological studies on the viral genome have since shown unequivocally that MCV is a poxvirus (see below). There are over 40 polypeptides detectable in purified virions (Oda *et al.*, 1982): two major structural proteins, present in the surface or sub-surface virion layers but absent from virion cores, show size variation between independent virus isolates. There is no antigenic cross-reactivity between MCV and other poxviruses, detectable by complement fixation (Mitchell, 1953) or by immunofluorescent staining of sectioned lesions (Shirodaria and Matthews, 1977). As such, MCV is distinct from other genera of vertebrate poxviruses, and so is unclassified.

The essential features of MCV replication within the infected keratinocyte are expected to be similar to those of the replication cycle of vaccinia virus, the prototypic poxvirus and a member of the orthopoxvirus genus (Fenner *et al.*, 1989). Infection with vaccinia virus is initiated by attachment and entry by endocytosis or fusion of the outer lipid coat with the cell membrane. Uncoating of the viral core releases the DNA and requires the combined action of host enzymes and newly synthesized virus proteins. Expression of viral genes is temporally regulated: the early class is expressed before the onset of viral DNA replication and transcription of the late class follows. Viral DNA replication and virion assembly occur within intracytoplasmic inclusions and do not require host nuclear functions. During the replicative cycle, host cell DNA replication and transcription are inhibited. Mature virions are released via the Golgi system or upon the death of the host cell.

8.7 ATTEMPTS AT VIRUS PROPAGATION

MCV has been transmitted by inoculation of human subjects (Juliusberg, 1905), but similar experiments with monkeys, apes, sheep, fowl, guinea-pigs and mice were unsuccessful, as were attempts to grow the virus on the chick embryo chorioallantoic membrane and in human embryonic skin organ cultures. Although for primary and continuous cultures of cells, limited transmission of a cytopathic effect (CPE: cell rounding and clumping) and induction of interferon were observed for a few passages, no culture system for the propagation of MCV has been established (Postlethwaite, 1970; McFadden et al., 1979). Attempts to rescue the virus by simultaneous infection of the chorioallantoic membrane with vaccinia, fowlpox or cowpox viruses were also unsuccessful (Mitchell, 1953). Propagation of MCV in human amnion cells has been reported (Francis and Bradford, 1976), but attempts to repeat this were unsuccessful.

MCV induces a CPE in several primate cell types (primary and continuous; simian and human fibroblasts or epithelial cells) and fresh human epidermis explants, but not in murine cells. The CPE persists for about 10 days, after which the cells return to normal and can respond to fresh challenge with MCV (McFadden et al., 1979). Immunofluorescence and electron microscopy were used to demonstrate viral attachment and release of virion cores into the cytoplasm. However, the cores are not uncoated, although transcription of MCV DNA from the intact cores does occur, and the infection cycle is thus aborted. This block to uncoating cannot be overcome by simultaneous infection with vaccinia virus, Shope rabbit fibroma virus or Yaba monkey tumour virus. The CPE can be abrogated by inhibitors of RNA and protein synthesis, showing it to be due to virus early gene expression. The failure to release core enzymes by completion of the uncoating process explains the inability of MCV non-genetically to reactivate heat-inactivated vaccinia virus (Brown et al., 1973).

As for the human papilloma viruses, the inability to culture MCV is thought to be due to the need for keratinocytes at a defined stage of differentiation which is difficult to obtain or sustain in vitro but which is necessary for completion of the uncoating process. In this respect, MCV behaves as a host-dependent, conditional lethal mutant (Postlethwaite, 1970).

8.8 GENOME STRUCTURE AND ORGANIZATION

The genome of *Molluscum contagiosum* virus is a linear double-stranded DNA molecule of length 180–200 kilobases. This has been determined by summation of restriction fragment sizes (Parr *et al.*, 1977; Darai *et al.*, 1986; Porter and Archard, 1987) and by direct measurement using field-inversion gel electrophoresis (Porter, 1989) and is typical of poxviral genomes, which range from 130 kb for the parapoxviruses to 280 kb for the avipoxviruses. Denaturation studies using electron microscopy of viral DNA demonstrated that the regions most easily denatured, and hence lowest in G+C content, are those near the ends of the genome (Parr *et al.*, 1977). In common with all poxvirus genomes and relating to their mode of replication (Wittek and Moss, 1980), the termini are covalently closed, the whole genome being a continuous strand of self-complementary DNA. This feature was confirmed by demonstrating the presence of restriction fragments, corresponding to the termini, that can rapidly renature following denaturation (Porter and Archard, 1987). The overall G+C content of MCV DNA is relatively high, estimated at 60% from the denaturation studies described above: that for nucleotide sequence data from a number of regions of the genome (totalling approximately 10 kb) is in close agreement with this figure (Porter, 1989; Blake *et al.*, 1991), which is different from that of the orthopoxviruses (G+C content 36%) and closer to that of the distinct genus of parapoxviruses (63%). This base composition reflects in the biased codon usage for MCV genes (Porter, 1989; Blake *et al.*, 1991) (see below).

A further property of poxvirus DNA is that the termini are inverted repeats and contain blocks of tandem direct repetitions, features important to the mechanism of replication (Wittek and Moss, 1980). For vaccinia virus strain WR, the terminal repeat is 10.3 kb long and contains two blocks of 13 and 17 repetitions of a 70 bp unit (Wittek and Moss, 1980). MCV DNA possesses similar structural features as determined by cross-hybridization of terminal restriction fragments and the observation that these fragments vary in size between independent isolates, indicative of polymorphism with respect to the number of tandem repetitions (Porter and Archard, 1987). For one subtype at least (MCV II; see below) these repetitions are organized into two blocks separated by a HindIII site. The number and size of the repetitions is unknown. The organization of the terminal region is discussed further below. The structural similarity of MCV DNA to vaccinia virus DNA implies a similar mechanism of DNA replication.

Comparison of DNA restriction endonuclease digest patterns of over 250 independent virus isolates for which data have been published (Darai *et al.*,

1986; Porter and Archard, 1987; Porter, 1989; Scholz *et al.*, 1989) has demonstrated three genetic subtypes of MCV, designated MCV I, II and III (Figure 8.1). Restriction maps of all three subtypes for the enzymes BamHI, ClaI and HindIII have been obtained by hybridization of isolated restriction fragments to digests of the whole genome, and differ extensively (Porter and Archard, 1987; Porter, 1989) (Figure 8.2). These data deny the possibility that one subtype may have arisen by recombination of the others, or that any subtype is a mutant of the copy-round type seen in orthopoxviruses (Archard *et al.*, 1984). Despite the independence suggested by a lack of restriction site conservation, all subtypes are closely related in nucleotide sequence as indicated by cross-hybridization of their genomes (Darai *et al.*, 1986; Porter and Archard, 1987; Porter, 1989). The MCV subtypes thus show greater variance than do the species of the orthopoxvirus genus, whose genomes show a high degree of restriction site conservation throughout the central region (Mackett and Archard, 1979). Parapoxvirus (Gassman *et al.*, 1985) and capripoxvirus (Gershon and Black, 1988) species also show restriction site conservation throughout large regions of their genomes. MCV DNA does not cross-hybridize with vaccinia virus DNA although comparative nucleotide sequence analysis for the MCV I and MCV II homologues of the vaccinia virus major envelope protein confirms their relatedness (94% homologous; see below). It is not known how variations in the sizes of structural proteins (Oda *et al.*, 1982) relates to MCV subtype.

The sizes of the genomes of MCV I, II and III used for the mapping studies (Porter and Archard, 1987; Porter, 1989) (Figure 8.2) are 190, 200 and 196 kb respectively. The sizes of their inverted terminal repeats are 3.9–5.4, 7.2–10.6 and 3.7–7.0 kb, i.e. notably larger for MCV II: that for MCV I has been more accurately determined to be 4.5 kb (see below).

In addition to the high degree of homology between the three subtype genomes, their overall organization is highly conserved. Mapping studies in which restriction fragments derived from one subtype were used to locate corresponding sequences in the other subtypes have shown that the genomes are essentially colinear, with two exceptions (Porter and Archard, 1987; Porter, 1989) (Figure 8.2). One of these is a 12 kb region of MCV I DNA, mapping around 150 kb from the left terminus, which is unique to this subtype. The significance of this unique sequence is unknown since there are no apparent biological differences between the viral subtypes. Alignment of the maps for MCV II and MCV III also requires a discontinuity at this position. The other exception is a discontinuity of 2–3 kb in the genome of MCV II near the left terminus, necessary for alignment of the terminal repeat with the other subtypes. However, the map position and direction of transcription of the homologue of the vaccinia virus major envelope protein in MCV I and MCV III are similar to that of the vaccinia gene (Porter, 1989; Blake *et al.*, 1991) (see below). If this similarity extends to other functions then the orientation of the MCV maps as shown in Figure

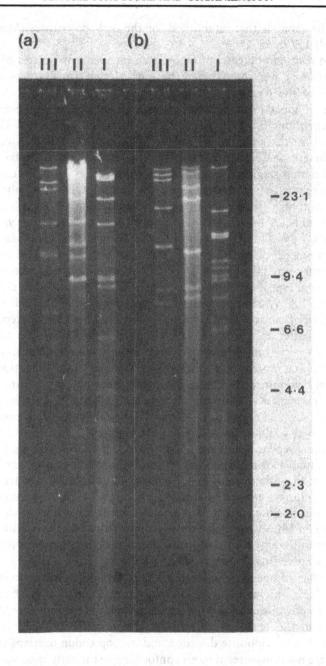

Figure 8.1 Comparative DNA restriction digests of MCV subtypes I, II and III with (a) ClaI and (b) BamHI. The positions of DNA size markers are shown (kb). Some background staining, due to residual host cell DNA, is seen. The three largest ClaI fragments for MCV I and II do not resolve well.

241

8.2 is the same as that established for the orthopoxviruses (Mackett and Archard, 1979).

When restriction digests of DNA from independent isolates of the same MCV subtype are compared, variations of two types are seen. One is the size of the terminal and near-terminal fragments that are assumed to contain the tandem repetitions, due to variations in the number of these repetitions (Porter *et al.*, 1987). Variations are always identical at both ends of the genome, due to the mode of poxviral DNA replication. The other occasional variation is due to loss or gain of restriction sites (i.e. restriction fragment length polymorphisms), as expected on the basis of some sequence heterogeneity between isolates. One such polymorphism occurs with high frequency: a BamHI site mapping to 76 kb from the left terminus of MCV I DNA was missing in 33% of isolates in one study group of 80 MCV I isolates (Porter *et al.*, 1987; Porter, 1989) and has also been reported independently for five of 142 isolates (Scholz *et al.*, 1989).

The near-terminal region of MCV I DNA containing the junction between the terminal repeat and unique sequences has been studied in detail (N.W. Blake and L.C. Archard, unpublished data). The terminal HindIII fragment O and part of the adjacent fragment K_L comprise the terminal repeat of this subtype (Porter and Archard, 1987; Porter, 1989). However, the tandem repetitions occur only within the former as only this fragment shows length polymorphism (Porter *et al.*, 1987). The sequence of the repetition unit has not been determined. Nucleotide sequence data of 2.4 kb from the distal end of fragment K_L, spanning the repeat/unique sequence junction, have been obtained and are 61% G+C rich. The junction has been positioned to within 1.6 kb of this sequence, giving a more accurate estimate of 4.5 kb for the terminal repeat size of the MCV I isolate used for mapping (see Figure 8.2). A number of repeats occur within this sequence on the terminal side of the junction: there are two occurrences of each of six repeats of 12–31 bp and a cluster of seven related sequences with a 14 bp consensus at the most distal position (Figure 8.3). These motifs have not previously been identified in MCV sequence data and differ from other poxvirus repetitive sequences, although a similar arrangement of repeats has been reported for orf, a parapoxvirus (Fraser *et al.*, 1990).

A number of open reading frames occur within this near-terminal region (Figure 8.3). The largest (with a potential product of at least 523 amino acids) extends throughout the repeats and into unique sequence, orientated away from the terminus. A second (greater than 192 amino acids) is orientated in the opposite direction and its stop codon overlaps that of the first: there is a downstream motif conforming to the early gene transcription termination signal. This tight clustering of open reading frames has been observed previously for MCV (Porter, 1989), and is reminiscent of that seen for vaccinia virus genes (Plucienniczak *et al.*, 1985). However, the identity of the products of these open reading frames is unknown; they

242

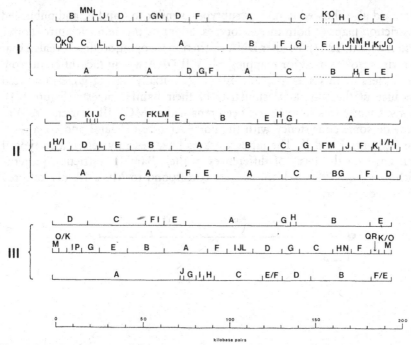

Figure 8.2 Comparison of the restriction maps of the genomes of MCV subtypes I, II and III. In each map the upper line is for BamHI, the middle for HindIII and the lower ClaI. A vertical line indicates a site for the relevant enzyme and the fragment between two sites is identified by a letter which refers to its size, 'A' being the largest. Two or more non-resolving fragments are given the same identification. The relative orientations and alignments were determined by cross-hybridization mapping. The three genomes are colinear. A region of MCV I DNA, encompassing BamHI fragments K and O, is unique to this subtype and is indicated with the broken line. There are two small regions of discontinuity marked in the MCV I and MCV II maps to allow for alignment: one of these is at the locus unique to MCV I whilst the other is due to the longer terminal repeat of MCV II.

are not homologous to genes adjacent to the tandem repetitions of vaccinia virus DNA (Wittek *et al.*, 1980).

8.9 *MOLLUSCUM CONTAGIOSUM* VIRUS SUBTYPE NOMENCLATURE

Parr and coworkers first demonstrated the existence of multiple MCV subtypes on the basis of different BamHI restriction digest patterns of viral DNA (Parr *et al.*, 1977). These authors described three subtypes, two of which were very similar and are probably variants of the same subtype. Darai *et al.* similarly described two subtypes, but used different nomenclature MCV I and II, corresponding to the subtypes II/III and I respectively of Parr and coworkers (Darai *et al.*, 1986). Porter and

Archard also showed two virus subtypes, similar to I and II, and published restriction maps of both these subtypes, adopting this nomenclature (Porter and Archard, 1987). Further DNA restriction mapping studies revealed that the virus isolate used for mapping MCV II DNA was in fact different from the others of this subtype, this difference initially having gone unnoticed because of the superficial similarity of their BamHI digests (Figure 8.1). Consequently this subtype has been renamed as MCV III (Porter, 1989) to provide some consistency with the nomenclature of Darai and coworkers for subtypes I and II. The latter workers have recently referred to type II 'variants' on the basis of differences in their BamHI restriction patterns (Scholz *et al.*, 1989) which, as they correspond to MCV III, are referred to as such below.

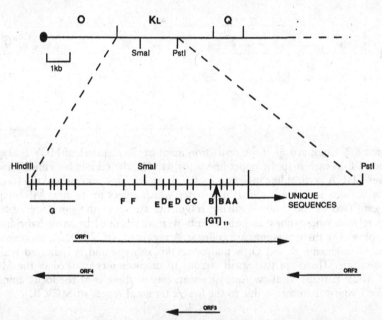

Figure 8.3 Above; HindIII map of the left terminus of MCV I DNA showing fragments O, K_L (the left-most of the two K fragments) and Q, with SmaI and PstI sites also shown. The terminal fragment (O) possesses the tandem repetitions, as deduced from its length polymorphism. Below; expanded HindIII-PstI fragment (2.4 kb), showing the junction between the left inverted terminal repeat and unique sequences, and the major open reading frames (ORF1-ORF4). The reading frame remains open upstream of the indicated start codon of ORF1, which could therefore initiate within HindIII fragment O. The stop codons for ORF1 and ORF2 overlap within the sequence TTAG. Also shown are an alternating dG-dT tract with potential Z-DNA structure, and repeats A-G: A, B and E are 12 bp, D is 16 bp, C is an imperfect 21 bp repeat, F is 31 bp, and G is a cluster of seven related sequences with the 14 bp consensus TTTTCTTCCGTGTGT. This arrangement of repeats is similar to the organization of the terminal region of the orf virus genome (Fraser *et al.*, 1990).

244

Thus, the existence of three distinct MCV subtypes has been demonstrated although some confusion over nomenclature resulted. The relative occurrences of the three subtypes are unequal: MCV I predominates and MCV III is rare (see below).

8.10 MOLECULAR EPIDEMIOLOGY

To date, molecular data relating to the virus subtype found in *Molluscum contagiosum* lesions are available for a total of 261 patients. The initial study was of eight patients (Parr *et al.*, 1977), whilst the two larger studies were of 147 patients (Scholz *et al.*, 1988, 1989) and 106 patients (Porter, 1989). All body sites have been represented, infection of the genitalia being common in adults due to sexual transmission (see below).

In cases where two or more members of a family presenting with *Molluscum contagiosum* have been studied (seven families), identical MCV DNA restriction patterns have been demonstrated, indicative of infection with the same viral isolate (Scholz *et al.*, 1988). Similarly, lesions removed from different sites on the same patient were shown to be due to the same virus type (Porter *et al.*, 1987) as is also the case, with one exception, of lesions removed from the same patients presenting on different occasions over periods of up to nine months (Scholz *et al.*, 1988; Porter *et al.*, 1987). Thus, although independent isolates show minor variations (e.g. with regard to the terminal repeat length), any given isolate is genetically stable, enhancing the value of molecular analysis for epidemiological studies. The exception mentioned was a woman with genital lesions shown to be due to MCV I at the initial presentation and due to MCV II seven weeks later (Porter *et al.*, 1987), presumably as a result of a new infection. Mixed infection has not been observed.

Table 8.1 presents these cumulative data, presented by MCV subtype and according to the geographical area from which the patients were drawn. The MCV II 'variants' (Scholz *et al.*, 1989) have been interpreted here as subtype III, as discussed above. The relative incidences of the MCV subtypes differ in the different patient study groups although the overall ratio of I:II:III is 226:32:3. There is no significant difference in occurrence with respect to the gender of the patient (Scholz *et al.*, 1988; 1989; Porter *et al.*, 1989; Porter, 1989). Whilst one study reported the absence of MCV II from young patients (<15 years) (Porter *et al.*, 1989), this is no longer the case following the analysis of lesions from a boy (<5 years) with this subtype (Scholz *et al.*, 1989). The three patients with MCV III were two males (<5 years and >15 years) (Scholz *et al.*, 1989) and a 16 year old girl (Porter, 1989). Thus, there is no association of any subtype with patient age.

Two other viruses that commonly infect the skin have multiple strains, with clinical features correlating with particular strains. It has been shown for human papilloma virus that certain strains are preferentially associated

Table 8.1 Cumulative data relating to the occurrence of MCV subtypes in a total of 261 patients. The geographical area from which patients were drawn is indicated. The relative incidence of the subtypes varies widely for the different studies

Reference	Study area	MCV subtype		
		I	II	III
Parr et al., 1977	Maryland, USA	4	4	0
Scholz et al., 1989	West Germany	32	1	0
Scholz et al., 1989	Hong Kong	6	0	0
Scholz et al., 1989	Scotland	104	2	2
Porter, 1989	South-east England	80	25	1
	Totals	226	32	3
		(86.6%)	(12.3%)	(1.1%)

with skin warts of particular morphology, site of infection or malignant potential (Jablonska and Orth, 1983; Orth and Favre, 1985). Of the two strains of herpes simplex virus, HSV1 is the predominant strain in the common recurring facial lesions, whilst HSV2 is most frequently associated with genital infections of adults (Chaney et al., 1983). In contrast, there is no preferential association of any MCV subtype with the site of infection. In particular, there is no predominance of either MCV I or MCV II in genital lesions (Porter et al., 1989; Porter, 1989; Scholz et al., 1988), despite earlier indications that this might have been so (Darai et al., 1986). There are no consistent morphological differences between lesions due to MCV I or MCV II (Porter, 1989; Scholz et al., 1988). Neither for the small number of cases of MCV III is there any feature peculiar to this subtype, although one of these three patients did have particularly large and numerous lesions (Porter, 1989).

As noted above, patients with impaired immunity may have numerous and persistent lesions. Amongst ten MCV patients with such conditions including eight with atopic dermatitis, one with icthyosis and one with pancytopaenia, there was no association with a particular virus subtype (Porter et al., 1989).

From the above analysis it seems that MCV subtype and genomic polymorphisms are not related to any clinical criteria. The only significant observation arising from the epidemiological data relates to the geographical distribution of each subtype (Table 8.1). All cases of MCV III were from patients presenting in Britain. The relative occurrence of MCV I to MCV II ranges from 1:1 in the small American study group, through 3:1 for England, to 50:1 for Scotland.

8.11 POSSIBLE MECHANISMS OF MCV-INDUCED TUMOURIGENESIS

Unlike the orthopoxviruses, a number of other poxviruses in addition to MCV are associated with proliferative responses in their natural hosts

(Baxby, 1984). There include fowlpoxvirus, an avipoxvirus causing epithelial cell hyperplasia, Yaba monkey tumour virus, an unclassified poxvirus inducing dermal or subcutaneous histiocytomas, and Shope rabbit fibroma virus, a leporipoxvirus responsible for subcutaneous fibromas. As for *Molluscum contagiosum* lesions, these tumours are benign and eventually resolve. A potential mechanism of poxvirus-induced tumourigenesis was suggested following the discovery that a vaccinia virus early gene product secreted from infected cells, vaccinia growth factor (VGF), is structurally and functionally homologous to the epidermal growth factor (EGF) family (Blomquist *et al.*, 1984; Brown *et al.*, 1985; Reisner, 1985; Stroobant *et al.*, 1985; Twardzik *et al.*, 1985). For vaccinia virus which grows lytically rather than causing tumour formation, this factor appears to determine its replicative efficiency (but not infectivity) by stimulating the growth of host cells early in infection (Buller *et al.*, 1988a). Vaccinia virus deleted for the VGF gene is attenuated *in vivo* and the limited cell proliferation of epidermal cells that normally occurs at the lesion periphery prior to their infection is reduced (Buller *et al.*, 1988a,b). Growth of vaccinia virus in A431 cells, which are highly responsive to EGF due to enhanced expression of its cell surface receptor, induces cell focus formation in contrast to plaque formation by the mutant virus, consistent with the concept of VGF-mediated cell proliferation (Buller *et al.*, 1988b)

A similar mechanism can be envisaged as being responsible for tumour formation in poxvirus infections, for which cell proliferation is not masked by cytolysis. For *Molluscum contagiosum*, this would involve secretion of a VGF homologue from MCV-infected keratinocytes which would stimulate proliferation of the underlying basal cells. It is the latter cells within the epidermis that are most responsive to EGF. A VGF homologue (Shope fibroma growth factor; SFGF) has been identified for Shope rabbit fibroma virus (Chang *et al.*, 1987), using an oligonucleotide probe directed against conserved elements of the EGF family of growth factors. A similar experimental approach was initially encouraging for MCV (Porter and Archard, 1987) but did not ultimately identify such a homologue, despite a thorough search with various oligonucleotides making use of sequence data derived from the p43K gene (see below) relating to codon usage in MCV genes (Porter, 1989). A direct test of this mechanism is not possible due to the inability to culture MCV *in vitro*. Another approach has also been taken, in which candidate fragments of MCV DNA were used to construct recombinants with a VGF-deleted vaccinia virus in an attempt to complement the mutant phenotype by assaying for focus formation following growth in A431 cells (N.W. Blake, M. Mackett and L.C. Archard, unpublished data). Limited regions of the MCV genome, corresponding to near-terminal fragments by analogy with the VGF and SFGF loci, were used but failed to restore the wild-type phenotype. Other experiments have demonstrated that MCV genes can be expressed in this recombinant system (see below). Thus it is possible that MCV induces tumour formation by a novel mechanism that does not involve an EGF-homologue.

8.12 THE *MOLLUSCUM CONTAGIOSUM* VIRUS HOMOLOGUE OF THE VACCINIA VIRUS MAJOR ENVELOPE PROTEIN: CONTROL OF MCV GENE EXPRESSION

Although a number of open reading frames have been identified in nucleotide sequence data derived from MCV DNA (Porter, 1989; N.W. Blake, unpublished data) that may encode viral proteins, only one of these has been identified, on the basis of homology with a previously studied vaccinia virus gene (Blake *et al.*, 1991). An open reading frame encoding a polypeptide of predicted molecular weight 43 kD (p43K) has been sequenced from a region close to the left terminus of MCV subtypes I and II. The amino acid sequence has a high degree of homology to an abundant late gene product of 372 amino acids, p37K, the major antigen present on the envelope of extracellular vaccinia virus (Hirt *et al.*, 1986). This antigen is acquired, along with a membrane of Golgi origin, by mature virions released prior to cytolysis and comprises 5–7% of their protein mass (Hiller and Weber, 1985). The p37K-containing membrane is not necessary for vaccinia virus infection but does enhance the process of cell adsorption (Payne and Norrby, 1978).

The p43K genes of MCV I and MCV II are 94% homologous at the nucleotide level and correspond to polypeptides of 388 amino acids (Blake *et al.*, 1991). The G+C content is 66% and reflects the strong bias for these nucleotides at the third 'wobble' position of codons: 82% and 87% for MCV I and II respectively. The gene products are 95% homologous. MCV p43K is 62% homologous (40% identical) to vaccinia virus p37K (Figure 8.4), despite the marked difference in nucleotide composition of MCV and vaccinia virus DNA. The greatest homology is between the C-termini, although the C-terminus of p43K extends 13 amino acids beyond that for p37K. Two insertions, of 7 and 8 amino acids, are necessary to optimize the sequence alignment. The MCV protein has two possible transmembrane regions, one near the N-terminus and one at the same position as a putative p37K transmembrane domain (amino acids 130–157) (Hirt *et al.*, 1986) (Figure 8.4). The level of amino acid identity in the N-terminal domain preceding this region is lower (26%) and may explain the lack of antigenic cross-reactivity of vaccinia virus and MCV.

The open reading frames flanking the p43K gene are not homologous between MCV subtypes or with vaccinia virus p37K flanking sequences. However, the positions of the p43K and p37K genes are similar within the genomes of these serologically unrelated viruses. The MCV gene is at approximately 24 kb from the junction of the left terminal repeat and unique

```
MCV I  p43K  .MGNLTSARPAGCKIVETLPATLPLALPTGSMLTYDCFDTLISQTQRELC 49
              :  | |::|||||   :  : :|::||: :|    :  ::
VACC   p37K  MWPFASVPAGAKCRLVETLPEN..MDFRSDHLTTFECFNEIITLAKKYIY 48

MCV I  p43K  IASYCCNLRSTPEGGHVLLRLLELARADVRVTIIVDEQSRDADATQLAGV 99
             |||:|||  ||  |: :: :| |    ::::  :::|| ::    :
VACC   p37K  IASFCCNPLSTTRGALIFDKLKEASEKGIKIIVLLDERGKRNLGELQSHC 98

MCV I  p43K  PNLRYLKLDVGELPGGKPGSLLSSFWVSDKRRF╔═══════════════╗149
             |:: :: ::::    | |||:::|||||  | ║YLGSASLTGGSISTIKS║
VACC   p37K  PDINFITVNIDK..KNNVGLLLGCFWVSDDERQ║YVGNASFTGGSIHTIKT║146
                                              ╚═════╦═══════════╝
MCV I  p43K  ╔════╗                                 ║146
             ║LGVY║SECEPLARDLRRRFRDYE........RLCARRCVRCLSLSTRFHLR 191
             ║||||║|::  |||      :|     || :|| :|::
VACC   p37K  ║LGVY║SDYPPLATDLRRRFDTFKAFNSAKNSWLNLCSAACCLPVSTAYHIK 196
             ╚════╝
MCV I  p43K  RHCENAFFSDAPESLIGSTRTFDADAVLAHVQAARSTIDMELLSLVPLVR 241
                :  || | || |:|  | :| | |:  :  |:  ||:| | :|| |
VACC   p37K  NPIGGVFFTDSPEHLLGYSRDLDTDVVIDKLKSAKTSIDIEHLAIVPTTR 246

MCV I  p43K  DEDSVQYWPRMHDALVRAALERNVRVRLLVGLWHRSDVFSLAAVKGLHEL 291
              ::    ||| : : :: ||::| |::||||| | :||:|:|  ::| |
VACC   p37K  VDGNSYYWPDIYNSIIEAAINRGVKIRLLVGNWDKNDVYSMATARSLDAL 296

MCV I  p43K  GVGHADISVRVFAIPGAKGEPLNNTKLLVVDDEYVHVTSADMDGTHYARH 341
              | :  |:||:||  |           ||||||:||||||||:|||::||||| |
VACC   p37K  CV.QNDLSVKVFTIQ.......NNTKLLIVDDEYVHITSANFDGTHYQNH 338

MCV I  p43K  AFVSFNCAERAFARALGALFERDWQSSFSSPLPRAPPPEPATLLPVN 388
             :|||||: :: :    :||||| || | |
VACC   p37K  GFVSFNSIDKQLVSEAKKIFERDWVSSHSKSLKI............ 372
```

Figure 8.4 Comparison of the amino acid sequences of the MCV I p43K and vaccinia virus p37K proteins. A vertical line represents an identity, whilst a colon represents a conservative amino acid change. Two gaps have been introduced to optimize the alignment. The regions boxed (p43K amino acids 133–153 and p37K amino acids 130–157) are hydrophobic, putative transmembrane domains.

sequences whereas the vaccinia virus gene is 28 kb from this junction. Both genes are transcribed towards the left terminus.

In addition to sequence conservation, the elements involved in regulation of expression of the p43K and p37K genes appear to be conserved. The p37K gene is expressed late, after the onset of DNA synthesis, and possesses the sequence motif TAAATG: this conserved feature of vaccinia virus late genes encompasses both the transcriptional start site and initiation codon (Hirt *et al.*, 1986). An upstream TTTTT motif is present and is another common feature of late genes. Despite the G/C-rich MCV sequence, similar motifs are present for the p43K gene which has the sequence TAAAATG encompassing the start codon and TTTTT at the corresponding position upstream (Blake *et al.*, 1991) (Figure 8.5). The MCV I and MCV II upstream regions are homologous until position -40 but then diverge, suggesting that all control elements occur within this region. The TAAAATG motif, with

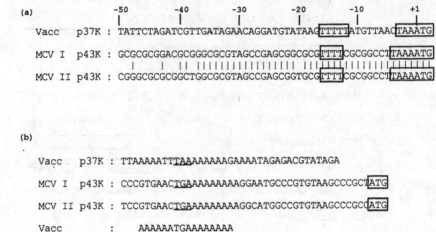

(a)

```
              -50        -40        -30        -20        -10        +1
               |          |          |          |          |          |
    Vacc  p37K : TATTCTAGATCGTTGATAGAACAGGATGTATAAGTTTTTATGTTAACTAAAATG

    MCV I p43K : GCGCGCGGACGCGGGCGCGTAGCCGAGCGGCGCGTTTTCGCGGCCTTAAAATG
                 |     |   ||  ||||||||||||||||||| ||| |||| ||||||||| |||||||
    MCV II p43K : CGGGCGCGCGGCTGGCGCGTAGCCGAGCGGTGCGTTTTCGCGGCCTTAAAATG
```

(b)

```
    Vacc  p37K : TTAAAAATTTAAAAAAAAAGAAAATAGAGACGTATAGA

    MCV I p43K : CCCGTGAACTGAAAAAAAAAGGAATGCCCGTGTAAGCCCGCTATG

    MCV II p43K : TCCGTGAACTGAAAAAAAAAGGCATGGCCGTGTAAGCCCGCCATG

    Vacc      :  AAAAAATGAAAAAAAA
    Concensus            T
```

Figure 8.5 (a) Comparison of nucleotides from −50 to +3 (numbered with the start codon as position 1–3) for the vaccinia virus p37K gene and the p43K genes for MCV I and MCV II. Boxed are the elements involved in initiation of transcription of the p37K gene and MCV counterparts, the latter being notable within the otherwise G/C-rich context. Evidence for initiation of p43K transcription within the TAAAATG motif is presented in Figure 7. (b) Comparison of sequence around the termination codons (underlined) of the p37K and p43K genes. The A-rich context in each case shows homology to the consensus sequence for the critical region of vaccinia virus early promoters, shown below (Davison and Moss, 1989). This implies early expression of the flanking open reading frames in each case, the initiation codons of which are boxed for the MCV sequences.

an additional base relative to the consensus, also occurs upstream from the vaccinia virus early gene ORF K3, which has some homology to the p37K gene (Boursnell *et al.*, 1988) (22% identity to both p37K and p43K), and for the vaccinia virus late gene ORF(N1L) which encodes a secreted protein (Kotwal *et al.*, 1989). The sequences surrounding the stop codons of the p37K and p43K genes are also A/T-rich and homologous to the consensus for vaccinia virus early gene promoters, implying adjacent early genes in each case (Figure 8.5).

The above analysis suggests that the genetic organization and the control of expression of MCV genes is similar in many respects to that of vaccinia virus. Similarity of poxvirus promoter elements for different genera is becoming established and initial experiments have shown that the transcriptional machinery of poxviruses is capable of transcribing heterologous poxvirus genes (Boyle and Couper, 1986; Macaulay and McFadden, 1989). These implications have been studied further by constructing a recombinant vaccinia virus containing a 64 kb BamHI fragment of MCV I DNA which includes the p43K gene (N.W. Blake, M. Mackett and L.C. Archard, unpublished data). The recombinant was constructed such that expression of MCV genes requires recognition of MCV promoters by vaccinia virus transcription factors. After infection, transcriptional analysis by Northern blotting demonstrated early transcripts specific to the MCV fragment, although their gene products have not been characterized (Figure 8.6). The p43K gene was not expressed at early times. S1 nuclease mapping was used to demonstrate expression of the p43K gene late in infection and to localize the transcriptional initiation site within the TAAAATG motif (Figure 8.7). Thus, MCV promoter elements can be recognized by the transcriptional machinery of the serologically-unrelated vaccinia virus, which has broad implications for the use of poxviruses with various host ranges as vectors for the expression of foreign genes.

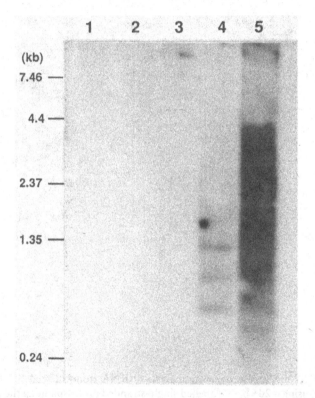

Figure 8.6 Northern blot of RNA made from CV-1 cells early (six hours post-infection in the presence of cycloheximide) and late (ten hours post-infection) after infection wih a recombinant vaccinia virus containing the 6.4 kb BamHI fragment J of MCV I, probed with this same fragment. 1; RNA from mock infected cells; 2 and 3; Early and late RNA respectively of vaccinia WR infected cells; 4 and 5; Early and late RNA respectively of cells infected with the vaccinia/MCV recombinant. The positions of RNA size markers are shown (kb). Discrete MCV- specific transcripts are demonstrated at early times. Late in infection the detected RNA is of heterogeneous length, a common feature of vaccinia virus late gene expression due to the lack of specific termination (Mahr and Roberts, 1984).

Figure 8.7 S1 nuclease protection analysis of RNA from infected CV-1 cells (see Figure 8.6) using a 246 bp 5'-labelled single-stranded probe spanning the p43K gene initiation codon. P; Untreated full-length probe; M; RNA from mock infected cells. E and L; Early and late RNA respectively from cells infected with the vaccinia/MCV recombinant. There is no p43K gene expression at early times after infection. Late RNA protects full-length probe, due to the presence of read-through transcripts (caption to Figure 8.6), and a cluster of fragments at 97–99 nucleotides that maps initiation of p43K gene transcription within the TAAAATG motif (Figure 8.5).

Two obvious and intriguing questions of MCV molecular biology remain to be resolved. The first is the nature of the constraints which confine viral replication to keratinocytes at a particular stage of differentiation: this may provide some insight into the process of cell differentiation as well as the molecular relationships between cellular and viral functions. The second is the molecular mechanism by which MCV infection stimulates the proliferation of a skin cell type which does not itself support viral infection: while this may well result from secretion by infected keratinocytes of a virus-coded growth factor, the nature of this factor and its putative receptor on basal cells remain unknown.

REFERENCES

Archard, L.C., Mackett, M., Barnes, D.E. and Dumbell, K.R. (1984) The genome structure of cowpox virus white pock mutants. *J. Gen. Virol.*, 65, 875–86.

Bateman, T. (1814) *A Practical Synopsis of Cutaneous Diseases*, 3rd edn. London, Longman Hurst Rees Orme & Brown.

Baxby, B. (1984) Poxviruses, in *Principles of Bacteriology, Virology and Immunity*, Vol. 4, 7th edn (eds F. Brown and G. Wilson) Edward Arnold, Kent, pp. 163–82.

Baxter, D.L. and Carson, W.E. (1966) *Molluscum contagiosum* of the sole. *Arch. Derm.*, 89, 471–2.

Becker, T.M., Blount, J.H., Douglas, J.D. and Judson, F.N. (1986) Trends in *Molluscum contagiosum* in the United States 1966–1983. *Sex. Trans. Dis.*, 13, 88–92.

Blake, N.W., Porter, C.D. and Archard, L.C. (1991) Characterisation of a *Molluscum contagiosum* virus homologue of the vaccinia virus p37K major envelope antigen. *J. Virol.*, 65, 3583–9.

Blank, H. and Rake, G. (1955) *Viral and Rickettsial Diseases of the Skin, Eye and Mucous Membranes of Man*. Boston, Little, Brown and Company.

Blomquist, M.C., Hunt, L.T. and Barker, W.C. (1984) Vaccina virus 19-kilodalton protein; relationship to several mammalian proteins, including two growth factors. *Proc. Natl Acad. Sci. USA*, 81, 7363–7.

Boursnell, M.E., Foulds, I.J., Campbell, J.I. and Binns, M.M. (1988) Non-essential genes in the vaccinia virus HindIII K fragment: a gene related to serine protease inhibitors and a gene related to the 37K vaccinia virus major envelope antigen. *J. Gen. Virol.*, 69, 2995–3003.

Boyle, D.B. and Coupar, B.E.H. (1986) Identification and cloning of the fowlpox virus thymidine kinase gene using vaccinia virus. *J. Gen. Virol.*, 67, 1591–600.

Brown, J.P., Twardzik, D.R., Marquardt, H. and Todaro, G.J. (1985) Vaccinia virus encodes a polypeptide homologous to epidermal growth factor and transforming growth factor. *Nature*, 313, 491–2.

Brown, S.T. and Weinberger, J. (1974) *Molluscum contagiosum*: sexually transmitted disease in 17 cases. *J. Am. Venereal Dis. Assoc.*, 1, 35–6.

Brown, S.T., Nalley, J.F. and Kraus, S.J. (1981) *Molluscum contagiosum. Sex. Trans. Dis.*, 8, 227–32.

Brown, T., Butler, P. and Postlethwaite, R. (1973) Nongenetic reactivation studies with the virus of *Molluscum contagiosum. J. Gen., Virol.* 19, 417–21.

Buller, R.M.L., Chakrabarti, S., Cooper, J.A. *et al.* (1988a) Deletion of the vaccinia virus growth factor gene reduces virus virulence. *J. Virol.*, **62**, 866–74.

Buller, R.M.L.; Chakrabarti, S., Moss, B. and Fredrickson, T. (1988b) Cell proliferative response to vaccinia virus is mediated by VGF. *Virology*, **164**, 182–92.

Chaney, S.M.J., Warren, K.G., Kettyls, J. *et al.* (1983) A comparative analysis of restriction enzyme digests of the DNA of herpes simplex virus isolated from genital and facial lesions. *J. Gen. Virol.*, **64**, 357–71.

Chang, W., Upton, C., Hu, S.-L. *et al.* (1987) The genome of shope fibroma virus, a tumorigenic poxvi rus, contains a growth factor gene with sequence similarity to those encoding epidermal growth factor and transforming growth factor alpha. *Mol. Cell. Biol.*, **7**, 535–40.

Cobbold, R.J.C. and Macdonald, A. (1970) *Molluscum contagiosum* as a sexually transmitted disease. *Practitioner*, **204**, 416–19.

Cotton, D.W.K., Cooper, C. Barrett, D.F. and Leppard, B.J. (1987) Severe atypical *molluscum contagiosum* in an immunocompromised host. *Brit. J. Derm.*, **116**, 871–6.

Darai, G., Reisner, H., Scholz, J. *et al.* (1986) Analysis of the genome of *Molluscum contagiosum* virus by restriction endonuclease analysis and molecular cloning. *J. Med. Virol.*, **18**, 29–39.

Davison, A.J. and Moss, B. (1989) Structure of vaccinia virus early promoters. *J. Mol. Biol.*, **210**, 749–69.

Department of Health and Social Security (1973) Sexually transmitted diseases. Extract from the annual report of the Chief Medical Officer to the Department of Health and Social Security for the year 1971. *Bri. J. Ven. Dis.*, **49**, 89–95.

Department of Health and Social Security (1985) Sexually transmitted diseases. Extract from the annual report of the Chief Medical Officer to the Department of Health and Social Security for the year 1983. *Genitourinary Med.*, **61**, 204–7.

Editorial BMJ (1968) *Molluscum contagiosum*. *Brit. Med. J.*, i, 459–60.

Epstein, W.L. and Fukuyama, K. (1973) Maturation of *Molluscum contagiosum* virus (MCV) in vivo: quantitative electron microscopic autoradiography. *J. Invest. Derm.*, **60**, 73–9.

Epstein, W.L., Conant, M.A. and Krasnobrod, H. (1966) *Molluscum contagiosum*: normal and virus infected epidermal cell kinetics. *J. Invest. Derm.*, **46**, 91–103.

Fenner, F., Wittek, R. and Dumbell, K.R. (1989) *The Orthopoxviruses*. Academic Press (San Diego).

Foulds, I.S. (1982) *Molluscum contagiosum*: an unusual complication of tatooing. *Brit. Med. J.*, **285**, 607.

Francis, R.D. and Bradford, H.B. (1976) Some biological and physical properties of *Molluscum contagiosum* virus propagated in cell culture. *J. Virol.*, **19**, 382–8.

Fraser, K.M., Hill, D.F., Mercer, A.A. and Robinson, A.J. (1990) Sequence analysis of the inverted terminal repetition in the genome of the Parapoxvirus, orf virus. *Virology*, **176**, 379–89.

Gassman, U., Wyler, R. and Wittek, R. (1985) Analysis of parapoxvirus genomes. *Arch. Virol.*, **83**, 17–31.

REFERENCES

Gershon, P.D. and Black, D.N. (1988) A comparison of the genomes of capripoxvirus isolates of sheep, goats and cattle. *Virology*, 164, 341–9.

Goldschmidt, H. and Kligman, A.M. (1958) Experimental inoculation of humans with ectodermotropic viruses. *J. Invest., Derm.*, 31, 175–82.

Goodman, D.S., Teplitz, E.D., Wishner, A. *et al.* (1987) Prevalence of cutaneous disease in patients with acquired immunodeficiency syndrome (AIDS) or AIDS-related complex. *J. Am. Acad. Derm.*, 17, 210–20.

Goodpasture, E.W. (1933) Borrelioses, fowlpox, *Molluscum contagiosum*, variola-vaccinia. *Science*, 77, 119–21.

Goodpasture, E.W. and Woodruff, C.E. (1931) A comparison of the inclusion bodies of fowlpox and *Molluscum contagiosum*. *Am. J. Pathol.*, 7, 1–7.

Gudgel, E.F. (1954) Can molluscum contagiosum be a venereal disease? *US Armed Forces Med. J.*, 5, 1207–8.

Henderson, W. (1841) Notice of the *Molluscum contagiosum*. *Edinburgh Med. Surg. J.*, 56, 213–18.

Heng, M.C.Y., Steuer, M.E., Levy, A. *et al.* (1989) Lack of host cellular immune response in eruptive *Molluscum contagiosum*. *Am. J. Dermatopathol.*, 11, 248–54.

Hiller, G. and Weber, K. (1985) Golgi-derived membranes that contain an acylated viral polypeptide are used for vaccinia virus envelopment. *J. Virol.*, 55, 651–9.

Hirt, P., Hiller, G. and Wittek, R. (1986) Localization and fine structure of a vaccinia virus gene encoding an envelope antigen. *J. Virol.*, 58, 757–64.

Ive, F.A. (1985) Follicular *Molluscum contagiosum*. *Brit. J. Derm.*, 113, 493–5.

Jablonska, S. and Orth, G. (1983) Human papovaviruses, in *Recent Advances in Dermatology* 6 (eds A.J. Rook and H.I. Maibach) Churchill Livingstone, (Edinburgh), pp. 1–36.

Jacob, P.H. (1970) *Molluscum contagiosum*. *Aerospace Med.*, 41, 1196.

Juliusberg, M. (1905) Zur Kenntnis des virus des *Molluscum contagiosum*. *Dtsch. Med. Wochenschr.*, 31, 1598–9.

Kotwal, G.J., Hugin, A.W. and Moss, B. (1989) Mapping and insertional mutagenesis of a vaccinia virus gene encoding a 13,800 Da secreted protein. *Virology*, 171, 579–87.

Laskaris, G. and Sklavounou, A. (1984) *Molluscum contagiosum* of the oral mucosa. *Oral Surg.*, 58, 688–91.

Lynch, P.J. and Minki, W. (1968) *Molluscum contagiosum* of the adult, probable venereal transmission. *Arch. Derm.*, 98, 141–3.

Macaulay, C. and McFadden, G. (1989) Tumourigenic poxviruses: characterisation of an early promoter from Shope fibroma virus. *Virology*, 172, 237–46.

Mackett, M. and Archard, L.C. (1979) Conservation and variation in orthopoxvirus genome structure. *J. Gen. Virol.*, 45, 683–701.

Mahr, A. and Roberts, B.E. (1984) Arrangement of late RNAs transcribed from a 7.1 kilobase EcoRI vaccinia virus DNA fragment. *J. Virol.*, 49, 510–20.

McFadden, G., Pace, W.E., Purres, J. and Dales, S. (1979) Biogenesis of poxviruses; transitory expression of *Molluscum contagiosum* early functions. *Virology*, 94, 297–313.

Mitchell, J.C. (1953) Observations on the virus of *Molluscum contagiosum*. *Brit. J. Exp. Path.*, 34, 44–9.

255

Oda, H., Ohayama, Y., Sameshima, T. and Hirakawa, K. (1982) Structural polypeptides of *Molluscum contagiosum* virus: their variability in various isolates and location within the virion. *J. Med. Virol.*, 9, 19–25.

Oriel, J.D. (1987) The increase in *Molluscum contagiosum*. *Brit. Med. J.*, 294, 74.

Orth, G. and Favre, M. (1985) Human papillomaviruses: biochemical and biological properties. *Clinics in Dermatol.*, 3, 27–42.

Parr, R.P., Burnett, J.W. and Garon, C.F. (1977) Structural characterisation of the *Molluscum contagiosum* virus. *Virology*, 81, 247–56.

Paterson, R. (1841) Causes and observations on the *Molluscum contagiosum* of Bateman with an account of the minute structure of the tumours. *Edinburgh Med. Surg. J.*, 56, 279–88.

Payne, L.G. and Norrby, E. (1978) Adsorption and penetration of enveloped and naked vaccinia virus particles. *J. Virol.*, 27, 19–27.

Pinkus, H. and Frisch, D. (1949) Inflammatory reactions to *Molluscum contagiosum* possibly of immunological nature. *J. Invest. Derm.*, 13, 289–94.

Pirie, G.D., Bishop, P.M., Burke, D.C. and Postlethwaite, R. (1971) Some properties of purified *Molluscum contagiosum* virus. *J. Gen. Virol.*, 13, 311–20.

Plucienniczak, A., Schroeder, E., Zettlmeissl, G. and Streeck, R.E. (1985) Nucleotide sequence of a cluster of early and late genes in a conserved segment of the vaccinia virus genome. *Nuc. Acids Res.*, 13, 985–98.

Porter, C.D. (1989) *Molluscum contagiosum* virus: genome structure and mechanism of tumourigenesis. PhD Thesis, University of London, UK.

Porter, C.D. and Archard, L.C. (1987) Characterisation and physical mapping of *Molluscum contagiosum* virus DNA and location of a sequence capable of encoding a conserved domain of epidermal growth factor. *J. Gen. Virol.*, 68, 673–82.

Porter, C.D., Muhlemann, M.F., Cream, J.J. and Archard, L.C. (1987) *Molluscum contagiosum*: characterization of viral DNA and clinical features. *Epidemiol. and Infect.*, 99, 563–7.

Porter, C.D., Blake, N.W., Archard, L.C. et al. (1989) *Molluscum contagiosum* virus types in genital and non-genital lesions. *Brit. J. Derm.*, 120, 37–41.

Postlethwaite, R. (1970) *Molluscum contagiosum*: a review. *Arch. Environ. Health*, 21, 432–52.

Postlethwaite, R., Watt, J.A., Hawley, T.G., Simpson, I. and Adam, H. (1967) Features of *Molluscum contagiosum* in the north-east of Scotland and in Fijian village settlements. *J. Hygiene*, 65, 281–91.

Reisner, A.H. (1985) Similarity between the vaccinia virus 19K early protein and epidermal growth factor. *Nature*, 313, 801–3.

Scholz, J., Rosen-Wolff, A., Bugert, J. et al. (1988) Molecular epidemiology of *Molluscum contagiosum*. *J. Infect. Dis.*, 158, 898–900.

Scholz, J., Rosen-Wolff, A., Bugert, J. et al. (1989) Epidemiology of *Molluscum contagiosum* using genetic analysis of the viral DNA. *J. Med. Virol.*, 27, 87–90.

Shelly, W.B. and Burmeister, V. (1986) Demonstration of a unique viral structure: the molluscum viral colony sac. *Brit. J. Derm.*, 115, 557–62.

Shirodaria, P.V. and Matthews, R.S. (1977) Observations on the antibody responses in *Molluscum contagiosum*. *Brit. J. Derm.*, 96, 29–34.

REFERENCES

Stroobant, P., Rice, A.R., Gullick, W.J. *et al.* (1985) Purification and characterisation of vaccinia virus growth factor. *Cell*, **42**, 383–93.

Sutton, J.S. and Burnett, J.W. (1969) Ultrastructural changes in dermal and epidermal cells of skin infected with *Molluscum contagiosum* virus. *J. Ultrastruct. Res.*, **26**, 177–96.

Twardzik, D.R., Brown, J.P., Ranchalis, J.E. (1985) Vaccinia virus-infected cells release a novel polypeptide functionally related to transforming and epidermal growth factors. *Proc. Natl Acad. Sci. USA*, **82**, 5300–4.

Van Rooyen, C.E. (1938) The micromanipulation and microdissection of the *Molluscum contagiosum* body. *J. Pathol.*, **46**, 425.

Wiles, U.J. and Kingery, L.B. (1919) Etiology of *Molluscum contagiosum*. *J. Cutan. Dis.*, **37**, 431–6.

Wilkin, J.K. (1977) *Molluscum contagiosum* venereum in a women's out-patient clinic: a venereally transmitted disease. *Am. J. Obst. Gynecol.*, **128**, 531–5.

Wittek, R. and Moss, B. (1980) Tandem repeats within the inverted terminal repetition of vaccinia virus DNA. *Cell*, **21**, 277–84.

Wittek, R., Cooper, J.A., Barbosa, E. and Moss, B. (1980) Expression of the vaccinia virus genome: analysis and mapping of mRNAs encoded within the inverted terminal repetition. *Cell*, **21**, 487–93.

Molecular biology of herpes simplex virus

DAVID E. BARKER, BERNARD ROIZMAN AND

MARJORIE B. KOVLER

> And nature must obey necessity
> Shakespeare, *Julius Caesar*,
> IV, iii

9.1 INTRODUCTION

Herpes . . . in the current post sexual-revolution era is certainly a familiar term yet the virus responsible is truly ancient. Indeed, herpes simplex virus has been with us for far longer than we have been *Homo sapiens*. Although oral cold sores were described by 100 AD (Wildy, 1973) and genital herpes was recognized as a sexually transmissible disease in the early eighteenth century (Astruc, 1736; Diday and Doyon, 1886) it was not until 1962 that two serologically distinct herpes simplex viruses were recognized; HSV-1 (primarily oral-labial) and HSV-2 (primarily genital) (Schnewiess, 1962; Nahmais and Dowdle, 1968). Detailed knowledge of HSV has emerged largely in the last 20 years, but it is accumulating at an accelerating pace in parallel with the sophistication of molecular biology.

9.2 TAXONOMY

Members of the family *herpesviridae* appear to have co-evolved with their hosts and have been diverging, one virus from another, throughout much of evolution (Roizman *et al.*, 1981). The herpes viruses currently are classified in three subfamilies (*alphaherpesvirinae, betaherpesvirinae* and

Molecular and Cell Biology of Sexually Transmitted Diseases
Edited by D. Wright and L. Archard
Published in 1992 by Chapman and Hall, London ISBN 0 412 36510 3

gammaherpesvirinae) defined initially according to biologic properties (Roizman *et al.*, 1981). Among human herpes viruses, HSV-1, HSV-2 and varicella-zoster virus (VZV) are members of *alphaherpesvirinae* whereas Epstein-Barr Virus (EBV) and cytomegalovirus (CMV) are members of the *gammaherpesvirnae* and *betaherpesvirinae* respectively. Recent studies suggest that in addition to biological properties, viruses within a subfamily share greater similarity in gene order, genome organization and protein homology than members of different subfamilies. However, viruses comprising the three subfamilies have diverged with respect to genetic content, sequence complexity and organization, and, for the conserved genes, the order in which they are arranged. Gene conservation and order is likely to become a major criterion by which herpes viruses are likely to be classified (Davison and Scott, 1986; Baer *et al.*, 1984; Lawrence *et al.*, 1990; McGeoch *et al.*, 1988). The two most recently isolated human pathogens, human herpes viruses 6 and 7 (Salhuddin *et al.*, 1986; Frenkel *et al.*, 1990) were discovered by cultivation of T lymphocytes *in vitro*.

All herpes viruses share at least four characteristics: (i) a morphologically identical virion (Figure 9.1); (ii) a viral chromosome consisting of a single large double-stranded DNA molecule; (iii) a highly ordered, intricately regulated cascade of gene expression during productive infection; and (iv) an ability to establish latent infections which can be maintained for the life of the host and which forms a crucial viral reservoir in the population (Roizman, 1990).

Although much of our understanding of HSV types comes from studies of a single serotype, HSV-1, some sequence comparisons (McGeoch *et al.*, 1988; McGeoch *et al.*, 1987) suggest that the two serotypes have collinear genomes in that every HSV-1 gene appears to have an identically positioned HSV-2 homologue. The colinearity of the HSV-1 and HSV-2 genomes has been established largely from studies of HSV-1 x HSV-2 recombinants (Morse *et al.*, 1977; 1978; Marsden *et al.*, 1978). While the matching of base pairs exceeds 50% (Kieff *et al.*, 1972), few restriction endonuclease cleavage sites are conserved in the genomes of both serotypes.

HSV exhibits two types of polymorphism at the DNA level. First the genome contains, at specific sites, variable numbers of copies of several repetitive elements (Roizman and Tognon, 1983). A second form of polymorphism is the variability in restriction endonuclease cleavage sites (Roizman and Tognon, 1983). As a general rule, the restriction endonuclease cleavage sites are conserved on serial passage of an isolate in culture and epidemiologically related individuals (mother, child, spouses, etc.) have yielded isolates which could not be differentiated with respect to their restriction endonuclease cleavage sites (Roizman and Tognon, 1983; Linnemann *et al.*, 1978; Buchman *et al.*, 1978). Thus, restriction fragment pattern analyses are a very useful epidemiologic tool in tracing the transmission of HSV from one individual to another. The significant findings to emerge from molecular epidemiologic studies carried out to date are that: (i) the increase in the incidence of genital herpes seen in the last few decades reflects a frenzy of transmission rather than an epidemic caused by one or a few strains (Roizman and Tognon, 1983); (ii) the increased incidence in neonatal herpes infections reflects an increase in the incidence of genital herpes (Sullivan-Bolyai et al., 1983; Rawls et al., 1971); and (iii) postnatal transmission resulting from exposure of newborns to individuals shedding virus has been documented (Linnemann et al., 1978; Buchman et al., 1978; Nahmais et al., 1983).

9.3 THE HERPES VIRUS VIRION

Herpes simplex virions contain the viral DNA, spermine, which may have a role in neutralizing the charge of the viral DNA, spermidine, lipids, and at least 33 different proteins which have been assigned Virion Protein (VP) numbers and which appear to be virally encoded (Spear and Roizman, 1972). These structural components are arranged in four architectural components (Figure 9.1).

The outer architectural component is the virion envelope (Figure 9.1). It consists of a lipid bilayer derived from the host cell and into which are inserted eight distinct glycoproteins on its outer surface and an as yet undefined number of proteins on its inner surface.

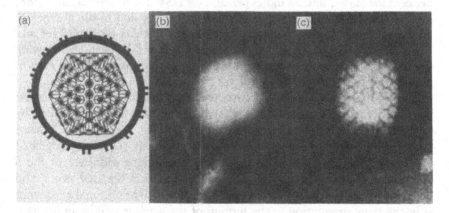

Figure 9.1 Schematic diagram and electron micrographs of HSV-1 virion and capsid. (a) Diagrammatic representation of the virion. The icosadeltahedral capsid is surrounded by a tegument and envelope with virally encoded glycoproteins projecting from its surface. The tegument is asymmetrically distributed between the capsid and the envelope (Roizman and Furlong, 1974); (b) a negatively stained HSV virion. Intact virions are impervious to negative stains; (c) a negatively stained HSV capsid exhibiting two-fold symmetry. A 4 nm wide channel running along the long axis of the capsomeres is made apparent by the negative stain (Roizman and Furlong, 1974).

A distinct, electron translucent architectural component, the tegument, is located between the envelope and the viral capsid. The tegument contains approximately 10 to 12 proteins (Roizman and Furlong, 1974).

The structure of the capsid, an icosadeltahedron consisting of 162 capsomeres (Figure 9.1), has been the subject of considerable research (Roizman and Furlong, 1974; Scrag et al., 1989). The capsids consist of approximately eight to ten proteins (Gibson and Roizman, 1972; 1974; Cohen et al., 1980). Immature capsids which do not contain DNA are not enveloped and remain in the nucleus (Vlazny et al., 1982). Capsids containing DNA become altered in protein structure, adhere, and become enveloped by patches containing viral tegument and membrane proteins in the inner lamellae of the nuclear membrane (Gibson and Roizman, 1974; Braun et al., 1984). The core consists of DNA, polyamines and proteins. The DNA is coiled in the form of a toroid (Furlong et al., 1972).

9.4 THE VIRAL GENOME

The genomes of herpesviruses vary in size, from approximately 120 to 250 Kbp (Davison and Scott, 1986; Baer et al., 1984; McGeoch et al., 1988; Kieff et al., 1971). The herpes simplex genome is a linear molecule of double-stranded DNA of greater than 152000 base pairs in length (McGeoch et al., 1988; Kieff et al., 1972). The G+C content of HSV DNA is 67 moles per cent for HSV-2 and 68 moles per cent for HSV-2 (McGeoch et al., 1988; Kieff et al., 1972). The genome circularizes rapidly and in the absence of de novo protein synthesis upon entry into the nucleus (Poffenberger and Roizman, 1985). A schematic representation of the HSV-1 genome showing the sequence and gene arrangements as well as the transcriptional map of the genome is illustrated in Figure 9.2.

The HSV genome consists of two covalently linked components, L and S. Each component consists of a stretch of unique sequences flanked by inverted repeats (Sheldrick and Berthelot, 1975). The inverted repeats flanking the unique sequences of the L component (U_L) are designated as ab and b'a' each 9000 bp in length. The unique sequence of the S component (U_S) are flanked by reiterated sequences designated as a'c' and ca each 6000 bp in length (Wadsworth et al., 1975). The terminal a sequence is in the same orientation at the termini of the genome and in an inverted orientation at the junction between the L and S components (Wadsworth et al., 1975). The a sequence flanking the L component may be present in one to as many as ten copies in tandem repeats whereas only one copy is present at the free terminus of the S component (Locker and Frenkel, 1979; Wagner and Summers, 1978). The a sequence consists of both unique and reiterated sequences and HSV strains differ with respect to the number of reiterated elements (Mocarski and Roizman, 1981; Davison and Wilkie, 1981). The a sequence of strain F [HSV-1(F)] is approximately 500 bp long (Mocarski

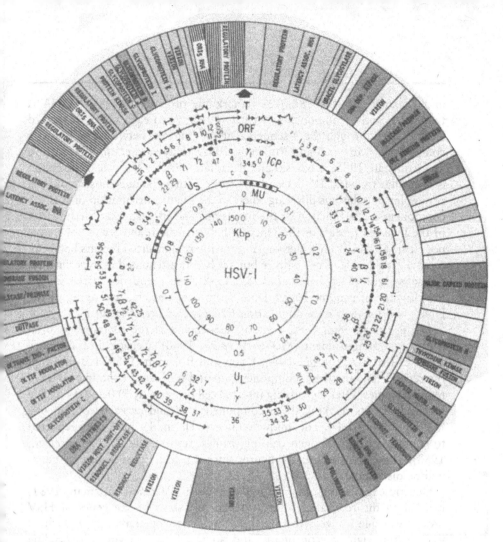

Figure 9.2 Schematic representation of the sequence and gene organization in HSV-1 genome. The viral genome is shown as a closed circle with the short and long components so arranged that linearization by cleavage at 12 o'clock would yield the Prototype (P) isomer of the DNA. The innermost circle correlates the reference genome Map Units (MU) (0.0 to 1.0) with the distances in thousands of base pairs (Kbp). The next circle shows the reiterated sequences, *a'b'* (dotted boxes) and *a'c'*, *ca* (open boxes) flanking the long (U$_L$) and short (U$_S$) unique regions (single lines). Infected Cell Protein (ICP) designations (ICP 0, ICP 34.5, etc.) and DNA transcriptional class (α, , 1, 2) are displayed beneath the HSV genes identified as Open Reading Frames (ORF), which are numbered sequentially (U$_S$ 1–12, U$_L$1–56). Some genes α0, , 34.5, or transcribed sequences (Latl) are contained within the repeats flanking the unique sequences; these are present in two copies per genome and are not numbered. ori$_S$, ori$_L$ are the locations of the short and long segment origins of DNA replication, respectively. Outside this circle are shown the domains and polarity of known HSV-1 transcripts (T). The infrequent spliced transcripts are identified by the arched lines at the sites of known introns. The outermost ring specifies the functions of the gene products. Dark shading – essential for replication in tissue culture; light gray – not essential for replication in tissue culture; striped – diploid gene, one copy of which is not required for replication in tissue culture; unshaded – requirement for replication in cell culture has not yet been examined. The schematic diagram is based on genetic and transcriptional studies, on the determined sequence of HSV-1 DNA, and on genetic engineering of novel genomes from which specific genes or sequences had been deleted.

and Roizman, 1981; 1982a). A remarkable property of HSV-1 DNA is that the L and S components can invert relative to each other (Hayward *et al.*, 1975) through the *a* sequence at the junction between the L and S components (Mocarski and Roizman, 1982a; Hayward *et al.*, 1975; Chou and Roizman, 1985; Mocarski and Roizman, 1982b). The rate of inversion is such that the viral DNA extracted from a single plaque consists of four equimolar populations differing solely in the relative orientation of L and S components. The function of the inversions and in particular the function of the inverted repeats at the junction between the L and S components is not clear. Viruses from which most of the internal inverted repeats had been removed are viable in cell culture but are highly attenuated in experimental animal systems (Meignier *et al.*, 1987; 1988). The *a* sequence carries signals for cleavage of genome length DNA from concatemers, packaging of the DNA, and for their own duplication (Vlazny and Frenkel, 1981).

Because of its high G+C content, sequencing of the genome has been a major accomplishment (McGeoch *et al.*, 1988). Genetic and sequence analyses (McGeoch *et al.*, 1988; Chou and Roizman, 1990) indicate that the repeats flanking the L component, *ab* and *b'a'*, each contain two open reading frames (ORF) specifying infected cell proteins (ICP) 0 and 34.5. The inverted repeats flanking the S component *c'a'* and *ca* each contain one ORF specifying ICP4. The unique sequences U_S and U_L have been reported to contain 12 and 56 open reading frames, respectively (McGeoch *et al.*, 1988). These are conservative figures since it is likely that additional genes will be discovered.

Genetic engineering of novel HSV genomes (Post and Roizman, 1981) has led to the realization that as much one third of the genes of HSV are dispensable for growth in cell culture. The dispensable genes include five of the eight glycoproteins and some of the enzymes (thymidine kinase, ribonucleotide reductase, dUTPase, etc.) involved in nucleic acid metabolism. In cell culture, the functions of some of the dispensable genes are complemented by cell functions (Sears *et al.*, 1985). However, most of these deletion mutants grow poorly in animals, suggesting that host tissues are unable fully to complement deletion mutants in missing functions (Meignier *et al.*, 1987a; 1988). In that sense, all viral genes are essential.

9.5 PRODUCTIVE INFECTION

The reproductive cycle of herpes simplex viruses schematically represented in Figure 9.3 differs in some details from that of other viruses. The key steps are as follows:

1. The initial cellular receptor to which the virus attaches is a moiety of heparin sulphate proteoglycan (Wudunn and Spear, 1989). The anti-receptors, i.e. the viral glycoproteins which mediate the initial attachments are glycoproteins B and C acting either in tandem or

independently (Wudunn and Spear, 1989; Campadelli-Fiume *et al.*, 1990). A subsequent step in the process of entry into cells involves glycoprotein D and H (Highlander *et al.*, 1987; 1988; Fuller *et al.*, 1989). These, with the involvement of other viral proteins probably mediate the fusion of the plasma membrane with the viral envelope. Viral glycoproteins B, D and H mediate essential functions as demonstrated by neutralization of HSV with specific antibody and other studies (Highlander *et al.*, 1987; 1988; Fuller *et al.*, 1989). The function of the other glycoproteins that are dispensable for growth in cell culture is not known; they may mediate attachment or penetration by interacting with different host surface proteins in highly differentiated cells.

2. Upon entry into cells, the capsid containing the DNA is transported to the nuclear pore (Batterson *et al.*, 1983). The DNA is released into the nucleus where it immediately circularizes in the absence of *de novo* protein synthesis (Poffenburger and Roizman, 1985). From here onwards, the expression of viral genes is coordinately regulated and sequentially ordered in a cascade fashion (Honess and Roizman, 1974; 1975).

3. Two tegument proteins brought into the infected cells play a significant role in the initial stages of infection. First, the *vhs* or virion host shut-off protein appears to mediate a RNAase activity which destabilizes all cytoplasmic mRNAs present in the infected cell during the early phases of infection (Read and Frenkel, 1983; Kwong and Frenkel, 1989). The specific function of *vhs* is not known, but the net effect of its activity is a general shortening of mRNA half life with the consequence that newly synthesized viral mRNA has access to the translational machinery of the cells.

   The second protein is the α gene trans-inducing factor (αTIF) (Campbell *et al.*, 1984; Pellet *et al.*, 1985). This protein is an abundant structural constituent of the virion; it is present in approximately 1000 molecules/virion. It also functions to induce the expression of the five α genes, the first set of genes expressed after infection (Post *et al.*, 1981). Studies at the non-permissive temperature with a mutant temperature sensitive in release of viral DNA from capsids at the nuclear pore, indicate that αTIF is released into the cytoplasm and makes its way into the nucleus independently of the viral DNA (Batterson and Roizman, 1983). Although it has been suggested that αTIF is not required for virus growth, the data do not exclude the possibility that the mutation is leaky (Steiner *et al.*, 1990).

4. All five α genes of HSV-1 and HSV-2 contain *cis*-acting sites required for their expression (Mackem and Roizman, 1982; Kristie and Roizman, 1984). The *cis*-acting site for induction of α genes by αTIF has the consensus sequence 5' GyATGnTAATGArATTCyTTGnGGGNC (Mackem and Roizman, 1982; Kristie and Roizman, 1984). At least

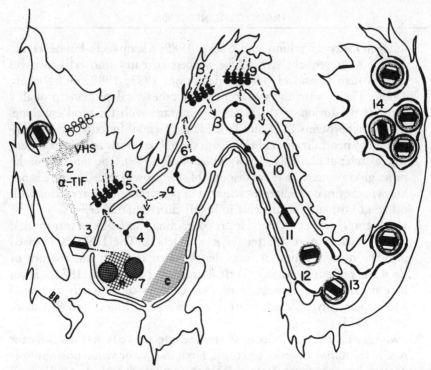

Figure 9.3 Schematic representation of the reproductive cycle in permissive cells. 1; Infectious enveloped virus attaches to cell receptors and viral membrane proteins fuse the envelope of the virus with the plasma membrane of the cell. 2; Viral tegument proteins are released into the cytoplasm; among these are the Virion Host Shut-off (*vhs*) factor which destabilize cytoplasmic mRNA (open polyribosomes on broken RNA) and the α-Trans Inducing Factor (αTIF) which is transported into the nucleus. 3; The capsid is transported to a nuclear pore, and viral DNA is released into the cell nucleus where it circularizes. 4; The promoter domains of the 5 α genes interact with cellular protein and αTIF and are transcribed. 5; α gene mRNAs are transported into the cytoplasm and are translated. Most α proteins are transported into the nucleus. 6: Functional α proteins, but especially ICP4 are essential for the expression of later (β and γ) genes. 7; Concurrent with expression of β genes, the cell undergoes major changes. These include margination and pulverization of the chromatin (*c*) and the nucleoli (*n*) disaggregate. 8; The β gene products include proteins responsible for maintenance of triphosphates pool, and repair and synthesis of viral DNA. The synthesis of viral DNA proceeds by a rolling circle model. 9; Concurrently with the onset of viral DNA replication (as early as 3–4 hours post infection), the last class of viral genes (γ) specifying largely the structural proteins of the virion are expressed. 10; Capsid proteins assemble into empty particles in the cell nucleus whereas membrane and tegument proteins aggregate in patches in all membranes of the cell. 11; Unit lengths of viral DNA are cleaved and packaged from head to tail concatamers. 12: Capsids containing viral DNA are modified. 13; The modified DNA containing capsids become attached to the tegument-envelope patches on the inner lamellae of the nuclear membrane and become enveloped by budding into the space between the inner and outer lamellae. 14; Enveloped virions probably exit the cell via the endoplasmic reticulum. See text for details and references.

some of the steps in the trans-activation of α genes are known. Thus a host protein designated by various names as αH1, OTF1, OCT1, binds to the sequence **ATGnTAAT** (Kristie and Roizman, 1987; 1988) αTIF and probably other cellular proteins then form higher order complexes which require (for their stability and efficiency) the sequence TAATGArAT and the downstream sequence of the consensus (McKnight *et al.*, 1987; Preston *et al.*, 1988; Kristie *et al.*, 1989).

5. The function of α genes and their products has been studied extensively. The product of αO, ICPO has been reported to be a promiscuous transactivator in transient expression systems (Everett, 1984; Gelman and Silverstein, 1988). Deletion mutants in αO replicate poorly, and no correlation can be made between *in vitro* and *in vivo* studies (Sacks and Schaffer, 1987; Stow and Stow, 1985). α27 is an essential gene required for both DNA synthesis and late gene expression (McCarthy *et al.*, 1989; Rice and Knipe, 1990). ICP22, the product of the α22 gene, appears to regulate late gene expression. In the case of deletion mutants, the function is complemented by host functions in some cells but not others (Sears *et al.*, 1985). The function of α47 is unknown and the gene is dispensable for growth in cells in culture (Longnecker and Roizmain, 1986). The last to be enumerated, α4, is the major regulatory protein of the virus. ICP4 autoregulates its own transcription and that of αO (DeLuca and Schaffer, 1988) and is required for the expression of virtually all of the genes tested to date.

ICP4 binds directly to DNA (Kristie and Roizman, 1986). Some of the binding sites are in the 5′ untranscribed domains whereas others are in 5′ transcribed non-coding domains (Michael *et al.*, 1988). Recent studies indicate that there is more than one DNA sequence to which the protein binds and the reported consensus sequence ATCGTCnnnnYCGRC is but one of the sites (Michael *et al.*, 1988).

6. The appearance of functional α proteins signals initiation of transcription of β genes. There are at least 15 β genes, and most are involved in nucleic acid metabolism. Among the β genes are seven whose function is essential for viral DNA synthesis and enzymes involved in maintenance of the deoxynucleoside triphosphate pools, and DNA repair (Challberg, 1986; Wu *et al.*, 1988; Crute *et al.*, 1989).

7. Viral DNA synthesis can be detected as early as 3–4 hours post infection. The viral genome has three origins of DNA synthesis, two in the inverted repeats flanking the S component and one in the middle of the L component (Mocarski and Roizman, 1982; Locker *et al.*, 1982). Precisely how they function is not clear since the earliest detectable viral DNA synthesis takes the form of a rolling circle (Jacob *et al.*, 1979). Deletions in individual origins are tolerated (Longnecker and Roizman, 1986) and an origin binding protein required for viral DNA synthesis has been identified and mapped (Elias and Lehman, 1988).

8. The transcription of late genes requires the synthesis of functional α and β proteins and, in many instances, concurrent viral DNA synthesis (Honess and Roizman, 1974; 1975; Wagner, 1984). Most late gene products appear to be virion structural proteins (Honess and Roizman, 1974). Late genes are heterogeneous. While some stringently require viral DNA synthesis for their expression, others are partially expressed in the presence of inhibitors which totally block DNA synthesis (e.g. Honess and Roizman, 1974; Mavromara–Nazos and Roizman, 1987).

9. Maturation of virions occurs in several steps. First, empty capsids assembled in the nucleus interact with specific sites in the *a* sequence of head-to-tail concatemers produced by the rolling circle viral DNA synthesis. The current model is that the packaging machinery of the virion cleaves the DNA at the *a* sequence, encapsidates the DNA from one end until a headful is packaged and an *a* sequence in the same orientation is encountered, and cleaves the concatemer again (Deiss *et al.*, 1986; Deiss and Frenkel, 1986). Second, the capsid containing the DNA undergoes modification in that a protein present in the empty capsids is cleaved rendering the particle competent for envelopment (Gibson and Roizman, 1974; Braun *et al.*, 1984; Preston *et al.*, 1983). Concurrent with capsid assembly, viral glycoproteins and tegument proteins assemble in patches in all membranes of the cells. These patches are electron opaque and prominent in the inner lamella of the nuclear membrane. Capsids containing DNA attach to the underside of these patches and bud into the space between the inner and outer lamellae. They are then transported by way of the endoplasmic reticulum to the extracellular fluid (Roizman and Furlong, 1974; Schwartz and Roizman, 1969).

10. The outcome of viral replication is cell death, a consequence of cumulative injuries that begin almost from the moment of entry of the virus in the susceptible cells. Among the events that contribute to cellular disfunction are disaggregation of the nucleolus concomitant with cessation in ribosomal RNA synthesis (Roizman and Furlong, 1974), margination and pulverization of the chromatin (Roizman and Furlong, 1974), modification of the structure and shape of the nucleus (Roizman and Furlong, 1974), and modification of the cellular membranes (Schwartz and Roizman, 1969). Some mutations in viral genes cause the infected cells to fuse with adjacent cells to form large polykaryocytes (Ejercito *et al.*, 1968). Modification of cellular membranes resulting from the incorporation of viral glycoproteins also cause the leakage of macromolecules from the infected cells (Wagner and Roizman, 1969).

9.6 LATENCY

A general property of all herpes viruses is their ability to maintain their genomes in a latent state for the life of their hosts. HSV, like other members of the *alphaherpesvirinae*, is maintained in the latent state in sensory neurons. In the natural course of infection, viral replication in susceptible cells at the portal of entry leads to infection of nerve endings, intra-axonal transport to sensory neurons, and release of viral DNA into the neuronal nucleus (Kristensson *et al.*, 1986; Lycke *et al.*, 1984). The events occurring in the neurons which ultimately maintain the virus in latent form are unknown. The available data were derived largely from studies done in experimental animal systems such as the mouse or the rabbit. The key observations are as follows: (1) The genome is maintained in an episomal form, i.e. circular or endless (Mellerick and Faser, 1987). (2) On the average, mouse ganglia contain 0.1 to 1 copies of the viral genome per ganglionic cell. As neurons constitute a small fraction of total cells and only a fraction of neurons become infected, there are approximately 10 to 100 viral genomes per neuron harbouring latent virus (Roizman and Sears, 1987). (3) Studies on deletion mutants strongly argue that replication at the portal of entry is not a prerequisite for the establishment of latency (Deluca *et al.*, 1985; Katz *et al.*, 1990). Furthermore, there is no evidence that expression of one or more viral genes is required to establish latency once the viral genome reaches the sensory neuron. (4) Sensory neurons harbouring latent virus accumulate large amounts of a nuclear transcript designated as latency associated transcript 1 (LAT1) derived by transcription of a small region of the viral genome located in the inverted repeats between the last ORF in the U_L (open reading frame 1 or 56) and a0 (Stevens *et al.*, 1987; Wagner *et al.*, 1988). LAT1 appears to be heterogeneous; the transcripts comprising this family differ in splice sites and a small fraction extend across the L- S component junction (Wagner *et al.*, 1988; Wechsler *et al.*, 1988). The function of LAT1 is not clear. Viral mutants from which both copies of the sequence had been deleted are capable of establishing latency and of reactivating spontaneously in the rabbit model (Javier *et al.*, 1988). Proponents of a function for LAT1 have reported that the induced reactivation rate for LAT1 mutants is reduced (Hill *et al.*, 1990). (5) Little is known of the mechanisms which induce the latent virus to initiate a productive infection. The studies done in animal models suggest that trauma to nerve endings as well as hormonal changes result in physiologic changes in neurons which trigger viral replication (Anderson *et al.*, 1961; Harbour *et al.*, 1983; Kwon *et al.*, 1982).

The mechanisms by which herpes simplex viruses establish, maintain and terminate the latent state are important issues that remain to be resolved.

9.7 VIRULENCE

Virulence in HSV is difficult to define and, until recently, attempts to define a molecular basis for the differences in lethality that have been observed between various isolates have been unrewarding. It has been noted for some time that strains sequentially passaged in culture tended to be less lethal when reintroduced to animal hosts, and that serial passage in a given animal host selected for more lethal strains upon rechallenge of that host. Recent work has demonstrated that mutants in glycoprotein D have altered ability to cause encephalitis (Izumi and Stevens, 1990). Work in this laboratory has found a gene termed ICP 34.5 whose deletion decreases the ability of HSV to cause encephalitis in an animal model by one million fold, while scarcely affecting the ability of the virus to grow in culture. In a mouse model this virus reactivates from latency with dramatically less efficiency. Whether ICP34.5 encodes a function required for viral growth or spread in the central nervous system has yet to be determined (Chou and Roizman, 1990; Chou et al., 1990). These discoveries provide logical targets for molecular genetic manipulation of the virus in attempts to cause attenuation.

9.8 ANTIVIRAL AGENTS

Most of the currently available antivirals active against herpes are nucleoside analogs, compounds which mimic the purine and pyrimidine bases of DNA. Acyclovir (Zovirax), DHPG (Ganciclovir) and BromoVinylDeoxyUridine (BVDU) are guanine nucleoside substitutes, while Vidarabine (Ara-A, Vira-A) is an adenine nucleoside cogener. Each of these drugs takes advantage of an enzyme encoded by Herpes simplex and VZV, called thymidine kinase (TK), which adds the initial phosphate group to free nucleosides. Cellular enzymes will readily add the subsequent two phosphate groups completely to 'charge' the nucleoside for DNA incorporation. It is presumed that the evolutionary advantage to HSV of incorporating a *tk* gene lies in the fact that HSV replicates a great deal of DNA in a very short period, enzymes like TK and ribonucleotide reductase (which converts RNA nucleosides to DNA ones) serve to expand the pool of available DNA building blocks and 'turbocharge' the host cell for its burst of herpes DNA replication. However, the herpes TK enzyme lacks the discrimination of its cellular counterpart and binds the false guanine nucleoside acyclovir 200 times more avidly than the cellular enzyme. Infected cell extracts are 30–120 times more efficient than uninfected cells in carrying out this conversion (Fyfe et al., 1978). Once phosphorylated, the nucleoside is now polar and unable to diffuse out of the cell: additional Acyclovir diffuses in, is trapped by the added PO_4 group and soon high levels of Acyclo-MP are reached. Cellular enzymes convert Acyclo-MP into the di and tri phosphate derivatives, Acyclo-TP being the active drug incorporated into the viral DNA. Cellular DNA polymerases are, like cellular TK, more stringent in their substrate requirements and reject Acyclo-TP as a substitute for guanidine, while the HSV DNA polymerase is inhibited by Acyclo-TP and incorporates it into the growing viral DNA.

This causes chain termination since Acyclovir lacks an appropriate hydroxyl group upon which to add additional bases. Acyclovir continues to bind the stalled viral DNA polymerase (St Clair *et al.*, 1980; Derse *et al.*, 1981; Furman *et al.*, 1984) by preventing the polymerase's 3'–5' exonuclease activity from functioning and locking the polymerase to the terminated strand.

It can be seen that the entire scheme described above depends on the function of the HSV TK gene product and mutants which inactivate the *tk* gene, and are therefore resistant to Acyclovir, were encountered almost as soon as the drug was introduced (Crumpacker *et al.*, 1982). Generally resistance to Acyclovir has taken the form of premature stop codons or missense mutations in the *tk* gene, although TK and polymerase mutants with altered substrate specificity have also been encountered (Crumpacker, 1989; Field *et al.*, 1981). Most of these strains are debilitated to some degree but can cause serious disease in immunocompromised hosts. HSV-2 strains are slightly less sensitive to Acyclovir than HSV-1 with VZV being less sensitive than HSV-2 (Crumpacker *et al.*, 1979).

Ganciclovir is related to acyclovir and can be phosphorylated similarly in HSV infected cells. It is active against CMV which lacks TK activity of its own but apparently induces sufficient host nucleoside kinase activity to activate Ganciclovir in infected cells. A recent report demonstrated that a number of Acyclovir resistant HSV strains were also Ganciclovir and Vidarabine resistant (Safrin *et al.*, 1990). Vidarabine has a lower therapeutic index than Acyclovir, being only 40 times more potent against viral than host DNA polymerases and high doses of Vidarabine are toxic to rapid dividing cells (Crumpacker, 1989). Although Vidarabine was the first drug shown to be useful in HSV encephalitis, subsequent trials demonstrated that Acyclovir was superior in treatment of this condition (Whitley *et al.*, 1980; 1986). BVDU is experimental in the US and appears to have potential utility against HSV-1 and VZV but is less active *in vitro* against HSV-2 than Acyclovir. Both TK- and polymerase-based Acyclovir resistance confers resistance also to BVDU.

Phosphonoformate (PFA) is a promising agent which was derived from PhosphonaceticAcid (PAA). PFA and PAA are analogs of pyrophosphate which is cleaved from a nucleoside triphosphate during nucleic acid synthesis. It is believed that PFA binds to the active site of DNA polymerase and inhibits binding of the true nucleoside substrate. PFA is not phosphorylated and TK-resistant isolates are susceptible to PFA, while the much more rarely encountered polymerase-based resistant strains are cross-resistant to PFA (Kern *et al.*, 1978; Safrin *et al.*, 1990).

9.9 IMMUNIZATION AGAINST INFECTION

As with all viral infections, the key to control is prevention rather than treatment: this is especially so with HSV as treatment does not affect the latent reservoir of virus, nor forestall future recurrences. Attempts at formulating a purified protein vaccine have met with mixed results, decreasing the severity of infection but not preventing it (Stanberry et al., 1987; Meignier et al., 1987b; Mertz et al., 1984). Despite the fact that a number of neutralizing antibodies and epitopes have been defined, the ability of HSV to grow despite effective humoral immunity leads to the conclusion that, without sufficient cell mediated reactivity, protection is likely to be incomplete (Mertz et al., 1984; Ashley et al., 1985; Zarling et al., 1988). A tissue culture high-passage vaccine strain of Varicella-Zoster Virus (Oka) appears to be effective in preventing acquisition of chicken-pox (Gershon et al., 1989). A live, genetically attenuated vaccine strain of Herpes simplex has been constructed and extensively tested in animals (Meignier et al., 1988; 1990). It is hoped that this strain, which expresses several HSV-2 glycoprotein genes in addition to HSV-1 genes, will stimulate the immune system against both HSV-1 and 2. Human trials of this prototype have begun recently. It has, however, been demonstrated that even wild-type infections sometimes fail to produce sufficient immune response to prevent exogenous reinfection (Buchman et al., 1979)

9.10 DETECTION OF HSV

Two crucial issues arise here; (1) being able to differentiate HSV 2 infected individuals from those who have HSV-1 infections, since women who have HSV-2 are more likely to have genital infections and therefore to transmit the virus to newborns who suffer from peripartum HSV infection (Johnson et al., 1989). Although monoclonal antibody kits are commercially available which can rapidly differentiate HSV isolates (Goldstein et al., 1983), only difficult research methods exist for serological differentiation of persons who are not actively shedding virus (Bernstein et al., 1984). HSV group-common seropositivity or seronegativity can be demonstrated easily. Methods to detect HSV-2 specific antibodies rapidly and efficiently as a surrogate marker for genital infection are being pursued. (2) For patients with suspected HSV encephalitis, a noninvasive, sensitive method of detecting HSV or reliably excluding its presence is needed to replace brain biopsy (Whitley et al., 1982). The in vitro amplification of DNA polymerase chain reaction (PCR) has shown promise as a method to detect HSV encephalitis using samples of cerebrospinal fluid (Rowley et al., 1990). PCR may also make it possible to screen large numbers of pregnant women near term with a highly sensitive inexpensive test, even if a positive test only mandates greater supervision of the at-risk infant by the pediatrician.

Discerning the serotype involved in Herpes simplex infection remains a necessity.

REFERENCES

Anderson, W.A., Magruder, B. and Kilbourne, E.D. (1961) Induced reactivation of herpes simplex virus in healed rabbit corneal lesions. *Proc. Soc. Exp. Biol. Med.*, 107, 628–32.

Ashley, R., Mertz, G., Clark, H. *et al.* (1985) Humoral immune response to herpes simplex virus Type 2 glycoproteins in patients receiving a glycoprotein subunit vaccine. *J. Virol.*, 56, 475–81.

Astruc, J. (1736) *De Morbis Veneris Libri Sex*, Paris.

Baer, R., *et al.* (1984) DNA sequence and expression of the B95-8 Epstein-Barr Virus genome. *Nature*, 310, 207.

Batterson, W., Furlong, D. and Roizman, B. (1983) Molecular genetics of herpes simplex virus. VII. Further characterization of a ts mutant defective in release of viral DNA and in other stages of viral reproductive cycle. *J. Virol.*, 45, 397–407.

Batterson, W. and Roizman, B. (1983) Characterization of the Herpes Simplex virion associated factor responsible for the induction of α genes. *J. Virol.*, 46, 371–7.

Bernstein, D.I., Lovett, M.A. and Bryson, Y.J. (1984) Serologic analysis of first episode nonprimary genital herpes simplex virus infection. *Am. J. Med.* 77, 1055–60.

Bernstein, D.I., Bryson, Y.J. and Lovett, M.A. (1985) Antibody response to type common and type unique epitopes of herpes simplex virus polypeptides. *J. Med. Virol.*, 15, 251–63.

Braun, D.K., Roizman, B. and Pereira, L. (1984) Characterization of post-translational products of herpes simplex virus gene 35 proteins binding to the surface of full but not empty capsids. *J. Virol.*, 49, 142–53.

Buchman, T.G., Roizman, B., Adams, G. and Stover, H. (1978) Restriction endonuclease footprinting of herpes simplex virus DNA: a novel epidemiologic tool applied to a nosocomial outbreak. *J. Inf. Dis.*, 138, 488–98.

Buchman, T.G., Roizman, B. and Nahmias, A.J. (1979) Demonstration of exogenous reinfection with herpes simplex virus type 2 by restriction endonuclease fingerprinting of viral DNA. *J. Inf. Dis.*, 140, 295–304.

Campbell, M.E.M., Palfreyman, J.W. and Preston, C.M. (1984) Identification of herpes simplex virus DNA sequences which encode a trans acting polypeptide responsible for stimulation of immediate early transcription. *J. Mol. Biol.*, 180, 1–19.

Campadelli-Fiume, G., Stirpe, D., Boscaro, A. *et al.* (1990) Glycoprotein C dependent attachment of attachment of herpes simplex virus to susceptible cells leading to productive infection. *J. Virol.*, 178, 213–22.

Challberg, M.D. (1986) A method for identifying the viral genes required for DNA replication. *Proc. Natl Acad. Sci. (USA)*, 83, 9094–8.

Chou, J., Kern, E.R., Whitley, R.J. and Roizman, B. (1990) Mapping of herpes simplex virus 1 eurovirulence to γ'34.5, a gene nonessential for growth in cell culture. *Science*, 250, 1202–6.

Chou, J. and Roizman, B. (1985) The isomerization of the herpes simplex virus 1 genome: Identification of the cis-acting and recombination sites within the domain of the a sequence. Cell, 41, 803–11.

Chou, J. and Roizman, B. (1990) The herpes simplex virus gene for ICP 34.5, which maps in inverted repeats, is conserved in several limited passage isolates but not in strain 17syn+. J. Virol., 64, 1014–20.

Cohen, G.S., Ponce De Leon, M., Deggelmann, H. et al. (1980) Structural analysis of the capsid proteins of herpes simplex virus types 1 and 2. J. Virol., 34, 521–31.

Crumpacker, C.S., Schnipper, L.E., Zaia, J.A. and Levine, M. (1979) Growth inhibition by acycloguanosine of herpesviruses isolated from human infections. Antimicrob. Agents Chemother., 15, 642–5.

Crumpacker, C.S. Schnipper, L.L., Marlowe, S.I. et al. (1982) Resistance to antiviral drugs of herpes simplex from a patient treated with Acyclovir. New. Eng. J. Med., 306, 343–6.

Crumpacker, C.S. (1989) Molecular targets of antiviral therapy. New. Eng. J. Med., 321, 163–72.

Crute, J.J., Tsurumi, T., Zhu, L. et al. (1989) Herpes simplex virus 1 helicase-primase: A complex of herpes-encoded gene products. Proc. Natl Acad. Sci. USA, 86, 2186–9.

Davison A.J. and Wilkie, N.M. (1981) Nucleotide sequences of the joint between the L and S segments of herpes simplex virus types 1 and 2. J. Gen. Virol., 55, 315–31.

Davison, A.J. and Scott, J.E. (1986) The complete DNA sequence of Varicella Zoster Virus genome. J. Gen. Virol., 67, 1759–816.

Deiss, L.P., Chou, J. and Frenkel, N. (1986) Functional domains within the a sequence involved in the cleavage-packaging of herpes simplex virus DNA. J. Virol., 59, 605–18.

Deiss, L.P. and Frenkel, N. (1986) Herpes simplex virus amplicon: cleavage of concatameric DNA is linked to packaging and involves amplification of the terminally reiterated a sequence. J. Virol., 57, 933–41.

DeLuca, N.A. and Schaffer, P.A. (1988) Physical and functional domains of the herpes simplex virus transcriptional regulatory protein ICP4. J. Virol., 62, 732–43.

DeLuca, N.A., McCathy, A.M. and Schaffer, P.A. (1985) Isolation and characterization of deletion mutants of herpes simplex type 1 in the gene encoding immediate-early regulatory protein ICP4. J. Virol., 56, 558–70.

Derse, D., Cheng, Y.-C., Furman, P.A. et al (1981) Inhibition of purified human and herpes simplex virus-induced DNA polymerases by 9–(2-hydroxymethyl) guanine triphosphate. J. Biol. Chem., 256, 11447–51.

Diday, P. and Doyon, A. (1886) Les Herpes Genitaux, Masson et Cie, Paris.

Ejercito, P.M., Kieff, E.D. and Roizman, B. (1968) Characterization of herpes simplex strains differing in their effects on social behavior of infected cells. J. Gen. Virol., 2, 357–64.

Elias, P. and Lehman, I.R. (1988) Interaction of origin binding protein with an origin of replication of herpes simplex virus 1. Proc. Natl Acad. Sci. USA, 85, 2959–63.

Everett, R.D. (1984) Trans-activation of transcription by herpes virus products:

requirements for two HSV-1 immediate early polypeptides for maximum activity. *EMBO J.*, 3, 3135–41.

Field, H., McMillan, A. and Darby, G. (1981) The sensitivity of acyclovir-resistant mutants of herpes simplex virus to other antiviral drugs. *J. Inf. Dis.*, 143, 281–5.

Frenkel, N., Schirmer, E.C., Wyatt, L.S. *et al.* (1990) Isolation of a new herpesvirus from human CD4 T cells. *Proc. Natl. Acad. Sci. USA*, 87, 748–52.

Fuller, A.O., Santos, R.E. and Spear, P.G. (1989) Neutralizing antibodies specific for glycoprotein H of herpes simplex virus permit viral attachment but prevent virion-cell fusion required for penetration. *J. Virol.*, 63, 3435–43.

Furlong, D., Swit, H. and Roizman, B. (1972) Arrangements of herpesvirus deoxyribonucleic acid in the core. *J. Virol.*, 10, 1071–4.

Furman, P.A., St Clair, M.H. and Spector, T. (1984) Acyclovir triphosphate is a suicide inhibitor of the herpes simplex virus DNA polymerase. *J. Biol. Chem.*, 259, 9575–9.

Fyfe, J.A., Keller, P.M., Furman, P.A. *et al.* (1978) Thymidine kinase from herpes simplex virus phosphorylates the new antiviral compound, 9–(2-hydroxymethyl)guanine. *J. Biol. Chem.*, 253, 8721–87.

Gelman, I.H. and Silverstein, S. (1988) Herpes simplex virus immediate early promoters are responsive to virus and cell trans-acting factors. *J. Virol.*, 61, 2286–96.

Gershon, A.A., Steinberg S.P. *et al.* (1989) Persistence of immunity to varicella in children with leukemia immunized with live attenuated varicella vaccine. *New Engl. J.Med.*, 320, 892–6.

Gibson, W. and Roizman, B. (1972) Proteins specified by herpes simplex virus VIII; Characterization and composition of multiple capsid forms of subtypes 1 and 2. *J. Virol.*, 10, 1044–52.

Gibson, W. and Roizman, B. (1974) Proteins specified by herpes simplex virus X; Staining and radiolabeling properties of β-capsid and virion proteins in polyacrylamide gels. *J. Virol.*, 13, 155–65.

Goldstein. L.C., Corey, L., McDougall, J.K. *et al.* (1983) Monoclonal antibodies to herpes simplex viruses: use in antigenic typing and rapid diagnosis. *J. Inf. Dis.*, 147, 829–37.

Harbour, D.A., Hill, T.J. and Blyth, W.A. (1983) Recurrent herpes simplex in the mouse: inflammation of the skin and reactivation of virus in the ganglia following peripheral stimuli. *J. Gen. Virol.*, 64, 1491–8.

Hayward. G.S. Jacob, R.J., Wadsworth, S.C. and Roizman, B. (1975) Anatomy of herpes simplex virus DNA: Evidence for populations of molecules that differ in the relative orientations of their long and short segments. *Proc. Natl Acad. Sci.*, 72, 4243–7.

Heine, J.W., Honess, R.W., Cassim, E. and Roizman B. (1974) Proteins specified by Herpes Simplex virus; XII the virion polypeptides of type 1 strains. *J. Virol.*, 14, 640–51.

Highlander, S.H., Sutherland, S.L., Gage, P.J., *et al.* (1987) Neutralizing monoclonal antibodies specific for herpes simplex virus glycoprotein D inhibit virus penetration. *J. Virol.*, 61, 3356.

Highlander, S.H. Cai, W., Person, S. *et al.* (1988) Monoclonal antibodies define a domain on herpes simplex virus glycoprotein B involved in virus penetration. *J. Virol.*, 62, 1881.

Hill, J.M., Sedarati, F., Javier, R.T. *et al* (1990) Herpes simplex virus latent phase transcription facilitates in-vivo reactivation. *Virology*, **174**, 117–25.

Honess, R.W. and Roizman, B. (1974) Regulation of herpesvirus macromolecular synthesis. I. Cascade regulation of the synthesis of the groups of viral proteins. *J. Virol.*, **14**, 8–19.

Honess, R.W. and Roizman, B. (1975) Regulation of herpesvirus macromolecular synthesis: sequential transition of polypeptide synthesis requires functional viral polypeptides. *Proc. Natl Acad. Sci. USA*, **72**, 1276–80.

Izumi, K.M. and Stevens, J.G. (1990) Molecular and biological characterization of a herpes simplex virus type 1 (HSV-1) neuroinvasiveness gene. *J. Exp. Med.*, **172**, 487–96.

Jacob, R.J., Morse, L.S. and Roizman, B. (1979) Anatomy of herpes simplex DNA XIII. Accumulation of head to tail concatamers in nuclei of infected cells and their role in the generation of the four isomeric arrangements of viral DNA. *J. Virol.*, **29**, 448–57.

Javier, R.T. and Stevens, J.G. (1988) A herpes simplex virus transcript abundant in latently infected neurons is dispensable for establishment of the infected state. *Virology*, **166**, 254.

Johnson, R.E., Nahmias, A.J., Magder, L.S. *et al.* (1989) A seroepidemiologic survey of the prevalence of herpes simplex virus Type 2 infection in the United States. *New Engl. J. Med.*, **321**, 7–12.

Katz, J.P., Bodin, E.T. and Coen, D.M. (1990) Quantitative polymerase chain reaction analysis of herpes simplex virus DNA in ganglia of mice infected with replication-incompetent mutants. *J. Virol.*, **64**, 4288–95.

Kern, E.R., Glasgow, L.A., Reno, J. and Balzi, A. (1978) Treatment of experimental herpesvirus infections with phosphonoformate and some comparisons with phosphonoacetate. *Antimicrob. Agents Chemother.*, **14**, 817–23.

Kieff, E.D., Bachenheimer, S.L. and Roizman, B. (1971) Size, composition, and structure of the DNA of subtypes 1 and 2 of herpes simplex virus. *J. Virol.*, **8**, 125–9.

Kieff, E.D., Hoyer, B., Bachenheimer, S. and Roizman, B. (1972) Relatedness of type 1 and type 2 herpes simplex viruses. *J. Virol.*, **9**, 738–45.

Kristie, T.M. and Roizman, B. (1984) Separation of sequences defining basal expression from those conferring α gene recognition within regulatory domains of herpes simplex virus 1 α genes. *Proc. Natl Acad. Sci. USA*, **81**, 4065–9.

Kristie, T.M. and Roizman, B. (1986) α4 the major regulatory protein of herpes simplex virus type 1 is stably and specifically associated with promoter regulatory domains of α genes and of selected other viral genes. *Proc. Natl Acad. Sci. USA*, **83**, 3218–22.

Kristie, T.M. and Roizman, B. (1987) Host cell proteins bind to the cis acting site required for virion mediated induction of herpes simplex virus 1 α genes. *Proc. Natl Acad. Sci. USA*, **84**, 71–5.

Kristie, T.M. and Roizman, B. (1988) Differentiation of DNA contact points of the host proteins binding at the *cis*-site for virion mediated induction of α genes of herpes simplex virus 1. *J. Virol.*, **62**, 1145–57.

Kristie, T.M., LeBowitz, J.H. and Sharp, P.A. (1989) The Octamer binding proteins form multi-protein-DNA complexes with the HSV αTIF regulatory protein. *EMBO J.*, **8**, 4229–38.

Kristensson, K., Lycke, E., Raytta, M. *et al.* (1986) Neuritic transport of herpes simplex virus in rat sensory neurons in vitro. Effects of substances interacting with microtubular function and axonal flow [Nocodazone, Taxol, and erythro-9–3–(2-hydoxynonyl)adenine]. *J. Gen. Virol.*, 67, 2023–8.

Kwon, B.S. Gangorosae, L.P., Green, K. and Hill, J. A. (1982) Kinetics of ocular herpes simplex virus shedding induced by epinephrine iontophoresis. *Invest. Opthalmol. Vis. Sci.*, 22, 818–21.

Kwong, A. and Frenkel, N. (1989) The herpes simplex virus virion host shutoff function. *J. Virol.*, 63, 4834–9.

Lawrence, G.L., Chee, M., Craxton, M.A. *et al.* (1990) Human Herpesvirus 6 is closely related to human cytomegalovirus. *J. Virol.*, 64, 287–99.

Lee, F.K., Coleman, M., Pereira, L. *et al.* (1985) Detection of herpes simplex virus type specific antibody with glycoprotein G. *J. Clin. Microbiol.*, 22, 641–4.

Linnemann, C.C., Jr, Buchman, T.G., Light, I.J., Ballard, J.L. and Roizman, B. (1978) Transmission of herpes-simplex virus type 1 in a nursery for the newborn: Identification of viral isolates by DNA fingerprinting. *Lancet* i, 964–6.

Locker, H. and Frenkel. F. (1979) Bam I, Kpn I and Sal I restriction enzyme maps of the DNAs of herpes simplex virus strains Justin and F: Occurrence of heterogeneities in defined regions of the viral DNA. *J. Virol.*, 32, 424–41.

Locker, H., Frenkel, H. and Halliburton, I. (1982) Structure and expression of class II defective herpes simplex virus genomes encoding infected cell polypeptide number 8. *J. Virol.*, 43 574–93.

Longnecker, R. and Roizman, B. (1986) Gerneration of an inverting herpes simplex virus mutant lacking the L-S junction a sequences, an origin of DNA synthesis incuding those specifying glycoprotein E, and α47. *J. Virol.*, 58, 583–91.

Lycke, E., Kristensson, K., Svennerholm, B. *et al.* (1984) Uptake and transport of herpes simplex virus in neurites of rat dorsal root ganglia cells in culture. *J. Gen. Virol.*, 65, 55–64.

Mackem, S. and Roizman, B. (1982) Structural features of the α genes 4, 0, and 27 promoter-regulatory sequences which confer α regulation on chimeric thymidine kinase genes. *J. Virol.*, 44, 939–49.

Mavromara-Nazos, P. and Roizman, B. (1987) Activation of Herpes simplex virus 2 genes by viral DNA replication. *Virology*, 161, 593–8.

McCarthy, A.M., McMahan, L. and Schaffer, P.A. (1989) Herpes simplex virus type 1 ICP27 deletion mutants exhibit altered patterns of transcription and are DNA deficient. *J. Virol.*, 63, 18–27.

McGeoch, D.J., Moss, H.W.M., McNab, D. and Frame, M.C. (1987) DNA sequence and genetic content of HindIII L region of the short unique component of the Herpes simplex virus type 2 genome: Identification of the gene encoding glycoprotein G and evolutionary comparisons. *J. Gen. Virol.*, 68, 19–38.

McGeoch, D., Dalrymple, M.A., Davison, A.J. *et al.* (1988) The complete sequence of the long unique region in the genome of herpes simplex virus type 1. *J. Gen. Virol.*, 69, 1531–74.

McKnight, J.L.C., Kristie, T.M. and Roizman, B. (1987) Binding of the virion protein mediating α gene induction in herpes simplex virus 1 infected cells to its cis site requires cellular proteins. *Proc. Natl Acad. Sci. USA*, 84, 7061.

Marsden, H.S., Stow, N.D., Preston, V.G. *et al.* (1978) Physical mapping of herpes simplex virus induced polypeptides. *J. Virol.*, 28, 624–42.

Meignier, B., Longnecker, R., Mavromara-Nazos, P. *et al.* (1987a) Virulence and establishment of latency by genetically engineered mutants of herpes simplex virus 1. *Virology*, 162, 251–4.

Meignier, B., Jourdier, T.M., Norrild, B. *et al.* (1987b) Immunization of experimental animals with reconstituted glycoprotein mixtures of herpes simplex virus 1 and 2: protection against challenge with virulent virus. *J. Inf. Dis.*, 155, 921–30.

Meignier, B., Longnecker, R. and Roizman, B. (1988) In vivo behavior of genetically engineered herpes simplex viruses R7017 and R7020: construction and evaluation in rodents. *J. Inf. Dis.*, 158, 602–14.

Meignier, B., Martin, B., Whitley, R. and Roizman, B. (1990) In vivo behavior of genetically engineered herpes simplex viruses R7017 and R7020. II. Studies in immunocompetent and immunosuppressed owl monkeys (*Aotus Trivirgatus*). *J. Inf. Dis.*, 162, 313–21.

Mellerick, D.M. and Faser, N.W. (1987) Physical state of the latent herpes simplex genome in a mouse model system: evidence suggesting an episomal state. *Virology*, 158, 265.

Mertz, G.J., Peterman, G., Ashley, R. *et al.* (1984) Herpes simplex virus type-2 glycoprotein-subunit vaccine: tolerance and humoral and cellular responses in humans. *J. Inf. Dis.*, 150, 242–9.

Micheal, N., Spector, D., Mavromana-Nazos, P. *et al.* (1988) The DNA binding properties of the major regulatory protein $\alpha 4$ of herpes simplex virus. *Science*, 239, 1531–4.

Mocarski, E., S. and Roizman, B. (1981) Site specific inversion sequence of herpes simplex virus genome: domain and structural features. *Proc. Natl Acad. Sci. USA*, 78, 7047–51.

Mocarski, E.S. and Roizman, B. (1982a) The structure and role of the herpes simplex virus DNA termini in inversion circularization and generation of virion DNA. *Cell*, 31, 89–97.

Mocarski, E.S. and Roizman, B. (1982b) Herpesvirus dependent amplification and inversion of a cell associated viral thymidine kinase gene flanked by viral *a* sequences and linked to an origin of viral DNA replication. *Proc. Natl Acad. Sci. USA*, 79, 5626–30.

Morse, L.S., Buchman, T.G., Roizman, B. and Schaffer, P.A. (1977) Anatomy of herpes simplex virus DNA IX; Apparent exclusion of some parental DNA arrangements in the generation of intertypic (HSV 1 × HSV 2) recombinants. *J. Virol.*, 24, 231–48.

Morse, L.S., Pereira, L., Roizman, B. and Schaffer, P.A. (1978) Anatomy of HSV DNA; XI Mapping of viral genes by analysis of polypeptides and functions specified by HSV 1 × HSV 2 recombinants. *J. Virol.*, 26, 389.

Nahmais, A.J. and Dowdle, W.R. (1968) Antigenic and biologic differences in herpesvirus himinis. *Prog. Med. Virol.*, 10, 110–59.

Nahmais, A.J., Keyserling, H.L. and Kerrick, C.M. (1983) Herpes simplex, in *Infectious Diseases of the Fetus and the Newborn Infant* (eds J.S. Remington and J.O. Klein), Saunders, Phila. PA., p.638.

Pellet, P.E., McKnight, J.L.C., Jenkins, F. and Roizman, B. (1985) Nucleotide sequence and predicted amino acid sequence of a protein encoded in a small herpes simplex virus DNA fragment capable of trans inducing α genes. *Proc. Natl Acad. Sci. USA*, 82, 5870–4.

Poffenberger, K. and Roizman, B. (1985) Studies on a non-inverting genome of a viable herpes simplex virus 1. Presence of head to tail linkages in packaged genomes and requirement for circularization after infection. *J. Virol.*, **53**, 589–95.

Post, L.E., Mackem, S. and Roizman, B. (1981) The regulation of genes of herpes simplex virus: expression of chimeric genes produced by fusion of thymidine kinase with α gene promoters. *Cell*, **24**, 555–65.

Post, L.E. and Roizman, B. (1981) A generalized technique for the deletion of specific genes in large genomes: α gene 22 of herpes simplex virus 1 is not essential for growth. *Cell*, **25**, 227–32.

Preston, C.M., Frame, M.C. and Campbell, M.E.M. (1988) A complex formed between cell components and an HSV structural polypeptide binds to a viral immediate early gene regulatory DNA sequence. *Cell*, **52**, 425.

Preston, V.G., Coates, A.M., and Rixon, F.J. (1983) Identification and characterization of a herpes simplex virus gene product required for encapsidation of viral DNA. *J. Virol.*, **45**, 1056–64.

Rawls, W.E., Gardner, H.L., Flanders, R.W. *et al.* (1971) Genital Herpes in 2 social groups. *Am. J. Obstet. Gynecol.*, **110**, 682.

Read, G.S. and Frenkel, N. (1983) Herpes simplex virus mutants defective in virion associated shut-off of host polypeptide synthesis and exhibiting abnormal synthesis of *a* (immediate early) viral polypeptides. *J. Virol.*, **46**, 498–512.

Rice, S.A. and Knipe, D.M. (1990) Genetic evidence for two distinct functions of the herpes simplex virus α protein ICP27. *J. Virol.*, **64**, 1704–15.

Roizman, B. and Furlong, D. (1974) The replication of herpesviruses, in *Comprehensive Virology vol 3* (eds H. Frenkel-Conrat and R.R. Wagner), Plenum Press, NY, pp. 229–403.

Roizman, B., Carmichel, L.E., Deinhardt, F. *et al.* (1981) Herpesviridae. Definition, provisional nomenclature and taxonomy. *Intervirol.*, **16**, 201–17.

Roizman, B. and Tognon, M. (1983) Restriction endonuclease patterns of herpes simplex virus DNA: Application to diagnosis and molecular epidemiology. Proc. Symp. on New Horizons in Diagnostic Virology. *Curr. Topics Microbiol. and Immunol.*, **104**, 275–86.

Roizman, B. and Sears, A.E. (1987) An inquiry into the mechanism of herpes simplex virus latency. *Annu. Rev. Microbiol.*, **41**, 543–71.

Roizman, B. (1990) Herpesviridae: A Brief Introduction, in *Virology* (eds B.N. Fields, D.M. Knipe, R.M. Channock *et al.*), Raven Press, New York, pp. 1787–93.

Rowley, A.H., Whitley, R.J., Lakeman, F.D. and Wolinsky, S. (1990) Rapid detection of herpes simplex virus DNA in cerebrospinal fluid of patients with herpes simplex encephalitis. *Lancet*, 440–1.

Sacks, W.R. and Schaffer, P.A. (1987) Deletion mutants in the gene encoding the herpes simplex virus type 1 immediate early protein ICP0 exhibit impaired growth in cell culture. *J. Virol.*, **61**, 829–39.

Safrin, S., Assaykeen, T., Follansbee, S. and Mills, J. (1990) Foscarnet therapy for acyclovir-resistant mucocutaneous herpes simplex virus infection in 26 AIDS patients: preliminary data. *J. Inf. Dis.*, **161**, 1078–84.

Salhuddin, S.Z., Ablashi, D.V., Markham, P.D. *et al.* (1986) Isolation of a new virus HBLV, in patients with lymphoproliferative disorders. *Science*, **234**, 596.

Scrag, J.D., Prasad, B.V.V., Rixon, F.J. and Chiu, W. (1989) Three dimensional structure of the HSV-1 nucleocapsid. *Cell*, 56, 651–60.

Schnewiess, K.E. (1962) Serologische Untersuchungen zur Typendifferenzierung des Herpesvirus Hominis. *Z. Immununitaesforsch Exp. Ther.*, 124, 24–8.

Schwartz, J. and Roizman, B. (1969) Concerning the egress of herpes simplex virus from infected cells: electron and light microscope observations. *Virology*, 38, 42–9.

Sears, A.E. Halliburton, I.W., Meignier, B. *et al.* (1985) Herpes simplex virus mutant deleted in the α 22 gene: growth and gene expression in permissive and restrictive cells, and establishment of latency in mice. *J. Virol.*, 55, 338–46.

Sheldrick, P. and Berthelot, N. (1975) Inverted repetitions in the chromosome of herpes simplex virus. *Cold Spring Harbor Symp. Quant. Biol.*, 39, 667–8.

Spear, P.G. and Roizman, B. (1972) Proteins specified by Herpes Simplex Virus; V Purification and structural proteins of the herpesvirion. *J. Virol.*, 9, 431–9.

Stanberry, L.R., Bernstein, D.I., Burke, R.L. *et al.* (1987) Vaccination with recombinant herpes simplex virus glycoproteins: Protection against initial and recurrent genital herpes. *J. Inf. Dis.*, 155, 914–20.

Steiner, I., Spivack, J.G., Deshmane, S.L. *et al.* (1990) A herpes simplex virus type 1 mutant containing a nontransinducing Vmw65 protein establishes latent infection in vivo in the absence of viral replication and reactivates efficiently from explanted trigeminal ganglia. *J. Virol.*, 64, 1630–8.

Stevens, J.G. Wagner, E., Dev. Rac. O. *et al.* (1987) RNA complimentary to a herpesvirus alpha gene mRNA is prominent in latently infected neurons. *Science*, 235, 1056.

St. Clair, M.H., Furman, P.A., Lubbers, C.A. and Elion, G.B. (1980) Inhibition of cellular and virally induced deoxyribonucleic acid polymerases by the triphosphate of acyclovir. *Antimicrob. Agents Chemother.*, 18, 741–5.

Stow, N.D. and Stow, E.C. (1985) Isolation and characterization of a herpes simplex type 1 mutant containing a deletion in the gene encoding the immediate early polypeptide Vmw110. *J. Gen. Virol.*, 67, 2571–85.

Sullivan-Bolyai, J., Hull, H.F., Wilson, C. and Corey, L. (1983) Neonatal Herpes simplex infections in King County Washington: Increasing incidence and epidemiological correlates. *J. Am. Med. Assoc.*, 250, 3059.

Vlazny, D.A. and Frenkel, N. (1981) Replication of herpes simplex virus DNA: location of replication recognition signals within defective virus genomes. *Proc. Natl Acad. Sci. USA*, 78, 742–6.

Vlazny, D.A., Kwong, A. and Frenkel, N. (1982) Site specific cleavage packaging of herpes simplex virus DNA and the selective maturation of nucleocapsids containing full length viral DNA. *Proc. Natl Acad. Sci. USA*, 79, 1423–7.

Wadsworth, S., Jacob, R.J. and Roizman, B. (1975) Anatomy of herpes simplex virus DNA II; Size composition and arrangement of the inverted terminal repetitions. *J. Virol.*, 15, 1487–97.

Wagner, E.K. and Roizman, B. (1969) RNA synthesis in cells infected with herpes simplex virus. I. The patterns of RNA synthesis in productively infected cells. *J. Virol.*, 4, 36–46.

Wagner, M.M. and Summers, W.C. (1978) Structure of the joint region and the termini of the DNA of herpes simplex virus type 1 *J. Virol.*, 27, 374–87.

Wagner, E.K. (1984) Individual HSV transcripts: characterization of specific genes, in *The Herpesviruses vol. 3*, (Ed. B. Roizman), New York, Plenum Press, pp. 45–104.

Wagner, E.K., Devi-Rao, G., Feldman, L.T. *et al.* (1988) Physical characterization of the herpes simplex virus latency-associated transcript in neurons. *J. Virol.*, 62, 1194.

Wechsler, S.L., Nesburn, A.B., Watson, R., *et al.* (1988) Fine mapping of the latency related gene of herpes simplex virus type 1: alternative splicing produces distinct latency related RNAs containing open reading frames. *J. Virol.*, 62, 4051–8.

Whitley, R.J., Nahmias, A.J., Soorng, S.J. *et al.* (1980) Vidarabine therapy of neonatal herpes simplex infection. *Pediatrics*, 66, 495–501.

Whitley, R.J., Alford, C.A., Hirsch, M.S. *et al.* (1986) Vidarabine versus Acyclovir in Herpes simplex encephalitis. *New Engl. J.Med.*, 314, 144–9.

Whitley, R.J. Soong S.-J., Linnemann, C. Jr., *et al.* (1982) Herpes simplex encephalitis: clinical assessment. *J. Am. Med. Assoc.*, 247, 317–20.

Wildy, P. (1973) Herpes history and classification, in *The Herpes Viruses* (Ed. A.S. Kaplan) Academic Press, NY, pp. 1–25.

Wudunn, D. and Spear. P.G. (1989) Initial interaction of herpes simplex virus with cells is binding to heparin sulfate. *J. Virol.*, 63, 52–8.

Wu, C.A., Nelson, N.I., McGeoch, D.J. and Challberg, M.D. (1988) Identification of the herpes simplex virus type 1 genes required for origin dependent DNA synthesis. *J. Virol.*, 62, 435–43.

Zarling, J.M., Moran, P.A., Brewer, L. *et al.* (1988) Herpes simplex virus (HSV) specific proliferative and cytotoxic T-cell responses in humans immunized with an HSV Type 2 glycoprotein subunit vaccine. *J. Virol.*, 62, 4481–5.

10

Anti-idiotypic therapeutic strategies in HIV infection

DAVID WILKS AND ANGUS G. DALGLEISH

10.1 INTRODUCTION – THE CD4/GP120 INTERACTION

The CD4 molecule has been shown to act as the major cellular receptor for the human immunodeficiency virus (HIV) (Dalgleish *et al.*, 1984; Klatzmann *et al.*, 1984; Maddon *et al.*, 1986), and viral tropism for CD4+ cells is mediated by the interaction of the HIV major envelope glycoprotein gp120 with CD4; this has been shown to occur with high affinity *in vitro* (Smith *et al.*, 1987). Monoclonal antibodies (MAbs) to CD4 can prevent the induction of syncytia by HIV in lymphocyte cultures *in vitro* (Dalgleish *et al.*, 1984; Sattentau *et al.*, 1986) and have also been shown to block the attachment of whole HIV particles or purified gp120 to CD4+ cells (McDougal *et al.*, 1986; Lundin *et al.*, 1987) These effects, which were central to the elucidation of CD4 as the virus receptor, have been presumed to be mediated either by steric hindrance or by direct competition for the gp120 binding site itself.

Individuals infected with the human immunodeficiency virus (HIV) develop antibodies against a variety of viral antigens, including the major envelope glycoprotein gp120 (Sarngadharan *et al.*, 1984). Neutralizing antibodies which are able to prevent infection and fusion of lymphocytes *in vitro* occur in the majority of patients (Weiss *et al.*, 1985; 1986; Groopman *et al.*, 1987). Neutralization titres correlate well with the total anti-envelope response although neutralization titres are relatively lower. Several studies have reported an association between the presence of neutralizing antibodies and a better clinical prognosis (Ho *et al.*, 1985; 1987; Robert-Guroff *et al.*, 1985; 1987; Ranki *et al.*, 1987; Weber *et al.*, 1987; Alesi *et al.*, 1989) although neutralizing and other HIV-specific antibodies do not prevent eventual clinical progression to AIDS.

Molecular and Cell Biology of Sexually Transmitted Diseases
Edited by D. Wright and L. Archard
Published in 1992 by Chapman and Hall, London ISBN 0 412 36510 3

Marked genetic variation, particularly in the gene encoding the viral envelope, has been demonstrated between HIV isolates from different patients, different tissues, and from the same patient at different times (Shaw et al., 1984; Hahn et al., 1986; Saag et al., 1988). It has been shown that viruses isolated from patients with advanced disease replicate faster and are more pathogenic than viruses isolated from patients with asymptomatic infection (reviewed in Fenyo et al., 1989). Experimental immunization of a variety of animal species with viral envelope proteins and peptides derived from the primary amino acid sequence of gp120 has generated neutralizing antibodies which are generally type specific (Weiss et al., 1986; Palker et al., 1988; Rusche et al., 1988; Nara et al., 1988). Chimpanzees infected productively with HIV also developed neutralizing antibodies, which were predominantly type specific (Nara et al., 1987; Goudsmit et al., 1988), as are most neutralizing antibodies present in the serum of infected patients (Cheng-Mayer et al., 1988; Looney et al., 1988). However, while neutralizing monoclonal antibodies can be effective at high dilution against specific isolates, human antisera exhibit the ability to broaden specificity with time (Weiss et al., 1986).

Since all isolates of HIV-1, HIV-2 and simian immunodeficiency virus use the high affinity interaction of gp120 with the CD4 antigen as the main mechanism of cell attachment, and since the CD4 binding site is highly conserved across all isolates, preventing this interaction is a candidate for therapeutic intervention, aimed at preventing this essential initial step in virus entry to cells (Dalgleish, 1986; Dalgleish et al., 1987; Sattentau and Weiss, 1988; Sattentau et al., 1988).

10.2 THE CD4 MOLECULE – STRUCTURE AND FUNCTION

The human differentiation antigen CD4 is expressed on a variety of cell types, in particular T lymphocytes, dendritic cells and cells of the macrophage/monocyte lineage. Its expression on T lymphocytes identifies a subset of cells which are predominantly associated with helper/inducer functions which are depleted in AIDS (Schroff et al., 1983). These cells show class II restriction, that is, they respond to antigen only when it is presented by other cells bearing homologous MHC class II molecules.

CD4 is a non-polymorphic 55 kD glycoprotein consisting of four tandem extracellular domains which contain significant sequence and structural homology with the variable and joining regions of immunoglobulin (Ig) supergene family members. There is a short hydrophobic transmembrane region and a cytoplasmic tail.

Members of the Ig supergene family have been defined by the presence of one or more regions homologous to the basic structural unit of immunoglobulin molecules, the Ig homology unit. These units are

284

characterized by a primary sequence about 70–110 residues in length with an essentially invariant disulphide bond spanning residues 50–70 and several other relatively conserved residues involved in establishing a tertiary structure referred to as an antibody fold. Ig homology units have been classified into V and C region units on the basis of primary sequence and X-ray crystallographic data. The tertiary structure of V and C regions consists of a series of antiparallel β strands connected by variable length loop sequences that assume a characteristic barrel or sandwich-like structure with two β sheets stabilized by the disulphide bond (reviewed in Hunkapiller and Hood, 1989; Williams, 1987). On the basis of the sequence homology between the amino-terminal domain of CD4 and the Ig (V region) homology unit, several authors have used the X-ray crystallographic structure of the myeloma protein REI to construct models of the putative tertiary structure of CD4 (Jameson et al., 1988; Sattentau et al., 1989; Kieber-Emmons et al., 1989) and X-ray crystallography studies of the two N-terminal domains of CD4 have recently confirmed that these domains do resemble immunoglobulin supergene family (IgSF) homology units in their tertiary structure (Wang et al., 1990; Ryu et al., 1990).

Polymorphic members of the Ig supergene family include immuno-globulins, MHC antigens and the T cell receptor for antigen. Non-polymorphic members of the family include CD4 and CD8 and a very large number of other cell surface proteins that have been shown to interact with each other to mediate cell adhesion and signalling, particularly between components of the immune system.

T cells respond to antigen in general only when it has been processed by antigen presenting cells (APC) and is presented as a peptide in association with MHC Class I or Class II molecules. Class I MHC presentation is usually associated with endogenous cellular products, including viral gene products in infected cells, whereas class II molecules are typically associated with presentation of peptides from molecules that have entered the cell by endocytosis. The T cell receptor (TCR) for antigen associates with the CD3 molecule during antigen recognition, and this complex interacts with the MHC/antigen complex on the APC.

Analysis of mature T cells indicates that they segregate into two classes: those that express CD4 and respond to antigen only in the context of MHC Class II – termed 'MHC class II restriction' – and those that express CD8 and show MHC class I restriction. It is now firmly established that CD4 binds to non-polymorphic parts of MHC Class II molecules (Sleckman et al., 1987; Gay et al., 1987; Doyle and Strominger, 1987); it is thought that CD4 associates with the TCR/CD3 complex conferring class II restriction by increasing the avidity of the T cell/APC interaction (reviewed in Janeway, 1989; Figure 10.1).

Whilst there is strong evidence that CD4 can bind directly to class II molecules and that this would enhance the avidity of the T cell/APC interaction, there is also evidence that CD4 may have other functions such as signal transduction. Anti-CD4 MAbs may, in some circumstances, suppress T cell activation in systems where accessory cells do not carry MHC Class II (Banks and Chess, 1985). Carrel *et al.* (1988) showed that CD4+ cells can be activated by the anti-CD4 Mab B66, and that conversely this Mab can inhibit the specific cytolytic activity of CD4+ effector cells.

Further evidence that CD4 plays a role in signal transduction during interaction with anti-CD4 Mabs or APCs has been provided by Veilette *et al.* (1988; 1989) who have shown that CD4 is associated with the internal membrane tyrosine-kinase p56^{lck}. Cross-linking of CD4 is associated with the rapid phosphorylation of the ζ-subunit of the TCR/CD3 complex and a rapid increase in tyrosine-specific protein kinase activity. Thus in addition to evidence that CD4 interacts directly with class II molecules, these data provide evidence for a specific p56^{lck}-mediated CD4 signal transduction pathway.

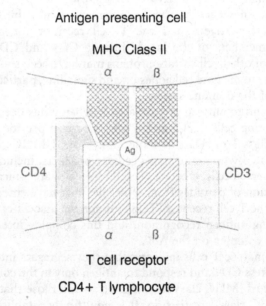

Antigen presenting cell

MHC Class II

T cell receptor

CD4+ T lymphocyte

Figure 10.1 A schematic representation of the putative interaction of an MHC Class II molecule, complexed with an antigenic peptide, with the T cell receptor on a CD4+T cell. CD4 binds non-polymorphic epitopes on MHC Class II, and may increase the avidity of the T cell: APC interaction, but there is also evidence that CD4 plays an active role in signal transduction (see text).

10.3 GP120 – STRUCTURE AND FUNCTION

The genome of HIV in its RNA form is about 9200 nucleotides. Like other retroviruses, it contains long terminal repeats (LTRs), that include signals for initiation and termination of transcription, and *gag*, *pol* and *env* genes organized in that order. In addition to these genes, HIV possesses three regulatory genes (*rev*, *tat* and *nef*) and three small genes (*vpu*, *vpr* and *vif*) which are believed to be involved in virus maturation and release (Cann and Karn, 1989). The LTR of HIV has a complex organization with several sites recognized to be *cis*-acting control elements, in addition to a sequence TAR which has been shown to be the major target for transactivation by the viral *tat* protein (reviewed in Hammarskjold and Rekosh, 1989).

The HIV *env* gene encodes a 160 kD precursor protein (gp160) which is cleaved to form the mature viral envelope proteins gp120 and gp41 (Robey *et al.*, 1985; Veronese *et al.*, 1985). The gp120 molecule has many potential sites for N-linked glycosylation; it is not known which of these sites are used, but gp120 is heavily glycosylated, with sugars accounting for half of its molecular weight (reviewed in Hammarskjold and Rekosh, 1989). Cleavage of gp160 is mediated by a cellular proteinase and is required for virus fusion and infectivity, but not for binding to CD4.

Gp120 is the external membrane glycoprotein. It is present on the membrane of infected cells and virus particles. It lacks a membrane anchoring domain and remains associated with the cell membrane by non-covalent interactions with gp41. gp41 contains several hydrophobic domains and is firmly anchored in the plasma membrane. Because of this loose association with gp41, gp120 is shed from infected cells and it has been postulated that soluble gp120 may contribute to the immunopathogenesis of AIDS (Habeshaw and Dalgleish, 1989). After binding to CD4, an allosteric rearrangement of the trimer spike is thought to occur, allowing the N-terminal of gp41 to insert into the target cell membrane and promote fusion (Hammarskjold and Rekesh, 1989).

The area of gp120 involved in binding to CD4 been found to lie in the C-terminal region of the molecule around residues 410–460 (Kowalski *et al.*, 1987; Lasky *et al.*, 1987; Nygren *et al.*, 1988). Although the CD4 binding site is well conserved across all isolates of HIV, it does not constitute a major neutralization epitope in infected individuals. Human antibody responses to gp120 do include neutralizing activity; the dominant neutralizing epitope identified in animal studies comprises residues 307–330, a disulphide-bonded loop with a highly variable sequence, which elicits type but not group specific neutralizing antibodies (Rusche *et al.*, 1988; Goudsmit *et al.*, 1988a; 1988b).

10.4 HOMOLOGY OF GP120 WITH MEMBERS OF THE IMMUNOGLOBULIN SUPERGENE FAMILY(IgSF)

In addition to the CD4 binding site, membrane fusion sequence and dominant neutralization epitope, the virus envelope contains several further areas of interest whose importance has yet to be fully delineated. Since gp120 and human MHC Class II molecules both bind to CD4, it is of particular interest that there are at least three areas of homology between the virus envelope and members of the human IgSF.

Young (1988) has pointed out that the sequence 261–270 of gp120 has significant homology to the consensus sequence for human HLA-DR, -DP and -DQ β-chain, residues 142–151:

```
        261         270
gp120   V V S T Q L L L N G
HLA-DR  V V S T * L I * N G
        142         151
```

Site-specific mutagenesis was used to abolish the N-linked glycosylation site at residue 269 and this resulted in a virus that was able to bind to CD4, and could mediate syncytium formation, but was non-infectious (Willey *et al.*, 1988). The function of this area has not been elucidated, but it is clear that in spite of its homology with HLA-DR, it is not involved in CD4 binding (Ho *et al.*, 1988).

Golding *et al.* (1989) have identified an area of homology between amino acids 837–843 of gp41 and amino acids 19–25 of the β-chain of human HLA:

```
            837        43
gp41        E G T D R V I
HLA βchain  N G T E R V R
            19         25
```

Murine MAbs against peptides derived from these sequences recognize both molecules. It was also shown that one third of sera from HIV-infected patients contained antibodies that recognized both epitopes and mediated *in vitro* suppression of T cell proliferative responses. These antibodies also mediated *in vitro*, antibody directed, cell mediated cytotoxicity. It was suggested that these antibodies might play a role in the loss of T cell function seen in patients before T cell numbers have started to decline.

A third area of homology between gp120 and HLA-DR molecules has been identified by Brinkworth (1989), who points out weak homology between amino acids 415–451 of gp120 with amino acids 15–63 of HLA-DQβ1. This area of gp120 lies adjacent to the CD4 binding site identified by Lasky *et al.* (1987). Brinkworth hypothesizes that HIV may have at some time in the past incorporated exon two of the gene for HLA-DQβ1 into its own genome and that this might explain its tropism for CD4.

The CD4 binding region of gp120 also shows significant homology to the IgSF, particularly to antibody complementarity determining regions (CDRs). This has been used to construct a molecular model of the site (Kieber-Emmons *et al.*, 1989). This model indicates that the residues 421–438 are potentially exposed for interaction with CD4. Furthermore, some neutralizing antibodies have been shown to be directed to this site (Sun *et al.*, 1989) and the peptide 421–438 has been shown to modulate CD4-dependent cellular function and to block virus binding (Kieber-Emmons *et al.*, 1989). Therefore it seems unlikely that the CD4 binding site is a 'canyon', inaccessible to antibodies – rather it appears that this site is immunosilent in the course of human infection. It is possible therefore that antibodies recognizing this site induced by, for example, an anti-idiotypic immunization could neutralize HIV *in vivo*.

It remains to be established whether these areas of IgSF homology are involved with interaction between gp120 and members of the IgSF other than CD4, or with epitopes on CD4 distinct from the established binding site for gp120.

10.5 IMMUNOSUPPRESSIVE EFFECTS OF GP120 *IN VITRO*

It has been postulated that gp120 shed from infected cells is itself responsible for the immunodeficiency characteristic of AIDS (Habeshaw and Dalgleish, 1989). Shalaby *et al.* (1987) showed that gp120 suppresses T cell proliferative responses to antigen at concentrations between 1 and 20 μg/ml. Mann *et al.* (1987) demonstrated suppression of phytohaemagglutinin-induced lymphocyte blastogenesis by gp120, and Diamond *et al.* (1988) have shown that gp120 bound to CD4+ cells impairs their ability to participate in class II -dependant immune functions. Manca *et al.*, (1990) have shown that gp120 inhibits T cell proliferation in an antigen-specific manner. Amadori *et al.* (1988) showed that whole HIV which had been inactivated by UV light was still able to mediate immunosuppressive effects *in vitro*. gp120 can directly inhibit CD4-Class II association (Clayton *et al.*, 1989; Lamarre *et al.*, 1989; Rosenstein *et al.*, 1990).

gp120 can activate resting B cells (Pahwa *et al.*, 1985). Polyclonal activation of B cells is likely to be due to a variety of factors (Fauci, 1988). Spickett *et al.* (1989) have shown that *in vitro* induction of immunoglobulin secretion by HIV varies between virus isolates, the cell line in which isolates are prepared and between B cell donors. The degree of purity of the B cell preparation is also important as some studies have reported that B cell activation occurs only in the presence of T cells. Cytomegalovirus and Epstein-Barr Virus, both potent activators of B cells may also contribute to this process (Martinez-Maza *et al.*, 1987). Polyclonal activation of B cells by gp120 may compromise the ability of the B cell population to respond to specific antigens.

In particular, gp120 appears to inhibit antigen specific T cell interactions which require correct recognition of self MHC class II by the T cell receptor. We have recently been able to demonstrate functional deletion of specific T cell responses in the presence of whole HIV and functional anergy in the presence of recombinant gp120 alone (Manca et al., 1990). These in vitro data fit well with the clinical spectrum of AIDS where the predominant problem is a defective response to recall antigens and not the ability to mount an immune response ab initio to a novel antigen.

This mechanism would be unlikely to explain the slow, almost inexorable decline of CD4+ cells as the number of deleted or anergic cells should still only represent a small percentage of the total CD4+ population. However, since gp120 binds CD4, has regions homologous to MHC class II and has the ability specifically to inhibit T cell receptor/MHC classII/antigen interactions, it is possible that it may alter the interpretation of self MHC class II. Such a scenario could induce a chronic allogeneic response (or, in this case, syngeneic as there are no 'foreign' MHC class II antigens) which would be like graft versus host disease (GVHD) (Manca et al., 1990). HIV infection and GVHD have many features in common, such as skin disease, diarrhoea and lymphadenopathy and the development of lymphomas. Thymic ontogeny of CD4+ cells involves selection of cells expressing T cell receptors that recognize antigen in association with the correct 'self' MHC class II antigens. In GVHD, it is likely that such cells are not selected for clonal expansion. The reprogramming or 'immunization' of the immune system to reject self MHC class II antigens is obviously a slow process and is, in theory, reversible. Indeed, GVHD induction can be modified by appropriate antisera which appear to recognize the epi-restriction sites on the T cell receptor specific for interaction with MHC molecules (DeGiorgi et al., 1989; 1990).

Masking of CD4 by gp120 may also interfere with the immune response to gp120 itself, which could lead to a clonally restricted response in comparison with the response to other virus components. There is experimental evidence to suggest that this occurs in vivo (Papadopoulos et al., 1988; Khalife et al., 1988). It follows that elimination of the CD4 binding site on gp120 might increase the immunogenicity of other gp120 epitopes. In preliminary studies, we have been able to show that certain forms of heat treatment of gp120 can abolish its ability to bind CD4 and its immunosuppressive effects, without loss of immunogenicity (Manca, unpublished data).

10.6 THE ANTI-IDIOTYPE STRATEGY

Some anti-CD4 MAbs bind to CD4 via a site close to that binding gp120 and prevent subsequent binding of HIV to CD4. If an anti-CD4 MAb competes with gp120 for a single binding site on CD4, it may possess in its variable region (or 'idiotype'), epitopes which resemble the CD4 binding site on gp120. Antibodies with specificity for these epitopes ('anti-idiotypic antibodies') might cross-react with a conserved site on gp120.

10.6.1 Idiotypes and anti-idiotypic antibodies

The description of a unique set of antigenic determinants present on human myeloma proteins was first reported by Slater *et al.* (1955); the term 'idiotype' was coined by Oudin (1974) to designate antigenic determinants that are unique to a small set of antibody molecules. Such unique antigenic determinants are referred to as private idiotypes. Idiotypes shared by larger sets of antibodies have since been observed in numerous species and are referred to as public idiotypes.

The network theory of immunoregulation, proposed by Jerne in 1974, stated that since lymphocytes can recognize virtually any molecular shape, they must be able to recognize the hypervariable regions of antibody molecules and receptors on other lymphocytes. Thus an idiotype on one antibody or lymphocyte could interact with an anti-idiotype on another antibody or cell. Such Id-anti-Id interactions could form a network that could function as a mechanism to control the immune response. Numerous studies have documented the manipulation of the murine immune response by injection of anti-idiotypic antibodies (reviewed in Dalgleish and Kennedy, 1988).

Anti-idiotypic antibodies may be classified by the way they bind to the variable region of the original antibody (Ab_1) (Bona and Kohler, 1984; Hiernaux, 1988). Alpha class anti-idiotypic antibodies ($Ab_2\alpha$) react with epitopes outside the antigen combining site (paratope) of the Ab_1 and do not interfere with binding of the Ab_1 to its antigen. $Ab_2\beta$ and $Ab_2\gamma$ class anti-idiotypic antibodies react with epitopes within the paratope of the Ab_1 and so prevent binding to the original antigen. $Ab_2\beta$ class anti-idiotypic antibodies bind with the paratope in the same way as the original antigen and thus bear conformational similarity to it, whereas $Ab_2\gamma$ class antibodies bind to different determinants and do not reproduce the structure of the original antigen. The classification of anti-idiotypic antibodies therefore depends upon an inhibition or competition assay to discriminate $Ab_2\alpha$s from other anti-idiotypes. The distinction between $Ab_2\beta$ and $Ab_2\gamma$ anti-idiotypic antibodies rests on the generation of third generation anti-idiotypic antibodies which reproduce the specificity of Ab_1, indicating that the Ab_2 structurally resembles the original antigen and must be of β class.

Figure 10.2 shows a simplified representation of these classes of anti-idiotypic antibody and a rudimentary anti-idiotypic network.

10.6.2 Anti-idiotypes as immunogens

The possibility that $Ab_2\beta$ anti-idiotypic antibodies might be used as immunogens against infectious organisms was conceived by Nisonoff

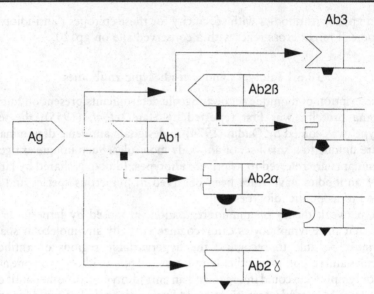

Figure 10.2 A rudimentary idiotypic network: antigen (Ag) is shown reacting with an antibody (Ab_1). Ab_1 in turn reacts with three different classes of anti-idiotypic antibody: $Ab_2\alpha$, which recognizes an epitope not associated with the antigen binding site, and $Ab_2\beta$ and $Ab_2\gamma$ which recognize antigen binding site associated epitopes. $Ab_2\beta$ reacts with Ab_1 in the same way as Ag and therefore resembles it structurally, whereas $Ab_2\gamma$ does not.

and Lamoyi (1981) and it is this class of anti-idiotype that has been examined in the case of HIV (Figure 10.3). The first example of a successful experimental anti-idiotypic immunization was reported in 1983 by Sacks *et al.* in a murine trypanosome system. It was subsequently established that the anti-idiotype was of $Ab_2\alpha$ class. Other studies have confirmed that $Ab_2\alpha$ anti-idiotypic antibodies can function as protective immunogens: it is presumed that $Ab_2\alpha$ anti-ids can induce protective immunity by effects on immunoregulation rather than by acting as a surrogate antigen *per se* (Schick *et al.*, 1987).

Since 1983, anti-idiotypic immunizations have been shown to induce protective immunity in a wide variety of experimental systems (Dalgleish and Kennedy, 1988), including many cases where the anti-idiotype used was of $Ab_2\beta$ class. Of particular interest were situations where the anti-idiotypic response was produced against virus specific T-cell clones (anti-clonotypic) (Finberg and Ertl, 1986), where it was used to identify a virus receptor (Gaulton and Greene, 1986) and where it was used to mimic non-protein antigens such as carbohydrate and phosphorylcholine (Stein and Soderstrom, 1984).

The use of anti-idiotypic immunization in human subjects has been restricted until recently to the experimental therapy of malignant disease.

Passive immunization with murine MAbs with specificity for tumour associated idiotypes has been used in leukaemia and lymphoma with variable results (reviewed in Brown *et al.*, 1989). Active immunization with anti-idiotypic antibody preparations raised against MAbs recognizing tumour associated antigens has been used in malignant melanoma (Kusama *et al.*, 1987; Ferrone and Kageshita, 1988) and colorectal carcinoma (Herlyn *et al.*, 1985) again with variable but encouraging results.

10.6.3 Anti-idiotypic immunization and HIV

A world-wide effort is aimed at producing a vaccine to protect against infection with HIV. Whilst changes in sexual behaviour have been effective in reducing transmission between homosexuals in the United States, there are limitations to the extent to which this will be successful on a world-wide scale, particularly in those areas such as Africa where HIV is transmitted predominantly between heterosexuals.

A number of AIDS vaccine strategies are in progress world-wide (reviewed in Koff and Fauci, 1989). Zagury *et al.* (1988) immunized healthy seronegative volunteers with a vaccinia HIV-*env* recombinant virus. Immunization was well tolerated but only low levels of neutralizing antibody and weak cellular responses were induced. In order to boost responses, some vacinees were also immunized with fixed autologous PBMCs infected with the same virus followed by boosting with recombinant

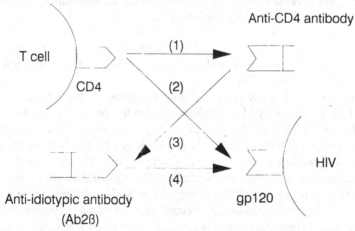

Figure 10.3 The theoretical basis of anti-idiotypic immunization against HIV: Anti-CD4 monoclonal antibodies and gp120 both bind to CD4 (arrows 1 and 2). $Ab_2\beta$ anti-idiotypic antibodies can reproduce the structure of antigens, in this case CD4 (arrow 3). If the anti-CD4 monoclonal antibody and gp120 bind to CD4 in identical ways, and if the anti-idiotypic antibody is $Ab_2\beta$ type, then the anti-idiotypic antibody may bind to the CD4-binding site on gp120 and prevent infection of the CD4+ cell (arrow 4).

gp160. This strategy is logistically unfeasible for large scale use, although it has demonstrated successfully that immunization can engender neutralizing responses. Naturally, no form of infectious challenge can be carried out.

Vaccinia HIV-*env* recombinant viruses have also been employed in USA, in healthy HIV-negative homosexual and bisexual men (Cooney *et al.*, 1991). Preliminary data indicate that individuals lacking pre-existing immunity to vaccinia mount strong T cell proliferative responses to HIV after exposure to this virus, whereas patients who are already vaccinia-immune mount low and transient HIV-specific T cell responses.

A number of HIV-negative volunteers from high and low risk groups have been immunized with recombinant gp160 expressed in a baculovirus vector-insect cell system in phase I dose ranging studies, the preliminary results of which were presented at the Vth International Conference on AIDS, at Montreal (Lane, 1989; Dolin *et al.*, 1989; Keefer *et al.*, 1989; Garrison *et al.*, 1989) Group specific T cell responses were elicited, but there was no neutralizing or enhancing antibody activity.

Salk (1987) recently proposed that by boosting the immune response against HIV during the asymptomatic period it might be possible to reduce the viral burden and delay the onset of AIDS. Levine *et al.* (1989) have immunized a number of subjects with ARC or asymptomatic HIV infection with irradiated, envelope-depleted HIV, which has been shown to be non-infectious in primates. Preliminary data indicate that this approach at least is safe.

There are several theoretical advantages to an anti-idiotype approach, targeted at the CD4 binding site on gp120:

1. The CD4 binding site on gp120 is conserved across all isolates, both in terms of primary sequence (Lasky *et al.*, 1987), and in functional terms (Sattentau *et al.*, 1988).
2. The CD4 binding site is not a dominant epitope in the context of native gp120; the majority of neutralizing antibodies produced in the course of infection by HIV are directed against type specific epitopes distant from the binding site (Rusche *et al.*, 1988). It is possible however that this epitope could be immunogenic if it were presented in a different context, that is, as part of the variable region of a monoclonal antibody.
3. A vaccine based on a MAb could be produced in large quantities using well developed technology. Recombinant protein antigens may not accurately reproduce conformational epitopes of the native molecule, and may differ in their pattern and degree of glycosylation. Theoretically this should be less of a problem with anti-idiotype based strategies since MAbs which recognize the tertiary structure of proteins or, for example, carbohydrate epitopes, should accurately reflect those structures as negative images within their binding sites.
4. Use of a MAb as an immunogen would avoid exposure of vaccinees

294

and personnel involved in vaccine production to live or attenuated virus strains.

5. There is evidence that gp120, reviewed above, can interfere with binding of CD4 to MHC Class II molecules (Clayton *et al.*, 1989; Lamarre *et al.*, 1989; Rosenstein *et al.*, 1990) thereby inhibiting Class II dependent immune functions of CD4+ cells. In view of these data, it may be preferable to avoid vaccine strategies based on immunization with whole gp120.

10.6.4 Early experimental evidence for the validity of an anti-idiotypic strategy in HIV infection

Preliminary results of anti-idiotypic immunization of experimental animals giving rise to anti-gp120 antibodies were reported at the Third International Conference on AIDS, in Washington DC in 1987, by two groups. Sattentau *et al.* (1987) immunized mice with the anti-CD4 MAbs MT151 and anti-Leu3a. Anti-idiotypic antisera to anti-Leu3a specifically stained the HIV infected cell line CEM, reacted with gp120 of two different isolates of HIV by radio-immunoprecipitation (RIPA) and inhibited syncytium formation between infected cells and a susceptible CD4+ cell line. These effects were not seen with antisera against MT151, although both antisera contained anti-idiotypic activity. Kennedy *et al.* (1987) immunized baboons with OKT4A and produced an anti-idiotype response that bound HIV and recombinant gp120 in solid phase assays, and reacted with an infected T-cell line. This antiserum inhibited the binding of both anti-Leu3a and OKT4A to CD4+ cells.

Chanh *et al.* (1987) raised a murine monoclonal IgM, designated HF1.7, against anti-Leu3a, which reacted with anti-Leu3a, but not a panel of irrelevant MAbs, on solid phase and inhibited the binding of anti-Leu3a to CD4+ cells. HF1.7 reacted with HIV antigens in commercial HIV ELISAs and recognized HIV infected cells but not uninfected cells when analysed by flow cytometry. It also reacted in Western blotting with recombinant gp160 and a 120 kD molecule from HIV-infected cell lysates. Unfortunately, this reagent was lost due to a laboratory accident and so has never been characterized in other laboratories.

Dalgleish *et al.* (1987) raised murine polyclonal antisera against anti-Leu3a; these were found to have anti-gp120 activity by solid phase assay, and were also shown to block weakly the formation of syncytia induced by a variety of HIV – 1 isolates.

Studies of the serological responses of HIV-infected individuals have suggested that infrequently anti-gp120 antibodies may arise which are anti-idiotypic to existing anti-CD4 MAbs. Lundin *et al.* (1988) screened 58 sera from HIV seropositive subjects and found three with anti-idiotypic activity against the anti-CD4 MAb T4.2, which is itself a potent inhibitor of syncytium formation. It was shown that affinity purified antibody bound both T4.2 and gp120. Forty-eight further samples were screened for anti-idiotypic antibodies to OKT4A and anti-Leu3a. One was positive for OKT4A, and none was positive for anti-Leu3a. We have recently screened a panel of 97 HIV seropositive individuals for anti-idiotypes to anti-Leu3a, MT151 and MT310 and found none positive (Wilks et al., 1990a). It seems likely therefore that if such antibodies do arise spontaneously, they are rare.

The CD4 binding site on gp120 is not the only conserved region. A site in gp41 has been reported by Zhou et al. (1987) as inducing cross-neutralizing antibodies and anti-idiotypic immunization against HIV, based on this epitope. Chimpanzees were immunized with a peptide equivalent to amino acids 735–752 of HTLV-III$_B$. Anti-peptide antibodies were purified and used to immunize rabbits, which developed an anti-idiotypic response which failed to recognize the original peptide. However, when the rabbit anti-idiotype preparation was used to immunize Balb/C mice the mice developed an Ab$_3$ response which not only recognized the rabbit anti-idiotype preparation, preventing it from binding to the chimpanzee anti-peptide antibodies, but also recognized the original peptide. The authors postulate that this result is due to activation of immunoregulatory mechanisms by Ab$_2\alpha$ antibodies in the rabbit anti-idiotype preparation.

10.6.5 Anti-idiotypic immunization in humans

In 1988, in the light of the animal evidence presented above, we carried out a pilot study of immunization of HIV infected individuals with anti-Leu3a (Wilks et al., 1990b). The objectives of this study were to examine the safety of administration of anti-Leu3a to HIV seropositive individuals, both in terms of acute toxic and allergic reactions and of long-term changes in clinical status. This should determine whether such individuals retain the ability to mount an antibody response to mouse immunoglobulin and, if so, whether that response includes any anti-idiotypic component and whether any anti-idiotypic antibodies produced bind to gp120. Anti-Leu3a was chosen as the immunizing MAb because of its ability to block syncytium formation at low concentrations, and because of the early success with animal immunizations. Four volunteers with the AIDS-related complex (ARC) were immunized with six injections of 1 mg of anti-Leu3a intramuscularly over a ten week period. No acute toxic or allergic reactions were seen, and no significant changes in clinical status, CD4 counts or p24 antigen levels were seen in the year following immunization. All four patients made an antibody response against mouse immunoglobulin which rendered FACS immunophenotyping of whole blood impossible, necessitating the use of washed lymphocytes (Wilks et al., 1990c). Anti-idiotypic responses were measured by ELISA and by the ability of serum to inhibit the binding of radiolabelled anti-Leu3a to CD4+ cells. All the patients made anti-idiotypic antibodies which blocked binding of anti-Leu3a to CD4 (i.e. $Ab_2\beta$ or $Ab_2\gamma$); titres varied between patients and did not correlate with indices of disease stage at the time of immunization. No significant change was seen in anti-gp120 titres by ELISA, or in the strength or pattern of neutralizing activity. Since these patients already have anti-gp120 antibodies by virtue of their infection, it was necessary to purify anti-idiotypic antibodies from serum by affinity chromatography, but no anti-gp120 activity was associated with these purified fractions. It appears therefore that, in humans at least, anti-idiotypes to anti-Leu3a do not have anti-gp120 activity.

However, this study did show that infected individuals retain the ability to mount anti-idiotypic responses to murine MAbs, and that this can be achieved without significant deleterious effects. In parallel with this study we have raised murine anti-idiotypic antibodies to a wide variety of anti-CDs, including anti-Leu3a. Conjugation of MAbs to the synthetic adjuvant muramyl dipeptide prior to immunization led to consistently higher titres than had previously been obtained, but in spite of this, no anti-gp120 activity was obtained. We have found that serum from immunized animals may inhibit syncytium formation two- to four-fold better than preimmune serum; this effect varies between MAbs, but is also seen with some MAbs which are not directed against CD4 (our unpublished data). It is possible that this phenomenon, which may be due to broadly cross-reactive components of the immune response such as IgM, might account for the positive results reported earlier. Reeves *et al.* (1991) have reported similar findings.

10.7 EPITOPE MAPPING OF CD4

In view of these results, it seems appropriate to reappraise the criteria by which MAbs might be selected as potential anti-idiotypic immunogens. Anti-Leu3a was selected for human immunization on the basis of its ability to inhibit syncytium formation at low concentrations and epitope mapping data suggesting that its binding site shares critical residues on CD4 with the gp120 binding site was only available in retrospect.

In order to raise an $Ab_2\beta$ anti-idiotypic response with anti-viral activity it is necessary that the anti-CD4 MAb should bind to CD4 at the same site as gp120. Accurate epitope mapping of monoclonal antibody binding sites would also be useful for confirming data about the gp120 and class II binding sites obtained by other means and to assist in the interpretation of studies aimed at elucidating the physiological functions of CD4 during interaction of CD4+ cells with other components of the immune system. We have used a radioimmunoassay of the ELISA inhibition method of Rath *et al.* (1988) to compare the relative functional affinities of a panel of anti-CD4 MAbs and have compared this with their ability to inhibit syncytium formation and to block gp120/CD4 binding (Wilks *et al.*, 1990d).

Differences in the *in vitro* functions of different anti-CD4 Mabs have been recognized since these Mabs were first raised. Much work has been published on the OKT4 series of antibodies which have been considered to recognize independent epitopes of CD4 (Rao *et al.*, 1983). Many groups have examined the effect of these Mabs on class II dependent *in vitro* immune functions using a variety of different techniques aimed at establishing whether they inhibit CD4-class II binding (Biddison *et al.*, 1982; 1984; Rogozinski *et al.*, 1984; Sleckman *et al.*, 1987; Gay *et al.*, 1987; Doyle and Strominger, 1987; Lamarre *et al.*, 1989; Clayton *et al.*, 1989)

There are consistencies and anomalies within these data, but it is interesting to note that antibodies consistently found to be 'inhibitors' are high affinity Mabs such as OKT4A and anti-Leu3a, whereas those giving anomalous results are lower affinity Mabs such as OKT4C, OKT4E and OKT4F. It is possible therefore that affinity, as well as epitope specificity, plays an important role in these experiments.

Soon after the discovery that CD4 functions as a receptor for HIV, McDougal *et al.* (1986) showed that some but not all anti-CD4 Mabs blocked binding of HIV to CD4+ cells. In order to map those epitopes of CD4 required for viral attachment, Sattentau *et al.* (1986) examined a large panel of Mabs for their ability to inhibit syncytium formation. However, a formal comparison of efficacy on a molar basis was not made. The ability of excess unlabelled Mab to cross-compete for CD4+ cells with small quantities of a limited range of high affinity labelled Mabs was also examined (Sattentau *et al.*, 1986; 1989). An important consideration in the interpretation of such results is the extent to which low affinity Mabs would be able to compete with higher affinity Mabs. There are many anomalies within these data, but as was noted for CD4-class II interactions, most of these anomalies relate to antibodies which we have found to rank low for functional affinity and so can be explained on that basis.

A variety of methods has been used to map epitopes of CD4 to critical residues within the primary sequence of the molecule, including the synthesis of truncated molecules and peptides (Berger *et al.*, 1988; Chao *et al.*, 1989; Jameson *et al.*, 1988), and CD4 constructs incorporating mutations, substitutions, deletions and insertions (Petersen and Seed, 1988; Clayton *et al.*, 1988; Mizukami *et al.*, 1988; Arthos *et al.*, 1989; Sattentau *et al.*, 1989). Although specific criticisms can be brought against each of these methods, there is general agreement on the location of Mab epitopes on CD4. However, the instances in which they differ tend again to involve the lower affinity Mabs referred to above such as OKT4E. A knowledge of relative functional antibody affinity is therefore important in the correct interpretation of such studies since it explains apparent anomalies and may suggest limitations on the value of the information obtained. A summary of the results of these studies, dealing with the commonly available MAbs, is

Table 10.1 Critical residues for binding of monoclonal antibodies and gp120, as determined by a variety of methods. Residues are numbered such that the cysteines at either end of the V_1 domain number 16 and 84 respectively

Monoclonal antibody				
Anti-Leu3a		35–49[2]		43[4]
MT151	130–159[1]	94,165	164[3]	94–96
OKT4A	9–38	60	57	64–66
OKT4B	130–159	165	164	94–96
OKT4C	9–38			24, 25, 88, 89
OKT4D	31–63	47	44–52	34, 50, 55
OKT4E	49–84	122	21, 91	24, 25, 42, 43, 88, 89
OKT4F	130–159			102–105
gp120	37–53	45–47	31, 44–47	42–55

[1]Jameson et al., 1988; [2]Peterson et al., 1988; [3]Mizukami et al., 1988; [4]Arthos et al., 1989; Sattentau et al., 1989.

presented in Table 10.1. By these criteria, critical residues for gp120 binding have been located in the region of amino acids 46–47.

The ability of antibodies and antisera to inhibit the formation of syncytia in infected lymphocyte cultures has been regarded as a measure of their ability to block the interaction of gp120 with CD4. We have shown that the ability to inhibit syncytium formation correlates strongly with antibody affinity (Wilks et al., 1990d). There are also antibodies of moderately high affinity such as L218 that block syncytium formation efficiently but only partially prevent the binding of gp120 to CD4+ cells. These discrepancies between gp120 blocking and inhibition of syncytium formation suggest that these two functions may be mediated by distinct regions of the CD4 molecule. Camerini and Seed (1990) have shown that substitution of residue 87 in human CD4 with the equivalent amino acid from chimpanzee CD4 abolishes the ability of cells to form syncytia without affecting gp120 binding. MAbs that block syncytium formation but do not block the binding of gp120 to CD4 may exert their effect by binding to CD4 in such a way as to prevent contact of the intact virion, or those parts of the virus envelope concerned with fusion, with the cell membrane. An analogy may be drawn with certain anti-gp120 MAbs which neutralize HIV effectively without blocking binding of gp120 to CD4 (Linsley et al., 1988). It is unlikely that the syncytium formation assay will be useful in identifying candidate Ab_1s for anti-idiotypic immunization.

Lifson et al. (1988) have synthesized a series of short peptides analogous to regions of the CD4 V_1 domain, and examined their ability to inhibit syncytium formation. Significant activity was found to be associated with the peptide 76–94; this activity was not associated with the main product of peptide synthesis but in a fraction containing benzylated peptide products generated during automated synthesis. The authors postulate that benzylation preserves the tertiary structure of the peptide, allowing it to compete with native CD4 for gp120. More recently, a peptide comprising amino acids 84–101 of the CD4 sequence has been shown to inhibit syncytium formation at 50 μg/ml, albeit 100-fold less efficiently than whole soluble CD4 (Jameson, personal communication). Although special methods are used to ensure conservation of the tertiary structure of these peptides, their activity is sequence specific. It is possible that these areas of the CD4 molecule are involved in fusion events taking place after gp120 binding.

X-ray crystallography studies of the two N-terminal domains of CD4 have confirmed that these domains resemble IgSF homology units in their tertiary structure (Wang et al., 1990; Ryu et al., 1990). These studies have also demonstrated high thermal parameters for the C'' ridge of the V_1 domain of CD4, the area thought to be involved in contact with gp120, predicting a degree of flexibility for that region. Allosteric effects may therefore underlie the ability of MAbs to prevent fusion of HIV with the cell membrane without inhibiting binding of gp120 to CD4. Conformational change following antibody binding has been demonstrated previously for histocompatibility antigens (Diamond et al., 1984; Parham, 1984). A better understanding of how gp120 and CD4 interact might allow a more rational choice of Ab_1 or the modification of an appropriate anti-CD4 reagent by recombinant DNA techniques (Dalgleish and Kennedy, 1988).

CONCLUSIONS

Anti-idiotypic strategies have been shown to induce protective immunity against infectious agents in a variety of animal systems, and have been used with partial success in the therapy of human malignancies. Theoretically, there are significant advantages to be gained from this approach in HIV infection but recent results have been disappointing, mainly because the choice of immunogen remains essentially empirical. A better understanding of the gp120/CD4 interaction, which may itself be responsible for the immunopathogenesis of AIDS, is required and anti-idiotypes themselves may help in this respect.

REFERENCES

Alesi, D.R., Ajello, F., Lupo, G. et al. (1989) Neutralizing antibody and clinical status of human immunodeficiency virus (HIV)-infected individuals. J. Med. Virol., 27, 7–12.

Amadori, A., Faulkner-Valle, G.P., De Rossi, A. et al. (1988) HIV mediated immunodepression: In vitro inhibition of T-lymphocyte proliferative response by ultraviolet-inactivated virus. Clin. Imm. Immunopath., 46, 37–54.

Arthos, J., Deen, K.C., Chaikin, M.A. et al. (1989) Identification of the residues in human CD4 critical for the binding of HIV. *Cell*, 57, 469–81.

Banks, I. and Chess, L. (1985) Perturbation of the T4 molecule transmits a negative signal to T cells. *J. Exp. Med.*, 162, 1294–303.

Berger, E.A., Fuerst, T.R. and Moss, B. (1988) A soluble recombinant polypeptide comprising the amino-terminal half of the extracellular region of the CD4 molecule contains an active binding site for the human immunodeficiency virus. *Proc. Natl Acad. Sci. USA*, 85, 2357–61.

Biddison, W.E., Rao, P.E., Talle, M.A. et al. (1982) Possible involvement of the OKT4 molecule in T cell recognition of class II HLA antigens: evidence from studies of cytotoxic T lymphocytes specific for SB antigens. *J. Exp. Med.*, 156, 1065–76.

Biddison, W.E., Rao, P.E., Talle, M.A. et al. (1984) Possible involvement of the T4 molecule in T cell recognition of class II HLA antigens: Evidence from studies of CTL-target cell binding. *J. Exp. Med.*, 159, 783–97.

Bona, C.A. and Kohler, H. (1984) Anti-idiotypic antibodies and internal images, in *Monoclonal and Anti-idiotypic Antibodies: Probes for receptor structure and function* (eds J.C. Ventner, C.M. Fraser and J. Lindstrom), in *Receptor biochemistry and methodology*, Vol 4. Alan Liss Inc., New York, pp. 141–9.

Brinkworth, R.I. (1989) The envelope glycoprotein of HIV-1 may have incorporated the CD4 binding site from HLA-DQ beta 1. *Life Sciences*, 45, iii–ix.

Brown, S.L., Miller, R.A. and Levy, R. (1989) Antiidiotype therapy of B cell lymphoma. *Semin. Oncol.*, 16, 199–210.

Camerini, D. and Seed, B. (1990) A CD4 domain important for HIV-mediated syncytium formation lies outside the virus binding site *Cell*, 60, 747–54.

Cann, A.J. and Karn, J. (1989) Molecular biology of HIV: new insights into the virus life-cycle. *AIDS*, 3 (suppl. 1), S19–34.

Carrel, S., Moretta, A., Pantaleo, G. et al. (1988) Stimulation and proliferation of CD4+ peripheral blood T lymphocytes induced by an anti-CD4 monoclonal antibody. *Eur. J. Immunol.*, 18, 333–9.

Chanh, T.C., Dreesman, G.R. and Kennedy, R.C. (1987) Monoclonal anti-idiotypic antibody mimics the CD4 receptor and binds human immunodeficiency virus. *Proc. Natl Acad. Sci. USA*, 84, 3891–5.

Chao, B.H., Costpoulos, D.S., Curiel, T. et al. (1989) A 113 amino acid fragment of CD4 produced in *Escherichia coli* blocks human immunodeficiency virus-induced cell fusion. *J. Biol. Chem.*, 264, 5812–17.

Cheng-Mayer, C., Homsy, J., Evans, L.A. and Levy, J.A. (1988) Identification of human immunodeficiency virus subtypes with distinct patterns of sensitivity to serum neutralisation. *Proc. Natl Acad. Sci. USA*, 85, 2815–19.

Clayton, L.K., Hussey, R.E., Steinbrich, R. et al. (1988) Substitution of murine for human CD4 residues identifies amino acids critical for HIV-gp120 binding. *Nature*, 335, 363–6.

Clayton, L.K., Sieh, M., Pious, D.A. and Reinherz, E.L. (1989) Identification of human CD4 residues affecting class II MHC versus HIV-1 gp120 binding. *Nature*, 339, 548–51.

Cooney, E.L., Collier, A.C., Greenberg, P.D. et al. (1991) Safety of and immunological response to a recombinant vaccinia virus vaccine expressing HIV envelope glycoprotein. *Lancet*, 337, 567–72.

Dalgleish, A.G. (1986) Antiviral strategies and vaccines against HTLV III/LAV. *J. Royal Coll. Phys. Lon.*, **20**, 258–67.

Dalgleish, A.G., Beverley, P.C.L., Clapham, P.R. *et al.* (1984) The CD4(T4) antigen is an essential component of the receptor for the AIDS retrovirus. *Nature*, **312**, 763–7.

Dalgleish, A.G. and Kennedy, R.C. (1988) Anti-idiotypic antibodies as immunogens: idiotype based vaccines. *Vaccine*, **6**, 215–20.

Dalgleish, A.G., Thomson, B.J., Chanh, T.C. *et al.* (1987) Neutralization of HIV isolates by anti-idiotypic antibodies which mimic the T4(CD4) epitope: a potential AIDS vaccine. *Lancet*, **ii**, 1047.

DeGiorgi, L., Habeshaw, J.A., Povey, S. and Matossian-Rogers, A. (1990) Reduction of graft-versus-host disease in neonatal F1 hybrid mice. *Clin. Exp. Imm.*, **79**, 130–4.

DeGiorgi, L., Povey, S., Habeshaw, J.A. and Davies, A.J.S. (1989) Prevention of graft-versus-host disease in neonatal F1 hybrid mice by pre-immunization of their mother with paternal spleen cells. *Trans. Proc.*, **21**(1), 3045–9.

Diamond, A.G., Butcher, G.W. and Howard, J.C. (1984) Localised conformational changes induced in a class I major histocompatability antigen by the binding of monoclonal antibodies. *J. Immunol.*, **132**, 1169–75.

Diamond, D.C., Sleckman, B., Gregory, T. *et al.* (1988) Inhibition of CD4+ T cell function by the HIV envelope protein, gp120 *J. Imm.*, **141**, 3715–17.

Dolin, R., Graham, B, Greenberg, S. *et al.* (1989) Safety and immunogenicity of HIV-1 recombinant gp160 vaccine candidate in normal volunteers. *Vth Int. Conf. AIDS*, Montreal 1989, Abstract ThCO 33.

Doyle, C. and Strominger, J.L. (1987) Interaction between CD4 and class II MHC molecules mediates cell adhesion. *Nature*, **330**, 256–9.

Fauci, A.S. (1988) The human immunodeficiency virus: Infectivity and mechanisms of pathogenesis. *Science*, **239**, 617–22.

Fenyo, E., Albert, J. and Asjo, B. (1989) Replicative capacity, cytopathic effect and cell tropism of HIV. *AIDS*, **3**(suppl 1), S5–12.

Ferrone, S. and Kageshita, T. (1988) Human high molecular weight melanoma associated antigen as a target for active specific immunotherapy: A phase I clinical trial with murine anti-idiotypic monoclonal antibodies *J. Dermatol.*, **15**, 457–65.

Finberg, R.W. and Ertl, H.C.J. (1986) Use of T-cell specific anti-idiotypes to immunize against viral infections. *Immunol. Rev.*, **90**, 129–55.

Garrison, L., Midthun, K., Fernie, B. *et al.* (1989) Western blots: Intermediate results and lot-to-lot variability in adults at low risk for HIV-1 infection. *Vth Int. Conf. AIDS*, Montreal 1989, Abstract TBP 123.

Gaulton, G.N. and Greene, M.I. (1986) Idiotypic mimicry of biological receptors. *Ann. Rev. Immunol.*, **4**, 253–80.

Gay, D., Maddon, P., Sekaly, R. *et al.* (1987) Functional interaction between human T-cell protein CD4 and the major histocompatibility complex HLA-DR antigen. *Nature*, **328**, 626–9.

Golding, H., Shearer, G.M., Hillman, K. *et al.* (1989) Common epitope in human immunodeficiency virus (HIV) I-gp41 and HLA Class II elicits immunosuppressive autoantibodies capable of contributing to immune dysfunction in HIV-1 infected individuals. *J. Clin. Inv.*, **83**, 1430–5.

Goudsmit, J., Debouck, C., Meloen, R.H. *et al.* (1988a) Human immunodeficiency virus type 1 neutralization epitope with conserved architecture elicits early type-specific antibodies in experimentally infected chimpanzees. *Proc. Natl Acad. Sci. USA*, 85, 4478–82.

Goudsmit, J., Thiriart, C. Smit, L. *et al.* (1988b) Temporal development of cross-neutralization between HTLV-III B and HTLV-III RF in experimentally infected chimpanzees. *Vaccine*, 6, 229–32.

Groopman, J.E., Benz, P.M., Ferriani, R. *et al.* (1987) Characterization of serum neutralization response to the human immunodeficiency virus (HIV). *AIDS Research and Human Retroviruses*, 3(1), 71–85.

Habeshaw, J.A. and Dalgleish, A.G. (1989) The relevance of the HIV env/CD4 interaction to the pathogenesis of acquired immune deficiency syndrome. *J. Acq. Immun. Def. Syn.*, 2, 457–8.

Hahn, B.H., Shaw, G.M., Taylor, M.E. *et al.* (1986) Genetic variation in HTLV-III/LAV over time in patients with AIDS or at risk for AIDS. *Science*, 232, 1548–53.

Hammarskjold, M. and Rekosh, D. (1989) The molecular biology of the human immunodeficiency virus. *Biochim. Biophys. Acta*, 989, 269–80.

Herlyn, D., Lubeck, M., Sears, H. and Koprowski, H. (1985) Specific detection of anti-idiotypic immune responses in cancer patients treated with murine monoclonal antibody. *J. Imm. Meth.*, 85, 27–38.

Hiernaux, J.R. (1988) Idiotypic vaccines and infectious diseases. *Infect. and Immun.*, 56, 1407–13.

Ho, D.D., Kaplan, J.C., Rackauskas, I.E. and Gurney, M.E. (1988) Second conserved domain of gp120 is important for HIV infectivity and antibody neutralization. *Science*, 239, 1021–3.

Ho, D.D., Rota, T.R. and Hirsch, M.S. (1985) Antibody to lymphadenopathy-associated virus in AIDS. *N. Eng. J. Med.* 312, 649–50.

Ho, D.D., Sarngadharan, M.G., Hirsch, M.S. *et al.* (1987) Human immunodeficiency virus neutralizing antibodies recognise several conserved domains on the envelope glycoproteins. *J. Virol.*, 61(6), 2024–8.

Hunkapiller, T. and Hood, L. (1989) Diversity of the immunoglobulin supergene family. *Adv. Immunol.*, 44, 1–63.

Hussey, R.E., Richardson, N.E., Kowalski, M. *et al.* (1988) A soluble CD4 protein selectively inhibits HIV replication and syncytium formation. *Nature*, 331, 78–81.

Jameson, B.A., Rao, P.E., Kong, L.I. *et al.* (1988) Location and chemical synthesis of a binding site for HIV-1 on the CD4 protein. *Science*, 240, 1335–9.

Janeway, C.A. (1989) The role of CD4 in T-cell activation: accessory molecule or co-receptor? *Immunol. Today*, 10(7), 234–8.

Jerne, N.K. (1974) Towards a network theory of the immune system. *Ann. Rev. Immunol.*, 125C, 373–89.

Keefer, M., Bonnez, W., Roberts, N. *et al.* (1989) HIV-1 recombinant gp160 vaccine recipients demonstrate gp160 specific lymphocyte proliferation prior to Western Blot reactivity. *Vth Int. Conf. AIDS*, Montreal 1989, Abstract MCP 8.

Kennedy, R.C., Eichberg, J.W., Lanford, R.E. and Dreesman, G.R. (1986) Anti-idiotypic antibody vaccine for type B viral hepatitis in chimpanzees. *Science*, 232, 220–3.

Kennedy, R.C., Zhou, E.-M., Dreesman, G.R. and Chanh, T.C. (1987) Internal image anti-idiotypes representative of homobodies mimic the CD4 molecule and bind human immunodeficiency virus. *III Int. Conf. AIDS*, p. 160, abstract TH.9.5.

Khalife, J. Guy, B. Capron, M. *et al.* (1988) Isotypic restriction of the antibody response to human immunodeficiency virus. *Aids Res. Hum. Retro.*, 4, 3–9.

Kieber-Emmons, T., Jameson, B. and Morrow, W. (1989) The gp120/CD4 interface: structural, immunological and pathological considerations. *Biochim. Biophys. Acta*, 989, 281–300.

Klatzmann, D., Champagne, E., Chamaret, S, *et al.* (1984) T-lymphocyte T4 molecule behaves as the receptor for human retrovirus LAV. *Nature*, 312, 767–8.

Koff, W.C. and Fauci, A.S. (1989) Human trials of AIDS vaccines: current status and future directions. *AIDS*, 3(suppl 1), S125–30.

Kowalski, M., Potz, J., Basiripour, L. *et al.* (1987) Functional regions of the envelope glycoprotein of human immunodeficiency virus type I. *Science*, 237, 1351–5.

Kusama, M., Kageshita, T., Tsugisaki, M. *et al.* (1987) Syngeneic antiidiotypic antisera to murine anti-human high-molecular-weight-melanoma-associated antigen monoclonal antibodies. *Cancer Res.*, 47, 4312–17.

Lamarre, D., Capon, D.J., Karp, D.R. *et al.* (1989) Class II MHC molecules and the HIV gp120 envelope protein interact with functionally distinct regions of the CD4 molecule. *EMBO J.*, 8(11), 3271–7.

Landau, N.R., Warton, M. and Littman, D.R. (1988) The envelope glycoprotein of the human imunodeficiency virus binds to the immunoglobulin-like domain of CD4. *Nature*, 334, 159–62.

Lane, C. (1989) Evaluation of a recombinant HIV-1 envelope protein as an immunogen in humans. *Vth Int. Conf. AIDS*, Montreal 1989, Abstract WCO 20.

Lasky, L.A., Nakamura, G., Smith, D.H. *et al.* (1987) Delineation of a region of the human immunodeficiency virus type 1 gp120 glycoprotein critical for interaction with the CD4 receptor. *Cell*, 50, 975–85.

Levine, A., Henderson, B., Groshen, S. *et al.* (1989) Immunization with inactivated, envelope depleted HIV immunogen in HIV infected men with ARC: preliminary report on exploratory studies in progress. *Vth Int. Conf. AIDS*, Montreal 1989, Abstract ThBO 44.

Lifson, J.D., Hwang, K.M., Nara P.L. *et al.* (1988) Synthetic CD4 peptide derivatives that inhibit HIV infection and cytopathicity. *Science*, 241, 712–16.

Linsley, P.S., Ledbatter, J.A., Kinney-Thomas, E. and Hu, S.-L. (1988) Effects of anti-gp120 monoclonal antibodies on CD4 receptor binding by the env protein of human immunodeficiency virus type I. *J. Virol.*, 62, 3695–702.

Looney, D.J., Fisher, A.G., Putney, S.D., *et al.* (1988) Type restricted neutralisation of molecular clones of human immunodeficiency virus. *Science*, 241, 357–9.

Lundin, K., Karlsson, A., Nygren, A. *et al.* (1988) Certain human gp120-HIV antibodies react with anti-CD4 antibodies. *Scand. J. Immunol.*, 27, 113–17.

Lundin, K., Nygren, A., Arthur, L.O., *et al.* (1987) A specific assay measuring binding of I125-gp120 from HIV to T4+/CD4+ cells. *J. Imm. Meth.*, 97, 93–100.

Maddon, P.J., Dalgleish, A.G., McDougal, J.S. *et al.* (1986) The T4 gene encodes the AIDS virus receptor and is expressed in the immune system and the brain. *Cell*, 47, 333–48.

Manca, F., Habeshaw, J.A. and Dalgleish, A.G. (1990) HIV envelope glycoprotein, antigen-specific T cell responses and soluble CD4. *Lancet*, 335, 811–15.

Mann, D.L., Lasane, F., Popovic, M., *et al.* (1987) HTLV-III large envelope protein (gp120) suppresses PHA-induced lymphocyte blastogenesis. *J. Immunol.*, 138, 2640–4.

Martinez-Maza, O., Crabb, E., Mitsuyasu, T.R. *et al.* (1987) Infection with the human immunodeficiency virus (HIV) is associated with an in vivo increase in B lymphocyte activation and immaturity. *J. Imm.*, 138, 3720–4.

McDougal, J.S., Nicholson, J.K.A., Cross, G.D. *et al.* (1986) Binding of the human retrovirus HTLV-III/LAV/ARV/HIV to the CD4 molecule: conformation dependance, epitope mapping, antibody inhibition, and potential for idiotypic mimicry. *J. Immunol.*, 137, 2937–44.

Mizukami, T., Fuerst, T.R., Berger, E.A. and Moss, B. (1988) Binding region for human immunodeficiency virus (HIV) and epitopes for HIV-blocking monoclonal antibodies of the CD4 molecule defined by site-directed mutagenesis. *Proc. Natl Acad. Sci. USA*, 85, 9273–7.

Nara, P.L., Robey, W.G., Arthur, L.O. *et al.* (1987) Persistent infection of chimpanzees with human immunodeficiency virus: Serological responses and properties of reisolated viruses. *J. Virol.*, 61, 3173–80.

Nara, P.L., Robey, W.G., Pyle, S.W. *et al.* (1988) Purified envelope glycoproteins from human immunodeficiency virus type 1 variants induce individual, type-specific neutralizing antibodies. *J. Virol.*, 62(8), 2622–8.

Nisonoff, A. and Lamyoi, E. (1981) Implications of the presence of an internal image of the antigen in anti-idiotypic antibodies: Possible application in vaccine production. *Clin. Immunol. Immunopath.*, 21, 397–406.

Nygren, A., Bergman, T., Matthews, T. *et al.* (1988) 95- and 25-kDa fragments of the human immunodeficiency virus envelope glycoprotein gp120 bind to the CD4 receptor. *PNAS*, 85, 6543–6.

Oudin, J. (1974) L'idiotypie des anticorps. *Ann. Immunol.*, 125C, 309–37.

Pahwa, S., Pahwa, R., Saxinger, C. *et al.* (1985) Influence of the human T-lymphotropic virus/lymphadenopathy-associated virus on functions of human lymphocytes: Evidence for immunosuppresive effects and polyclonal B-cell activation by banded viral preparations. *Proc. Natl Acad. Sci. USA*, 82, 8198–202.

Palker, T.J., Clark, M.E., Langlois, A.J. *et al.* (1988) Type-specific neutralization of the human immunodeficiency virus with antibodies to env-encoded synthetic peptides. *Proc. Natl Acad. Sci. USA*, 85, 1932–6.

Papadopoulos, N.M., Costello, R., Ceroni, M. and Moutsopoulos, H.M. (1988) Identification of HIV-specific oligoclonal bands in serum of carriers of HIV antibody. *Clin. Chem.*, 34, 973–5.

Parham, P. (1984) Changes in conformation with loss of alloantigenic determinants of a histocompatability antigen (HLA-B7) induced by monoclonal antibodies. *J. Immunol.*, 132, 2975–83.

Peterson, A. and Seed, B. (1988) Genetic analysis of monoclonal antibody and HIV binding sites on the human lymphocyte antigen CD4. *Cell*, 54, 65–72.

Ranki, A., Weiss, S.H., Valle, S.-L. *et al.* (1987) Neutralizing antibodies in HIV

(HTLV-III) infection: correlation with clinical outcome and antibody response against different viral proteins. *Clin. Exp. Immunol.*, **69**, 231–9.

Rao, P.E., Talle, M.A., Kung, P.C. and Goldstein, G. (1983) Five epitopes of a differentiation antigen on human inducer T cells distinguished by monoclonal antibodies. *Cell. Imm.*, **80**, 310–19.

Rath, S., Stanley, C.M. and Steward, M.W. (1988) An inhibition enzyme immunoassay for estimating relative antibody affinity and affinity heterogeneity. *J. Imm. Meth.*, **106**, 245–9.

Reeves, J.P., Buck, D., Berkower, I., *et al.* (1991) Anti-Leu3a induces combining site-related anti-idiotypic antibody without inducing anti-HIV activity. *AIDS Research and Human Retroviruses*, **7**(1), 55–63.

Robert-Guroff, M., Brown, M. and Gallo, R.C. (1985) HTLV-III neutralising antibodies in patients with AIDS and AIDS-related complex. *Nature*, **316**, 72–4.

Robert-Guroff, M., Giardina, P.J., Robey, W.G. *et al.* (1987) HTLV-III neutralizing antibody development in transfusion dependant seropositive patients with B-thalassemia. *J. Immunol.*, **138**, 3731–6.

Robey, W., Safai, B., Orozslan, S. *et al.* (1985) Characterization of envelope and core structural gene products of HTLV-III with sera from AIDS patients. *Science*, **225**, 593–5.

Rogozinski, L., Bass, A., Glickman, E. *et al.* (1984) The T4 surface antigen is involved in the induction of helper function. *J. Imm*, **132**, 735–9.

Rosenstein, Y., Burakoff, S.J. and Herrmann, S.H. (1990) HIV-gp120 can block CD4-class II MHC-mediated adhesion. *J. Imm.*, **144**, 526–31.

Rusche, J.R., Javaherian, K., McDanal, C. *et al.* (1988) Antibodies that inhibit fusion of human immunodeficiency virus-infected cells bind a 24 amino acid sequence of the viral envelope, gp120. *Proc. Natl Acad. Sci. USA*, **85**, 3198–202.

Ryu, S.E. Kwong, P., Truneh, A. *et al.* (1990) Crystal structure of an HIV-binding recombinant fragment of human CD4. *Nature*, **348**, 419–26.

Saag, M.S., Hahn, B.H., Gibbons, J. *et al.* (1988) Extensive variation of human immunodeficiency virus type 1 in vivo. *Nature*, **334**, 440–4.

Sacks, D.L. and Sher, A. (1983) Evidence that anti-idiotype induced immunity to experimental African Trypanosomiasis is genetically restricted and requires recognition of combining-site related idiotopes. *J. Imm.*, **131**(3), 1511–15.

Salk, J. (1987) Prospects for the control of AIDS by immunization of seropositive individuals. *Nature*, **327**, 473–6.

Sarngadharan, M.G., Popovic, M., Bruch, L. *et al.* (1984) Antibodies reactive with human T-lymphotropic retroviruses (HTLV-III) in the serum of patients with AIDS. *Science*, **224**, 506–8.

Sattentau, Q. and Weiss, R.A. (1988) The CD4 antigen: physiological ligand and HIV receptor. *Cell*, **52**, 631–3.

Sattentau, Q.J., Arthos, J., Deen, K. *et al.* (1989) A mutational analysis of epitopes of the HIV-binding domain of CD4 using CD4 antibodies and anti-idiotypes. *J. Exp. Med.*, **170**, 1319–34.

Sattentau, Q.J., Dalgleish, A.G., Weiss, R.A. and Beverley, P.C.L. (1986) Epitopes of the CD4 antigen and HIV infection. *Science*, **234**, 1120–3.

Sattentau, Q.J., Weber, J.N., Weiss, R.A. and Beverley, P.C.L. (1987) Antisera to

Leu3a with anti-idiotypic activity react with gp110/130 of HIV-1 and LAV-2. *III Int. Conf. AIDS*, p. 160, abstract TH.9.4.

Sattentau, Q., Clapham, P.R., Weiss R.A. *et al.* (1988) The human and simian immunodeficiency viruses HIV-1 HIV-2 and SIV interact with similar epitopes on their cellular receptor the CD4 molecule. *AIDS*, 2, 101–5.

Schick, M.R., Dreesman, G.R. and Kennedy, R.C. (1987) Induction of an anti-hepatitis B surface antigen response in mice by noninternal image (Ab2) anti-idiotypic antibodies. *J. Immunol.*, 138, 3419–25.

Schnittman, S.M., Lane, H.C, Higgins, S.E. *et al.* (1986) Direct polyclonal activation of human B lymphocytes by the acquired immune deficiency virus. *Science*, 233, 1084–6.

Schroff, R.W., Gottlieb, M.S., Prince, H.E. *et al.* (1983) Immunological studies of homosexual men with immunodeficiency and Kaposi's sarcoma. *Clin. Imm. Immunopath.*, 27, 300–14.

Shalaby, M.R., Krowka, J.E., Gregory, T.J. *et al.* (1987) The effects of human immunodeficiency virus recombinant envelope glycoprotein on immune cell functions in vitro. *Cell. Immunol.*, 110, 140–8.

Shaw, G.M., Hahn, B.H., Arya, S.K. *et al.* (1984) Molecular characterization of human T-cell leukaemia (lymphotropic) virus type III in the acquired immune deficiency syndrome. *Science*, 226, 1165–71.

Slater, R.J., Ward, S.M. and Kunkel, H.G. (1955) Immunological relationships among the myeloma proteins. *J. Exp. Med.*, 101, 85–108.

Sleckman, B.P., Peterson, A., Jones, W.K., *et al.* (1987) Expression and function of CD4 in a murine T-cell hybridoma. *Nature*, 328, 351–3.

Smith, D.H., Byrn, R.A., Masters, S.A. *et al.* (1987) Blocking of HIV-1 infectivity by a soluble secreted form of the CD4 antigen. *Science*, 238, 1704–7.

Spickett, G., Beattie, R., Farrant, J. *et al.* (1989) Assessment of responses of normal human B lymphocytes to different isolates of human immunodeficiency virus: role of normal donor and of cell line used to prepare viral isolates. *Aids Res. Hum. Retro.*, 5(3), 355–66.

Stein, K.E. and Soderstrom, T. (1984) Neonatal administration of idiotypic or anti-idiotypic antibodies primers for protection against *Escherichia coli* K13 infection in mice. *J. Exp. Med.*, 160, 1001–11.

Sun, N., Ho, D., Sun, C. *et al.* (1989) Generation and characterization of monoclonal antibodies to the putative CD4 binding domain of human immunodeficiency virus type 1 gp120. *J. Virol.*, 63, 3579–85.

Veillete, A., Bookman, M.A., Horak, E.M. and Bolen, J.B. (1988) The CD4 and CD8 T cell surface antigens are associated with the internal membrane tyrosine-protein kinase p56-lck. *Cell*, 55, 301–8.

Veillette, A., Bookman, M.A., Horak, E. *et al.* (1989) Signal transduction through the CD4 receptor involves the activation of the internal membrane tyrosine-protein kinase p56-lck. *Nature*, 338, 257–9.

Veronese, F., DeVico, A., Copeland, T. *et al.* (1985) Characterization of gp41 as the transmembrane protein coded by the HTLV-III/LAV envelope gene. *Science*, 229, 1402–5.

Wang, J., Yan, Y., Garrett, T. *et al.* (1990) Atomic structure of a fragment of human CD4 containing two immunoglobulin-like domains. *Naure*, 348, 411–18.

Weber, J.N., Clapham, P.R., Weiss, R.A. *et al.* (1987) Human immunodeficiency virus infection in two cohorts of homosexual men: neutralising sera and association of anti-gag antibody with prognosis. *Lancet*, i, 119–22.

Weiss, R.A., Clapham, P.R., Weber, J.N. *et al.* (1986) Variable and conserved neutralisation antigens of the human immunodeficiency virus. *Nature*, 324, 572–5.

Weiss, R.A., Clapham, P.R., Cheingsong-Popov, R. *et al.* (1985) Neutralisation of human T-lymphotropic virus type III by sera of AIDS and AIDS-risk patients. *Nature*, 316, 69–72.

Wilks, D., Byrom, N., Walker, L. *et al.* (1990a) Characteristic immunophenotyping artefact seen in patients with anti-mouse immunoglobulin antibodies. *Cytometry*, 11, 318–19.

Wilks, D., Walker, L.C., Habeshaw, J.A. *et al.* (1990b) Anti-CD4 autoantibodies and screening for anti-idiotypic antibodies to anti-CD4 monoclonal antibodies in HIV seropositive individuals. *AIDS*, 4, 113–18.

Wilks, D., Walker, L.C., O'Brien, J. *et al.* (1990c) Differences in affinity of anti-CD4 monoclonal antibodies predict their effects on syncytium induction by human immunodeficiency virus (HIV). *Immunology*, 71, 10–15.

Wilks, D., Walker, L., Habeshaw, J.A. *et al.* (1991) Anti-idiotypic responses to immunization with anti-Leu3a in HIV seropositive individuals. *J. Infect. Diseases*, 163, 389–92.

Willey, R., Bonifacino, J., Potts, B. *et al.* (1988) Biosynthesis, cleavage and degradation of the human immunodeficiency virus envelope glycoprotein gp160. *Proc. Natl Acad. Sci. USA*, 85, 9580–4.

Williams, A.F. (1987) A year in the life of the immunoglobulin superfamily. *Immunol. Today*, 8 (10), 298–303.

Young, J.A.T. (1988) HIV and HLA similarity (letter). *Nature*, 333, 215.

Zagury, D., Bernard, J., Cheynier, R. *et al.* (1988) A group specific anamnestic immune reaction against HIV-1 induced by a candidate vaccine against AIDS. *Nature*, 332, 728–31.

Zhou, E.M., Chanh, T.C., Dreesman, G.R. *et al.* (1987) Immune response to human immunodeficiency virus: In vivo administration of anti-idiotype induces an anti-gp160 response specific for a synthetic peptide, *J. Imm.*, 139, 2950–6.

Index

All entries in *italics* represent diagrams, those in **bold** are tables.